An Introduction to Generalized Linear Models

Fourth Edition

CHAPMAN & HALL/CRC
Texts in Statistical Science Series

Series Editors
Joseph K. Blitzstein, *Harvard University, USA*
Julian J. Faraway, *University of Bath, UK*
Martin Tanner, *Northwestern University, USA*
Jim Zidek, *University of British Columbia, Canada*

An Introduction to Generalized Linear Models

Fourth Edition

By

Annette J. Dobson

and

Adrian G. Barnett

CRC Press
Taylor & Francis Group
Boca Raton London New York

CRC Press is an imprint of the
Taylor & Francis Group, an **informa** business

CRC Press
Taylor & Francis Group
6000 Broken Sound Parkway NW, Suite 300
Boca Raton, FL 33487-2742

International Standard Book Number-13: 978-1-138-74168-3 (Hardback)
International Standard Book Number-13: 978-1-138-74151-5 (Paperback)

Library of Congress Cataloging-in-Publication Data

Names: Dobson, Annette J., 1945- author. | Barnett, Adrian G., author.
Title: An introduction to generalized linear models / by Annette J. Dobson, Adrian G. Barnett.
Other titles: Generalized linear models
Description: Fourth edition. | Boca Raton : CRC Press, 2018. | Includes bibliographical references and index.
Identifiers: LCCN 2018002845| ISBN 9781138741683 (hardback : alk. paper) | ISBN 9781138741515 (pbk. : alk. paper) | ISBN 9781315182780 (e-book : alk. paper)
Subjects: LCSH: Linear models (Statistics)
Classification: LCC QA276 .D589 2018 | DDC 519.5--dc23
LC record available at https://lccn.loc.gov/2018002845

Visit the Taylor & Francis Web site at
http://www.taylorandfrancis.com

and the CRC Press Web site at
http://www.crcpress.com

Printed and bound in Great Britain by
TJ International Ltd, Padstow, Cornwall

To Beth.

Contents

Preface

The original purpose of the book was to present a unified theoretical and conceptual framework for statistical modelling in a way that was accessible to undergraduate students and researchers in other fields.

The second edition was expanded to include nominal and ordinal logistic regression, survival analysis and analysis of longitudinal and clustered data. It relied more on numerical methods, visualizing numerical optimization and graphical methods for exploratory data analysis and checking model fit.

The third edition added three chapters on Bayesian analysis for generalized linear models. To help with the practical application of generalized linear models, Stata, R and WinBUGS code were added.

This fourth edition includes new sections on the common problems of model selection and non-linear associations. Non-linear associations have a long history in statistics as the first application of the least squares method was when Gauss correctly predicted the non-linear orbit of an asteroid in 1801.

Statistical methods are essential for many fields of research, but a widespread lack of knowledge of their correct application is creating inaccurate results. Untrustworthy results undermine the scientific process of using data to make inferences and inform decisions. There are established practices for creating reproducible results which are covered in a new Postface to this edition.

The data sets and outline solutions of the exercises are available on the publisher's website: http://www.crcpress.com/9781138741515. We also thank Thomas Haslwanter for providing a set of solutions using Python: https://github.com/thomas-haslwanter/dobson.

We are grateful to colleagues and students at the Universities of Queensland and Newcastle, Australia, and those taking postgraduate courses through the Biostatistics Collaboration of Australia for their helpful suggestions and comments about the material.

Annette J. Dobson and Adrian G. Barnett
Brisbane, Australia

Chapter 1

Introduction

1.1 Background

This book is designed to introduce the reader to generalized linear models, these provide a unifying framework for many commonly used statistical techniques. They also illustrate the ideas of statistical modelling.

The reader is assumed to have some familiarity with classical statistical principles and methods. In particular, understanding the concepts of estimation, sampling distributions and hypothesis testing is necessary. Experience in the use of t-tests, analysis of variance, simple linear regression and chisquared tests of independence for two-dimensional contingency tables is assumed. In addition, some knowledge of matrix algebra and calculus is required.

The reader will find it necessary to have access to statistical computing facilities. Many statistical programs, languages or packages can now perform the analyses discussed in this book. Often, however, they do so with a different program or procedure for each type of analysis so that the unifying structure is not apparent.

Some programs or languages which have procedures consistent with the approach used in this book are **Stata**, **R**, **S-PLUS**, **SAS** and **Genstat**. For Chapters 13 to 14, programs to conduct Markov chain Monte Carlo methods are needed and **WinBUGS** has been used here. This list is not comprehensive as appropriate modules are continually being added to other programs.

In addition, anyone working through this book may find it helpful to be able to use mathematical software that can perform matrix algebra, differentiation and iterative calculations.

1.2 Scope

The statistical methods considered in this book all involve the analysis of relationships between measurements made on groups of subjects or objects.

For example, the measurements might be the heights or weights and the ages of boys and girls, or the yield of plants under various growing conditions. We use the terms **response, outcome** or **dependent variable** for measurements that are free to vary in response to other variables called **explanatory variables** or **predictor variables** or **independent variables**—although this last term can sometimes be misleading. Responses are regarded as random variables. Explanatory variables are usually treated as though they are non-random measurements or observations; for example, they may be fixed by the experimental design.

Responses and explanatory variables are measured on one of the following scales.

1. **Nominal** classifications: e.g., red, green, blue; yes, no, do not know, not applicable. In particular, for **binary, dichotomous** or **binomial** variables there are only two categories: male, female; dead, alive; smooth leaves, serrated leaves. If there are more than two categories the variable is called **polychotomous, polytomous** or **multinomial**.

2. **Ordinal** classifications in which there is some natural order or ranking between the categories: e.g., young, middle aged, old; diastolic blood pressures grouped as ≤ 70, 71–90, 91–110, 111–130, ≥ 131 mmHg.

3. **Continuous** measurements where observations may, at least in theory, fall anywhere on a continuum: e.g., weight, length or time. This scale includes both **interval scale** and **ratio scale** measurements—the latter have a well-defined zero. A particular example of a continuous measurement is the time until a specific event occurs, such as the failure of an electronic component; the length of time from a known starting point is called the **failure time**.

Nominal and ordinal data are sometimes called **categorical** or **discrete variables** and the numbers of observations, **counts** or **frequencies** in each category are usually recorded. For continuous data the individual measurements are recorded. The term **quantitative** is often used for a variable measured on a continuous scale and the term **qualitative** for nominal and sometimes for ordinal measurements. A qualitative, explanatory variable is called a **factor** and its categories are called the **levels** for the factor. A quantitative explanatory variable is sometimes called a **covariate**.

Methods of statistical analysis depend on the measurement scales of the response and explanatory variables.

This book is mainly concerned with those statistical methods which are relevant when there is just *one response variable* although there will usually be several explanatory variables. The responses measured on different subjects are usually assumed to be statistically independent random variables although this requirement is dropped in Chapter 11, which is about correlated data, and in subsequent chapters. Table 1.1 shows the main methods of statistical analysis for various combinations of response and explanatory variables and the chapters in which these are described. The last three chapters are devoted to Bayesian methods which substantially extend these analyses.

The present chapter summarizes some of the statistical theory used throughout the book. Chapters 2 through 5 cover the theoretical framework that is common to the subsequent chapters. Later chapters focus on methods for analyzing particular kinds of data.

Chapter 2 develops the main ideas of classical or frequentist statistical modelling. The modelling process involves four steps:

1. Specifying models in two parts: equations linking the response and explanatory variables, and the probability distribution of the response variable.
2. Estimating fixed but unknown parameters used in the models.
3. Checking how well the models fit the actual data.
4. Making inferences; for example, calculating confidence intervals and testing hypotheses about the parameters.

The next three chapters provide the theoretical background. Chapter 3 is about the **exponential family of distributions**, which includes the Normal, Poisson and Binomial distributions. It also covers **generalized linear models** (as defined by Nelder and Wedderburn (1972)). Linear regression and many other models are special cases of generalized linear models. In Chapter 4 methods of classical estimation and model fitting are described.

Chapter 5 outlines frequentist methods of statistical inference for generalized linear models. Most of these methods are based on how well a model describes the set of data. For example, **hypothesis testing** is carried out by first specifying alternative models (one corresponding to the null hypothesis and the other to a more general hypothesis). Then test statistics are calculated which measure the "goodness of fit" of each model and these are compared. Typically the model corresponding to the null hypothesis is simpler, so if it fits the data about as well as a more complex model it is usually preferred on the grounds of parsimony (i.e., we retain the null hypothesis).

Chapter 6 is about **multiple linear regression** and **analysis of variance** (ANOVA). Regression is the standard method for relating a continuous response variable to several continuous explanatory (or predictor) variables. ANOVA is used for a continuous response variable and categorical or qualitative explanatory variables (factors). **Analysis of covariance** (ANCOVA) is used when at least one of the explanatory variables is continuous. Nowa-

Table 1.1 *Major methods of statistical analysis for response and explanatory variables measured on various scales and chapter references for this book. Extensions of these methods from a Bayesian perspective are illustrated in Chapters 12–14.*

Response (chapter)	Explanatory variables	Methods
Continuous (Chapter 6)	Binary	t-test
	Nominal, >2 categories	Analysis of variance
	Ordinal	Analysis of variance
	Continuous	Multiple regression
	Nominal & some continuous	Analysis of covariance
	Categorical & continuous	Multiple regression
Binary (Chapter 7)	Categorical	Contingency tables Logistic regression
	Continuous	Logistic, probit & other dose-response models
	Categorical & continuous	Logistic regression
Nominal with >2 categories (Chapters 8 & 9)	Nominal	Contingency tables
	Categorical & continuous	Nominal logistic regression
Ordinal (Chapter 8)	Categorical & continuous	Ordinal logistic regression
Counts (Chapter 9)	Categorical	Log-linear models
	Categorical & continuous	Poisson regression
Failure times (Chapter 10)	Categorical & continuous	Survival analysis (parametric)
Correlated responses (Chapter 11)	Categorical & continuous	Generalized estimating equations Multilevel models

days it is common to use the same computational tools for all such situations. The terms **multiple regression** or **general linear model** are used to cover the range of methods for analyzing one continuous response variable and multiple explanatory variables. This chapter also includes a section on **model selection** that is also applicable for other types of generalized linear models

Chapter 7 is about methods for analyzing binary response data. The most common one is **logistic regression** which is used to model associations between the response variable and several explanatory variables which may be categorical or continuous. Methods for relating the response to a single continuous variable, the dose, are also considered; these include **probit analysis** which was originally developed for analyzing dose-response data from bioassays. Logistic regression has been generalized to include responses with more than two nominal categories (**nominal, multinomial, polytomous** or **polychotomous logistic regression**) or ordinal categories (**ordinal logistic regression**). These methods are discussed in Chapter 8.

Chapter 9 concerns **count** data. The counts may be frequencies displayed in a **contingency table** or numbers of events, such as traffic accidents, which need to be analyzed in relation to some "exposure" variable such as the number of motor vehicles registered or the distances travelled by the drivers. Modelling methods are based on assuming that the distribution of counts can be described by the Poisson distribution, at least approximately. These methods include **Poisson regression** and **log-linear models**.

Survival analysis is the usual term for methods of analyzing failure time data. The parametric methods described in Chapter 10 fit into the framework of generalized linear models although the probability distribution assumed for the failure times may not belong to the exponential family.

Generalized linear models have been extended to situations where the responses are correlated rather than independent random variables. This may occur, for instance, if they are **repeated measurements** on the same subject or measurements on a group of related subjects obtained, for example, from **clustered sampling**. The method of **generalized estimating equations** (GEEs) has been developed for analyzing such data using techniques analogous to those for generalized linear models. This method is outlined in Chapter 11 together with a different approach to correlated data, namely **multilevel modelling** in which some parameters are treated as random variables rather than fixed but unknown constants. Multilevel modelling involves both fixed and random effects (mixed models) and relates more closely to the Bayesian approach to statistical analysis.

The main concepts and methods of Bayesian analysis are introduced in Chapter 12. In this chapter the relationships between classical or frequentist

methods and Bayesian methods are outlined. In addition the software Win-
BUGS which is used to fit Bayesian models is introduced.

Bayesian models are usually fitted using computer-intensive methods
based on Markov chains simulated using techniques based on random numbers. These methods are described in Chapter 13. This chapter uses some
examples from earlier chapters to illustrate the mechanics of Markov chain
Monte Carlo (MCMC) calculations and to demonstrate how the results allow
much richer statistical inferences than are possible using classical methods.

Chapter 14 comprises several examples, introduced in earlier chapters,
which are reworked using Bayesian analysis. These examples are used to illustrate both conceptual issues and practical approaches to estimation, model
fitting and model comparisons using WinBUGS.

Finally there is a Postscript that summarizes the principles of good
statistical practice that should always be used in order to address the
"**reproducibility crisis**" that plagues science with daily reports of "break-
throughs" that turn out to be useless or untrue.

Further examples of generalized linear models are discussed in the books
by McCullagh and Nelder (1989), Aitkin et al. (2005) and Myers et al. (2010).
Also there are many books about specific generalized linear models such as
Agresti (2007, 2013), Collett (2003, 2014), Diggle et al. (2002), Goldstein
(2011), Hilbe (2015) and Hosmer et al. (2013).

1.3   Notation

Generally we follow the convention of denoting random variables by uppercase italic letters and observed values by the corresponding lowercase letters.
For example, the observations $y_1, y_2, ..., y_n$ are regarded as realizations of the
random variables $Y_1, Y_2, ..., Y_n$. Greek letters are used to denote parameters
and the corresponding lowercase Roman letters are used to denote estimators
and estimates; occasionally the symbol $\hat{}$ is used for estimators or estimates.
For example, the parameter β is estimated by $\hat{\beta}$ or b. Sometimes these con-
ventions are not strictly adhered to, either to avoid excessive notation in cases
where the meaning should be apparent from the context, or when there is a
strong tradition of alternative notation (e.g., e or ε for random error terms).

Vectors and matrices, whether random or not, are denoted by boldface
lower- and uppercase letters, respectively. Thus, **y** represents a vector of observations

$$\begin{bmatrix} y_1 \\ \vdots \\ y_n \end{bmatrix}$$

or a vector of random variables

$$\begin{bmatrix} Y_1 \\ \vdots \\ Y_n \end{bmatrix},$$

$\boldsymbol{\beta}$ denotes a vector of parameters and $\mathbf{X}$ is a matrix. The superscript ^{T} is used for a matrix transpose or when a column vector is written as a row, e.g., $y = [Y_1, \ldots, Y_n]^T$.

The probability density function of a continuous random variable Y (or the probability mass function if Y is discrete) is referred to simply as a **probability distribution** and denoted by

$$f(y; \boldsymbol{\theta})$$

where $\boldsymbol{\theta}$ represents the parameters of the distribution.

We use dot ($\cdot$) subscripts for summation and bars ($^-$) for means; thus,

$$\bar{y} = \frac{1}{N} \sum_{i=1}^{N} y_i = \frac{1}{N} y_{\cdot}.$$

The expected value and variance of a random variable Y are denoted by $E(Y)$ and $\text{var}(Y)$, respectively. Suppose random variables $Y_1, \ldots, Y_N$ are independent with $E(Y_i) = \mu_i$ and $\text{var}(Y_i) = \sigma_i^2$ for $i = 1, \ldots, n$. Let the random variable W be a **linear combination** of the Y_i's

$$W = a_1 Y_1 + a_2 Y_2 + \ldots + a_n Y_n, \tag{1.1}$$

where the a_i's are constants. Then the expected value of W is

$$E(W) = a_1 \mu_1 + a_2 \mu_2 + \ldots + a_n \mu_n \tag{1.2}$$

and its variance is

$$\text{var}(W) = a_1^2 \sigma_1^2 + a_2^2 \sigma_2^2 + \ldots + a_n^2 \sigma_n^2. \tag{1.3}$$

1.4   Distributions related to the Normal distribution

The sampling distributions of many of the estimators and test statistics used in this book depend on the Normal distribution. They do so either directly because they are derived from Normally distributed random variables or asymptotically, via the Central Limit Theorem for large samples. In this section we give definitions and notation for these distributions and summarize the relationships between them. The exercises at the end of the chapter provide practice in using these results which are employed extensively in subsequent chapters.

1.4.1   Normal distributions

1. If the random variable Y has the Normal distribution with mean μ and variance σ^2, its probability density function is

$$f(y; \mu, \sigma^2) = \frac{1}{\sqrt{2\pi\sigma^2}} \exp\left[-\frac{1}{2}\left(\frac{y-\mu}{\sigma}\right)^2\right].$$

   We denote this by $Y \sim N(\mu, \sigma^2)$.

2. The Normal distribution with $\mu = 0$ and $\sigma^2 = 1$, $Y \sim N(0, 1)$, is called the **standard Normal distribution**.

3. Let $Y_1, \ldots, Y_n$ denote Normally distributed random variables with $Y_i \sim N(\mu_i, \sigma_i^2)$ for $i = 1, \ldots, n$ and let the covariance of Y_i and Y_j be denoted by

$$\text{cov}(Y_i, Y_j) = \rho_{ij}\sigma_i\sigma_j,$$

   where ρ_{ij} is the correlation coefficient for Y_i and Y_j. Then the joint distribution of the Y_i's is the **multivariate Normal distribution** with mean vector $\boldsymbol{\mu} = [\mu_1, \ldots, \mu_n]^T$ and variance-covariance matrix $\mathbf{V}$ with diagonal elements σ_i^2 and non-diagonal elements $\rho_{ij}\sigma_i\sigma_j$ for $i \neq j$. We write this as $\mathbf{y} \sim \text{MVN}(\boldsymbol{\mu}, \mathbf{V})$, where $\mathbf{y} = [Y_1, \ldots, Y_n]^T$.

4. Suppose the random variables $Y_1, \ldots, Y_n$ are independent and Normally distributed with the distributions $Y_i \sim N(\mu_i, \sigma_i^2)$ for $i = 1, \ldots, n$. If

$$W = a_1 Y_1 + a_2 Y_2 + \ldots + a_n Y_n,$$

   where the a_i's are constants, then W is also Normally distributed, so that

$$W = \sum_{i=1}^{n} a_i Y_i \sim N\left(\sum_{i=1}^{n} a_i\mu_i, \sum_{i=1}^{n} a_i^2\sigma_i^2\right)$$

   by Equations (1.2) and (1.3).

1.4.2   Chi-squared distribution

1. The **central chi-squared distribution** with n degrees of freedom is defined as the sum of squares of n independent random variables $Z_1, \ldots, Z_n$ each with the standard Normal distribution. It is denoted by

$$X^2 = \sum_{i=1}^{n} Z_i^2 \sim \chi^2(n).$$

In matrix notation, if $\mathbf{z} = [Z_1, \ldots, Z_n]^T$, then $\mathbf{z}^T \mathbf{z} = \sum_{i=1}^{n} Z_i^2$ so that $X^2 = \mathbf{z}^T \mathbf{z} \sim \chi^2(n)$.

2. If X^2 has the distribution $\chi^2(n)$, then its expected value is $E(X^2) = n$ and its variance is $\mathrm{var}(X^2) = 2n$.

3. If $Y_1, \ldots, Y_n$ are independent, Normally distributed random variables, each with the distribution $Y_i \sim N(\mu_i, \sigma_i^2)$, then

$$X^2 = \sum_{i=1}^{n} \left(\frac{Y_i - \mu_i}{\sigma_i} \right)^2 \sim \chi^2(n) \tag{1.4}$$

because each of the variables $Z_i = (Y_i - \mu_i)/\sigma_i$ has the standard Normal distribution $N(0, 1)$.

4. Let $Z_1, \ldots, Z_n$ be independent random variables each with the distribution $N(0, 1)$ and let $Y_i = Z_i + \mu_i$, where at least one of the μ_i's is non-zero. Then the distribution of

$$\sum Y_i^2 = \sum (Z_i + \mu_i)^2 = \sum Z_i^2 + 2 \sum Z_i \mu_i + \sum \mu_i^2$$

has larger mean $n + \lambda$ and larger variance $2n + 4\lambda$ than $\chi^2(n)$ where $\lambda = \sum \mu_i^2$. This is called the **non-central chi-squared distribution** with n degrees of freedom and **non-centrality parameter** λ. It is denoted by $\chi^2(n, \lambda)$.

5. Suppose that the Y_i's are not necessarily independent and the vector $\mathbf{y} = [Y_1, \ldots, Y_n]^T$ has the multivariate Normal distribution $\mathbf{y} \sim \mathrm{MVN}(\boldsymbol{\mu}, \mathbf{V})$ where the variance–covariance matrix $\mathbf{V}$ is non-singular and its inverse is $\mathbf{V}^{-1}$. Then

$$X^2 = (\mathbf{y} - \boldsymbol{\mu})^T \mathbf{V}^{-1} (\mathbf{y} - \boldsymbol{\mu}) \sim \chi^2(n). \tag{1.5}$$

6. More generally if $\mathbf{y} \sim \mathrm{MVN}(\boldsymbol{\mu}, \mathbf{V})$, then the random variable $\mathbf{y}^T \mathbf{V}^{-1} \mathbf{y}$ has the non-central chi-squared distribution $\chi^2(n, \lambda)$ where $\lambda = \boldsymbol{\mu}^T \mathbf{V}^{-1} \boldsymbol{\mu}$.

7. If $X_1^2, \ldots, X_m^2$ are m independent random variables with the chi-squared distributions $X_i^2 \sim \chi^2(n_i, \lambda_i)$, which may or may not be central, then their sum

also has a chi-squared distribution with $\sum n_i$ degrees of freedom and non-centrality parameter $\sum \lambda_i$, that is,

$$\sum_{i=1}^{m} X_i^2 \sim \chi^2 \left(\sum_{i=1}^{m} n_i, \sum_{i=1}^{m} \lambda_i \right).$$

This is called the **reproductive property** of the chi-squared distribution.

8. Let $\mathbf{y} \sim \text{MVN}(\boldsymbol{\mu}, \mathbf{V})$, where $\mathbf{y}$ has n elements but the Y_i's are not independent so that the number k of linearly independent rows (or columns) of $\mathbf{V}$ (that is, the rank of $\mathbf{V}$) is less than n and so $\mathbf{V}$ is singular and its inverse is not uniquely defined. Let $\mathbf{V}^-$ denote a generalized inverse of $\mathbf{V}$ (that is a matrix with the property that $\mathbf{V}\mathbf{V}^-\mathbf{V} = \mathbf{V}$). Then the random variable $\mathbf{y}^T\mathbf{V}^-\mathbf{y}$ has the non-central chi-squared distribution with k degrees of freedom and non-centrality parameter $\lambda = \boldsymbol{\mu}^T\mathbf{V}^-\boldsymbol{\mu}$.

   For further details about properties of the chi-squared distribution see Forbes et al. (2010).

9. Let $\mathbf{y}_1, \ldots, \mathbf{y}_n$ be n independent random vectors each of length p and $\mathbf{y}_n \sim \text{MVN}(\mathbf{0}, \mathbf{V})$. Then $\mathbf{S} = \sum_{i=i}^{n} \mathbf{y}_i\mathbf{y}_i^T$ is a $p \times p$ random matrix which has the Wishart distribution $\text{W}(\mathbf{V}, n)$. This distribution can be used to make inferences about the covariance matrix $\mathbf{V}$ because $\mathbf{S}$ is proportional to $\mathbf{V}$. In the case $p = 1$ the Y_i's are independent random variables with $Y_i \sim \text{N}(0, \sigma^2)$, so $Z_i = Y_i/\sigma \sim \text{N}(0,1)$. Hence, $\mathbf{S} = \sum_{i=1}^{n} Y_i^2 = \sigma^2 \sum_{i=1}^{n} Z_i^2$ and therefore $\mathbf{S}/\sigma^2 \sim \chi^2(n)$. Thus, the Wishart distribution can be regarded as a generalisation of the chi-squared distribution.

1.4.3   t-distribution

The **t-distribution** with n degrees of freedom is defined as the ratio of two independent random variables. The numerator has the standard Normal distribution and the denominator is the square root of a central chi-squared random variable divided by its degrees of freedom; that is,

$$T = \frac{Z}{(X^2/n)^{1/2}} \tag{1.6}$$

where $Z \sim \text{N}(0,1)$, $X^2 \sim \chi^2(n)$ and Z and X^2 are independent. This is denoted by $T \sim \text{t}(n)$.

1.4.4   F-distribution

1. The **central F-distribution** with n and m degrees of freedom is defined as the ratio of two independent central chi-squared random variables, each

divided by its degrees of freedom,

$$F = \frac{X_1^2}{n} \bigg/ \frac{X_2^2}{m} , \tag{1.7}$$

where $X_1^2 \sim \chi^2(n), X_2^2 \sim \chi^2(m)$ and X_1^2 and X_2^2 are independent. This is denoted by $F \sim \mathrm{F}(n,m)$.

2. The relationship between the t-distribution and the F-distribution can be derived by squaring the terms in Equation (1.6) and using definition (1.7) to obtain

$$T^2 = \frac{Z^2}{1} \bigg/ \frac{X^2}{n} \sim \mathrm{F}(1,n) , \tag{1.8}$$

that is, the square of a random variable with the t-distribution, $t(n)$, has the F-distribution, $F(1,n)$.

3. The **non-central F-distribution** is defined as the ratio of two independent random variables, each divided by its degrees of freedom, where the numerator has a non-central chi-squared distribution and the denominator has a central chi-squared distribution, that is,

$$F = \frac{X_1^2}{n} \bigg/ \frac{X_2^2}{m} ,$$

where $X_1^2 \sim \chi^2(n, \lambda)$ with $\lambda = \boldsymbol{\mu}^T \mathbf{V}^{-1} \boldsymbol{\mu}$, $X_2^2 \sim \chi^2(m)$, and X_1^2 and X_2^2 are independent. The mean of a non-central F-distribution is larger than the mean of central F-distribution with the same degrees of freedom.

1.4.5   Some relationships between distributions

We summarize the above relationships in Figure 1.1. In later chapters we add to this diagram and a more extensive diagram involving most of the distributions used in this book is given in the Appendix. Asymptotic relationships are shown using dotted lines and transformations using solid lines. For more details see Leemis (1986) from which this diagram was developed.

1.5   Quadratic forms

1. A **quadratic form** is a polynomial expression in which each term has degree 2. Thus, $y_1^2 + y_2^2$ and $2y_1^2 + y_2^2 + 3y_1 y_2$ are quadratic forms in y_1 and y_2, but $y_1^2 + y_2^2 + 2y_1$ or $y_1^2 + 3y_2^2 + 2$ are not.

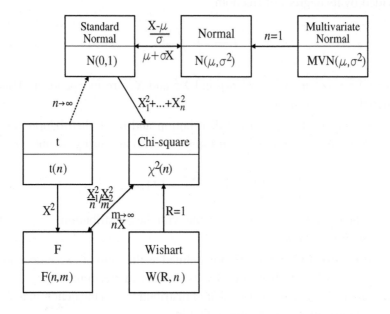

Figure 1.1 *Some relationships between common distributions related to the Normal distribution, adapted from Leemis (1986). Dotted line indicates an asymptotic relationship and solid lines a transformation.*

2. Let $\mathbf{A}$ be a symmetric matrix

$$
\begin{bmatrix}
a_{11} & a_{12} & \cdots & a_{1n} \\
a_{21} & a_{22} & \cdots & a_{2n} \\
\vdots & & \ddots & \vdots \\
a_{n1} & a_{n2} & \cdots & a_{nn}
\end{bmatrix},
$$

where $a_{ij} = a_{ji}$; then the expression $\mathbf{y}^T \mathbf{A} \mathbf{y} = \sum_i \sum_j a_{ij} y_i y_j$ is a quadratic form in the y_i's. The expression $(\mathbf{y} - \boldsymbol{\mu})^T \mathbf{V}^{-1} (\mathbf{y} - \boldsymbol{\mu})$ is a quadratic form in the terms $(y_i - \mu_i)$ but not in the y_i's.

3. The quadratic form $\mathbf{y}^T \mathbf{A} \mathbf{y}$ and the matrix $\mathbf{A}$ are said to be **positive definite** if $\mathbf{y}^T \mathbf{A} \mathbf{y} > 0$ whenever the elements of $\mathbf{y}$ are not all zero. A necessary and sufficient condition for positive definiteness is that all the determinants

$$
|A_1| = a_{11}, |A_2| = \begin{vmatrix} a_{11} & a_{12} \\ a_{21} & a_{22} \end{vmatrix}, |A_3| = \begin{vmatrix} a_{11} & a_{12} & a_{13} \\ a_{21} & a_{22} & a_{23} \\ a_{31} & a_{32} & a_{33} \end{vmatrix}, \ldots, \text{ and}
$$

$|A_n| = \det \mathbf{A}$ are positive. If a matrix is positive definite, then it can be inverted and also it has a square root matrix $\mathbf{A}^*$ such that $\mathbf{A}^* \mathbf{A} = \mathbf{A}$. These properties are useful for the derivation of several theoretical results related to estimation and the probability distributions of estimators.

4. The rank of the matrix $\mathbf{A}$ is also called the degrees of freedom of the quadratic form $Q = \mathbf{y}^T \mathbf{A} \mathbf{y}$.

5. Suppose $Y_1, \ldots, Y_n$ are independent random variables each with the Normal distribution $N(0, \sigma^2)$. Let $Q = \sum_{i=1}^{n} Y_i^2$ and let $Q_1, \ldots, Q_k$ be quadratic forms in the Y_i's such that

$$Q = Q_1 + \ldots + Q_k,$$

where Q_i has m_i degrees of freedom $(i = 1, \ldots, k)$. Then $Q_1, \ldots, Q_k$ are independent random variables and $Q_1/\sigma^2 \sim \chi^2(m_1)$, $Q_2/\sigma^2 \sim \chi^2(m_2), \ldots$, and $Q_k/\sigma^2 \sim \chi^2(m_k)$, if and only if

$$m_1 + m_2 + \ldots + m_k = n.$$

This is Cochran's theorem. A similar result also holds for non-central distributions. For more details see Forbes et al. (2010).

6. A consequence of Cochran's theorem is that the difference of two independent random variables, $X_1^2 \sim \chi^2(m)$ and $X_2^2 \sim \chi^2(k)$, also has a chi-squared distribution

$$X^2 = X_1^2 - X_2^2 \sim \chi^2(m - k)$$

provided that $X^2 \geq 0$ and $m > k$.

1.6   Estimation

1.6.1   Maximum likelihood estimation

Let $\mathbf{y} = [Y_1, \ldots, Y_n]^T$ denote a random vector and let the joint probability density function of the Y_i's be

$$f(\mathbf{y}; \boldsymbol{\theta})$$

which depends on the vector of parameters $\boldsymbol{\theta} = [\theta_1, \ldots, \theta_p]^T$.

The **likelihood function** $L(\boldsymbol{\theta}; \mathbf{y})$ is algebraically the same as the joint probability density function $f(\mathbf{y}; \boldsymbol{\theta})$ but the change in notation reflects a shift of emphasis from the random variables $\mathbf{y}$, with $\boldsymbol{\theta}$ fixed, to the parameters $\boldsymbol{\theta}$, with $\mathbf{y}$ fixed. Since L is defined in terms of the random vector $\mathbf{y}$, it is itself a random variable. Let Ω denote the set of all possible values of the parameter vector $\boldsymbol{\theta}$; Ω is called the **parameter space**. The **maximum likelihood estimator** of θ is the value $\widehat{\boldsymbol{\theta}}$ which maximizes the likelihood function, that is,

$$L(\widehat{\boldsymbol{\theta}}; \mathbf{y}) \geq L(\boldsymbol{\theta}; \mathbf{y}) \qquad \text{for all } \boldsymbol{\theta} \text{ in } \Omega.$$

Equivalently, $\widehat{\boldsymbol{\theta}}$ is the value which maximizes the **log-likelihood function** $l(\boldsymbol{\theta};\mathbf{y}) = \log L(\boldsymbol{\theta};\mathbf{y})$ since the logarithmic function is monotonic. Thus,

$$l(\widehat{\boldsymbol{\theta}};\mathbf{y}) \geq l(\boldsymbol{\theta};\mathbf{y}) \qquad \text{for all } \boldsymbol{\theta} \text{ in } \Omega.$$

Often it is easier to work with the log-likelihood function than with the likelihood function itself.

Usually the estimator $\widehat{\boldsymbol{\theta}}$ is obtained by differentiating the log-likelihood function with respect to each element θ_j of $\boldsymbol{\theta}$ and solving the simultaneous equations

$$\frac{\partial l(\boldsymbol{\theta};\mathbf{y})}{\partial \theta_j} = 0 \qquad \text{for } j = 1, \ldots, p. \qquad (1.9)$$

It is necessary to check that the solutions do correspond to maxima of $l(\boldsymbol{\theta};\mathbf{y})$ by verifying that the matrix of second derivatives

$$\frac{\partial^2 l(\boldsymbol{\theta};\mathbf{y})}{\partial \theta_j \partial \theta_k}$$

evaluated at $\boldsymbol{\theta} = \widehat{\boldsymbol{\theta}}$ is negative definite. For example, if $\boldsymbol{\theta}$ has only one element θ, this means it is necessary to check that

$$\left[\frac{\partial^2 l(\theta, y)}{\partial \theta^2} \right]_{\theta = \widehat{\theta}} < 0.$$

It is also necessary to check if there are any values of $\boldsymbol{\theta}$ at the edges of the parameter space Ω that give local maxima of $l(\boldsymbol{\theta};\mathbf{y})$. When all local maxima have been identified, the value of $\widehat{\boldsymbol{\theta}}$ corresponding to the largest one is the maximum likelihood estimator. (For most of the models considered in this book there is only one maximum and it corresponds to the solution of the equations $\partial l / \partial \theta_j = 0, \; j = 1, \ldots, p.$)

An important property of maximum likelihood estimators is that if $g(\boldsymbol{\theta})$ is any function of the parameters $\boldsymbol{\theta}$, then the maximum likelihood estimator of $g(\boldsymbol{\theta})$ is $g(\widehat{\boldsymbol{\theta}})$. This follows from the definition of $\widehat{\boldsymbol{\theta}}$. It is sometimes called the **invariance property** of maximum likelihood estimators. A consequence is that we can work with a function of the parameters that is convenient for maximum likelihood estimation and then use the invariance property to obtain maximum likelihood estimates for the required parameters.

In principle, it is not necessary to be able to find the derivatives of the likelihood or log-likelihood functions or to solve Equation (1.9) if $\widehat{\boldsymbol{\theta}}$ can be found numerically. In practice, numerical approximations are very important for generalized linear models.

Other properties of maximum likelihood estimators include consistency, sufficiency, asymptotic efficiency and asymptotic normality. These are discussed in books such as Cox and Hinkley (1974) or Forbes et al. (2010).

1.6.2   Example: Poisson distribution

Let $Y_1, \ldots, Y_n$ be independent random variables each with the Poisson distribution

$$f(y_i; \theta) = \frac{\theta^{y_i} e^{-\theta}}{y_i!}, \qquad y_i = 0, 1, 2, \ldots$$

with the same parameter θ. Their joint distribution is

$$f(y_1, \ldots, y_n; \theta) = \prod_{i=1}^{n} f(y_i; \theta) \quad = \quad \frac{\theta^{y_1} e^{-\theta}}{y_1!} \times \frac{\theta^{y_2} e^{-\theta}}{y_2!} \times \cdots \times \frac{\theta^{y_n} e^{-\theta}}{y_n!}$$

$$= \quad \frac{\theta^{\Sigma y_i} e^{-n\theta}}{y_1! y_2! \ldots y_n!}.$$

This is also the likelihood function $L(\theta; y_1, \ldots, y_n)$. It is easier to use the log-likelihood function

$$l(\theta; y_1, \ldots, y_n) = \log L(\theta; y_1, \ldots, y_n) = \left(\sum y_i\right) \log \theta - n\theta - \sum (\log y_i!).$$

To find the maximum likelihood estimate $\widehat{\theta}$, use

$$\frac{dl}{d\theta} = \frac{1}{\theta} \sum y_i - n.$$

Equate this to zero to obtain the solution

$$\widehat{\theta} = \sum y_i / n = \bar{y}.$$

Since $d^2 l / d\theta^2 = -\sum y_i / \theta^2 < 0$, l has its maximum value when $\theta = \widehat{\theta}$, confirming that $\bar{y}$ is the maximum likelihood estimate.

1.6.3   Least squares estimation

Let $Y_1, \ldots, Y_n$ be independent random variables with expected values $\mu_1, \ldots, \mu_n$, respectively. Suppose that the μ_i's are functions of the parameter vector that we want to estimate, $\boldsymbol{\beta} = [\beta_1, \ldots, \beta_p]^T$; $p < n$. Thus

$$E(Y_i) = \mu_i(\boldsymbol{\beta}).$$

The simplest form of the **method of least squares** consists of finding the estimator $\widehat{\boldsymbol{\beta}}$ that minimizes the sum of squares of the differences between Y_i's and their expected values

$$S = \sum [Y_i - \mu_i(\boldsymbol{\beta})]^2.$$

Usually $\widehat{\boldsymbol{\beta}}$ is obtained by differentiating S with respect to each element β_j of $\boldsymbol{\beta}$ and solving the simultaneous equations

$$\frac{\partial S}{\partial \beta_j} = 0, \qquad\qquad j = 1,\ldots,p.$$

Of course it is necessary to check that the solutions correspond to minima (i.e., the matrix of second derivatives is positive definite) and to identify the global minimum from among these solutions and any local minima at the boundary of the parameter space.

Now suppose that the Y_i's have variances σ_i^2 that are not all equal. Then it may be desirable to minimize the weighted sum of squared differences

$$S = \sum w_i [Y_i - \mu_i(\boldsymbol{\beta})]^2,$$

where the weights are $w_i = (\sigma_i^2)^{-1}$. In this way, the observations which are less reliable (i.e., the Y_i's with the larger variances) will have less influence on the estimates.

More generally, let $\mathbf{y} = [Y_1,\ldots,Y_n]^T$ denote a random vector with mean vector $\boldsymbol{\mu} = [\mu_1,\ldots,\mu_n]^T$ and variance–covariance matrix $\mathbf{V}$. Then the **weighted least squares estimator** is obtained by minimizing

$$S = (\mathbf{y} - \boldsymbol{\mu})^T \mathbf{V}^{-1}(\mathbf{y} - \boldsymbol{\mu}).$$

1.6.4   Comments on estimation

1. An important distinction between the methods of maximum likelihood and least squares is that the method of least squares can be used without making assumptions about the distributions of the response variables Y_i beyond specifying their expected values and possibly their variance–covariance structure. In contrast, to obtain maximum likelihood estimators we need to specify the joint probability distribution of the Y_i's.

2. For many situations maximum likelihood and least squares estimators are identical.

3. Often numerical methods rather than calculus may be needed to obtain parameter estimates that maximize the likelihood or log-likelihood function or minimize the sum of squares. The following example illustrates this approach.

1.6.5   Example: Tropical cyclones

Table 1.2 shows the number of tropical cyclones in northeastern Australia for the seasons 1956–7 (season 1) through 1968–9 (season 13), a period of fairly consistent conditions for the definition and tracking of cyclones (Dobson and Stewart 1974).

Table 1.2 *Numbers of tropical cyclones in 13 successive seasons.*

Season	1	2	3	4	5	6	7	8	9	10	11	12	13
No. of cyclones	6	5	4	6	6	3	12	7	4	2	6	7	4

Let Y_i denote the number of cyclones in season i, where $i = 1, \ldots, 13$. Suppose the Y_i's are independent random variables with the Poisson distribution with parameter θ. From Example 1.6.2, $\widehat{\theta} = \bar{y} = 72/13 = 5.538$. An alternative approach would be to find numerically the value of θ that maximizes the log-likelihood function. The component of the log-likelihood function due to y_i is

$$l_i = y_i \log \theta - \theta - \log y_i!.$$

The log-likelihood function is the sum of these terms

$$l = \sum_{i=1}^{13} l_i = \sum_{i=1}^{13} (y_i \log \theta - \theta - \log y_i!).$$

Only the first two terms in the brackets involve θ and so are relevant to the optimization calculation because the term $\sum_1^{13} \log y_i!$ is a constant. To plot the log-likelihood function (without the constant term) against θ, for various values of θ, calculate $(y_i \log \theta - \theta)$ for each y_i and add the results to obtain $l^* = \sum (y_i \log \theta - \theta)$. Figure 1.2 shows l^* plotted against θ.

Clearly the maximum value is between $\theta = 5$ and $\theta = 6$. This can provide a starting point for an iterative procedure for obtaining $\widehat{\theta}$. The results of a simple bisection calculation are shown in Table 1.3. The function l^* is first calculated for approximations $\theta^{(1)} = 5$ and $\theta^{(2)} = 6$. Then subsequent approximations $\theta^{(k)}$ for $k = 3, 4, \ldots$ are the average values of the two previous estimates of θ with the largest values of l^* (for example, $\theta^{(6)} = \frac{1}{2}(\theta^{(5)} + \theta^{(3)})$). After 7 steps, this process gives $\widehat{\theta} \simeq 5.54$ which is correct to 2 decimal places.

1.7   Exercises

1.1 Let Y_1 and Y_2 be independent random variables with $Y_1 \sim N(1,3)$ and $Y_2 \sim N(2,5)$. If $W_1 = Y_1 + 2Y_2$ and $W_2 = 4Y_1 - Y_2$, what is the joint distribution of W_1 and W_2?

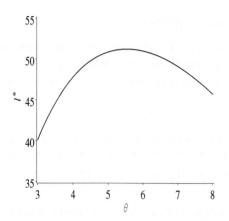

Figure 1.2 *Graph showing the location of the maximum likelihood estimate for the data in Table 1.2 on tropical cyclones.*

Table 1.3 *Successive approximations to the maximum likelihood estimate of the mean number of cyclones per season.*

k	$\theta^{(k)}$	l^*
1	5	50.878
2	6	51.007
3	5.5	51.242
4	5.75	51.192
5	5.625	51.235
6	5.5625	51.243
7	5.5313	51.24354
8	5.5469	51.24352
9	5.5391	51.24360
10	5.5352	51.24359

1.2 Let Y_1 and Y_2 be independent random variables with $Y_1 \sim N(0,1)$ and $Y_2 \sim N(3,4)$.

a. What is the distribution of Y_1^2?

b. If $\mathbf{y} = \begin{bmatrix} Y_1 \\ (Y_2 - 3)/2 \end{bmatrix}$, obtain an expression for $\mathbf{y}^T \mathbf{y}$. What is its distribution?

c. If $\mathbf{y} = \begin{pmatrix} Y_1 \\ Y_2 \end{pmatrix}$ and its distribution is $\mathbf{y} \sim \text{MVN}(\boldsymbol{\mu}, \mathbf{V})$, obtain an expression for $\mathbf{y}^T \mathbf{V}^{-1} \mathbf{y}$. What is its distribution?

1.3 Let the joint distribution of Y_1 and Y_2 be MVN$(\boldsymbol{\mu}, \mathbf{V})$ with

$$\boldsymbol{\mu} = \begin{pmatrix} 2 \\ 3 \end{pmatrix} \quad \text{and} \quad \mathbf{V} = \begin{pmatrix} 4 & 1 \\ 1 & 9 \end{pmatrix}.$$

a. Obtain an expression for $(\mathbf{y} - \boldsymbol{\mu})^T \mathbf{V}^{-1}(\mathbf{y} - \boldsymbol{\mu})$. What is its distribution?

b. Obtain an expression for $\mathbf{y}^T \mathbf{V}^{-1} \mathbf{y}$. What is its distribution?

1.4 Let $Y_1, \ldots, Y_n$ be independent random variables each with the distribution $N(\mu, \sigma^2)$. Let

$$\bar{Y} = \frac{1}{n} \sum_{i=1}^{n} Y_i \quad \text{and} \quad S^2 = \frac{1}{n-1} \sum_{i=1}^{n} (Y_i - \bar{Y})^2.$$

a. What is the distribution of $\bar{Y}$?

b. Show that $S^2 = \frac{1}{n-1} \left[\sum_{i=1}^{n} (Y_i - \mu)^2 - n(\bar{Y} - \mu)^2 \right]$.

c. From (b) it follows that $\sum (Y_i - \mu)^2 / \sigma^2 = (n-1)S^2/\sigma^2 + \left[(\bar{Y} - \mu)^2 n / \sigma^2 \right]$. How does this allow you to deduce that $\bar{Y}$ and S^2 are independent?

d. What is the distribution of $(n-1)S^2/\sigma^2$?

e. What is the distribution of $\frac{\bar{Y} - \mu}{S/\sqrt{n}}$?

1.5 This exercise is a continuation of the example in Section 1.6.2 in which $Y_1, \ldots, Y_n$ are independent Poisson random variables with the parameter θ.

a. Show that $E(Y_i) = \theta$ for $i = 1, \ldots, n$.

b. Suppose $\theta = e^\beta$. Find the maximum likelihood estimator of β.

c. Minimize $S = \sum (Y_i - e^\beta)^2$ to obtain a least squares estimator of β.

1.6 The data in Table 1.4 are the numbers of females and males in the progeny of 16 female light brown apple moths in Muswellbrook, New South Wales, Australia (from Lewis, 1987).

a. Calculate the proportion of females in each of the 16 groups of progeny.

b. Let Y_i denote the number of females and n_i the number of progeny in each group $(i = 1, \ldots, 16)$. Suppose the Y_i's are independent random variables each with the Binomial distribution

$$f(y_i; \theta) = \binom{n_i}{y_i} \theta^{y_i} (1 - \theta)^{n_i - y_i}.$$

Find the maximum likelihood estimator of θ using calculus and evaluate it for these data.

c. Use a numerical method to estimate $\widehat{\theta}$ and compare the answer with the one from (b).

Table 1.4 *Progeny of light brown apple moths.*

Progeny group	Females	Males
1	18	11
2	31	22
3	34	27
4	33	29
5	27	24
6	33	29
7	28	25
8	23	26
9	33	38
10	12	14
11	19	23
12	25	31
13	14	20
14	4	6
15	22	34
16	7	12

Chapter 2

Model Fitting

2.1 Introduction

The model fitting process described in this book involves four steps:

1. Model specification—a model is specified in two parts: an equation linking the response and explanatory variables and the probability distribution of the response variable.
2. Estimation of the parameters of the model.
3. Checking the adequacy of the model—how well it fits or summarizes the data.
4. Inference—for classical or frequentist inference this involves calculating confidence intervals, testing hypotheses about the parameters in the model and interpreting the results.

In this chapter these steps are first illustrated using two small examples. Then some general principles are discussed. Finally there are sections about notation and coding of explanatory variables which are needed in subsequent chapters.

2.2 Examples

2.2.1 Chronic medical conditions

Data from the Australian Longitudinal Study on Women's Health (Lee et al. 2005) show that women who live in country areas tend to have fewer consultations with general practitioners (family physicians) than women who live near a wider range of health services. It is not clear whether this is because they are healthier or because structural factors, such as shortage of doctors, higher costs of visits and longer distances to travel, act as barriers to the use of general practitioner (GP) services. Table 2.1 shows the numbers of chronic medical conditions (for example, high blood pressure or arthritis) reported

Table 2.1 *Number of chronic medical conditions of 26 town women and 23 country women with similar use of general practitioner services.*

Town
0 1 1 0 2 3 0 1 1 1 1 2 0 1 3 0 1 2 1 3 3 4 1 3 2 0
$n = 26$, mean = 1.423, standard deviation = 1.172, variance = 1.374
Country
2 0 3 0 0 1 1 1 1 0 0 2 2 0 1 2 0 0 1 1 1 0 2
$n = 23$, mean = 0.913, standard deviation = 0.900, variance = 0.810

by samples of women living in large country towns (town group) or in more rural areas (country group) in New South Wales, Australia. All the women were aged 70–75 years, had the same socio-economic status and had three or fewer GP visits during 1996. The question of interest is: Do women who have similar levels of use of GP services in the two groups have the same need as indicated by their number of chronic medical conditions?

The Poisson distribution provides a plausible way of modelling these data as they are count data and within each group the sample mean and variance are similar. Let Y_{jk} be a random variable representing the number of conditions for the kth woman in the jth group, where $j = 1$ for the town group and $j = 2$ for the country group and $k = 1, \ldots, K_j$ with $K_1 = 26$ and $K_2 = 23$. Suppose the Y_{jk}'s are all independent and have the Poisson distribution with parameter θ_j representing the expected number of conditions.

The question of interest can be formulated as a test of the null hypothesis $H_0 : \theta_1 = \theta_2 = \theta$ against the alternative hypothesis $H_1 : \theta_1 \neq \theta_2$. The model fitting approach to testing H_0 is to fit two models, one assuming H_0 is true, that is

$$E(Y_{jk}) = \theta; \quad Y_{jk} \sim \mathrm{Po}(\theta), \tag{2.1}$$

and the other assuming it is not, so that

$$E(Y_{jk}) = \theta_j; \quad Y_{jk} \sim \mathrm{Po}(\theta_j), \tag{2.2}$$

where $j = 1$ or 2. Testing H_0 against H_1 involves comparing how well Models (2.1) and (2.2) fit the data. If they are about equally good, then there is little reason for rejecting H_0. However, if Model (2.2) is clearly better, then H_0 would be rejected in favor of H_1.

If H_0 is true, then the log-likelihood function of the Y_{jk}'s is

$$l_0 = l(\theta; \mathbf{y}) = \sum_{j=1}^{J} \sum_{k=1}^{K_j} (y_{jk} \log \theta - \theta - \log y_{jk}!), \tag{2.3}$$

where $J = 2$ in this case. The maximum likelihood estimate, which can be obtained as shown in the example in Section 1.6.2, is

$$\widehat{\theta} = \sum\sum y_{jk}/N,$$

where $N = \sum_j K_j$. For these data the estimate is $\widehat{\theta} = 1.184$ and the maximum value of the log-likelihood function, obtained by substituting this value of $\widehat{\theta}$ and the data values y_{jk} into (2.3), is $\widehat{l}_0 = -68.3868$.

If H_1 is true, then the log-likelihood function is

$$l_1 = l(\theta_1, \theta_2; \mathbf{y}) = \sum_{k=1}^{K_1} (y_{1k}\log\theta_1 - \theta_1 - \log y_{1k}!)$$

$$+ \sum_{k=1}^{K_2} (y_{2k}\log\theta_2 - \theta_2 - \log y_{2k}!). \tag{2.4}$$

(The subscripts on l_0 and l_1 in (2.3) and (2.4) are used to emphasize the connections with the hypotheses H_0 and H_1, respectively). From (2.4) the maximum likelihood estimates are $\widehat{\theta}_j = \sum_k y_{jk}/K_j$ for $j = 1$ or 2. In this case $\widehat{\theta}_1 = 1.423, \widehat{\theta}_2 = 0.913$ and the maximum value of the log-likelihood function, obtained by substituting these values and the data into (2.4), is $\widehat{l}_1 = -67.0230$.

The maximum value of the log-likelihood function l_1 will always be greater than or equal to that of l_0 because one more parameter has been fitted. To decide whether the difference is statistically significant, we need to know the sampling distribution of the log-likelihood function. This is discussed in Chapter 4.

If $Y \sim Po(\theta)$ then $E(Y) = var(Y) = \theta$. The estimate $\widehat{\theta}$ of $E(Y)$ is called the **fitted value** of Y. The difference $Y - \widehat{\theta}$ is called a **residual** (other definitions of residuals are also possible, see Section 2.3.4). Residuals form the basis of many methods for examining the adequacy of a model. A residual is usually standardized by dividing by its standard error. For the Poisson distribution an approximate standardized residual is

$$r = \frac{Y - \widehat{\theta}}{\sqrt{\widehat{\theta}}}.$$

The standardized residuals for Models (2.1) and (2.2) are shown in Table 2.2 and Figure 2.1. Examination of individual residuals is useful for assessing certain features of a model such as the appropriateness of the probability distribution used for the responses or the inclusion of specific explanatory variables. For example, the residuals in Table 2.2 and Figure 2.1 exhibit some skewness, as might be expected for the Poisson distribution.

Table 2.2 *Observed values and standardized residuals for the data on chronic medical conditions (Table 2.1), with estimates obtained from Models (2.1) and (2.2).*

Value of Y	Frequency	Standardized residuals from (2.1); $\widehat{\theta} = 1.184$	Standardized residuals from (2.2); $\widehat{\theta}_1 = 1.423$ and $\widehat{\theta}_2 = 0.913$
		Town	
0	6	−1.088	−1.193
1	10	−0.169	−0.355
2	4	0.750	0.484
3	5	1.669	1.322
4	1	2.589	2.160
		Country	
0	9	−1.088	−0.956
1	8	−0.169	0.091
2	5	0.750	1.138
3	1	1.669	2.184

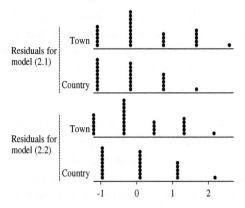

Figure 2.1 *Plots of residuals for Models (2.1) and (2.2) for the data in Table 2.2 on chronic medical conditions.*

The residuals can also be aggregated to produce summary statistics measuring the overall adequacy of the model. For example, for Poisson data denoted by independent random variables Y_i, provided that the expected values θ_i are not too small, the standardized residuals $r_i = (Y_i - \widehat{\theta}_i)/\sqrt{\widehat{\theta}_i}$ approximately have the standard Normal distribution $N(0, 1)$, although they are not usually independent. An intuitive argument is that, approximately,

$r_i \sim N(0,1)$ so $r_i^2 \sim \chi^2(1)$, and hence,

$$\sum r_i^2 = \sum \frac{(Y_i - \widehat{\theta}_i)^2}{\widehat{\theta}_i} \sim \chi^2(m). \tag{2.5}$$

In fact, it can be shown that for large samples, (2.5) is a good approximation with m equal to the number of observations minus the number of parameters estimated in order to calculate to fitted values $\widehat{\theta}_i$ (for example, see Agresti, 2013, page 479). Expression (2.5) is, in fact, the usual chi-squared goodness of fit statistic for count data which is often written as

$$X^2 = \sum \frac{(o_i - e_i)^2}{e_i} \sim \chi^2(m),$$

where o_i denotes the observed frequency and e_i denotes the corresponding expected frequency. In this case $o_i = Y_i$, $e_i = \widehat{\theta}_i$ and $\sum r_i^2 = X^2$.

For the data on chronic medical conditions, for Model (2.1)

$$\sum r_i^2 = 6 \times (-1.088)^2 + 10 \times (-0.169)^2 + \ldots + 1 \times 1.669^2 = 46.759.$$

This value is consistent with $\sum r_i^2$ being an observation from the central chi-squared distribution with $m = 23 + 26 - 1 = 48$ degrees of freedom. (Recall from Section 1.4.2 that if $X^2 \sim \chi^2(m)$, then $E(X^2) = m$, and notice that the calculated value $X^2 = \sum r_i^2 = 46.759$ is near the expected value of 48.)

Similarly, for Model (2.2)

$$\sum r_i^2 = 6 \times (-1.193)^2 + \ldots + 1 \times 2.184^2 = 43.659,$$

which is consistent with the central chi-squared distribution with $m = 49 - 2 = 47$ degrees of freedom. The difference between the values of $\sum r_i^2$ from Models (2.1) and (2.2) is small: $46.759 - 43.659 = 3.10$. This suggests that Model (2.2) with two parameters may not describe the data much better than the simpler Model (2.1). If this is so, then the data provide evidence supporting the null hypothesis H_0: $\theta_1 = \theta_2$. More formal testing of the hypothesis is discussed in Chapter 4.

The next example illustrates steps of the model fitting process with continuous data.

2.2.2   Example: Birthweight and gestational age

The data in Table 2.3 are the birthweights (in grams) and estimated gestational ages (in weeks) of 12 male and female babies born in a certain hospital. The

Table 2.3 *Birthweight (grams) and gestational age (weeks) for boys and girls.*

	Boys		Girls	
	Age	Birthweight	Age	Birthweight
	40	2968	40	3317
	38	2795	36	2729
	40	3163	40	2935
	35	2925	38	2754
	36	2625	42	3210
	37	2847	39	2817
	41	3292	40	3126
	40	3473	37	2539
	37	2628	36	2412
	38	3176	38	2991
	40	3421	39	2875
	38	2975	40	3231
Mean	38.33	3024.00	38.75	2911.33

mean ages are almost the same for both sexes but the mean birthweight for boys is higher than the mean birthweight for girls. The data are shown in the scatter plot in Figure 2.2. There is a linear trend of birthweight increasing with gestational age and the girls tend to weigh less than the boys of the same gestational age. The question of interest is whether the rate of increase of birthweight with gestational age is the same for boys and girls.

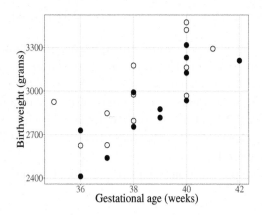

Figure 2.2 *Birthweight plotted against gestational age for boys (open circles) and girls (solid circles); data in Table 2.3.*

Let Y_{jk} be a random variable representing the birthweight of the kth baby in group j where $j = 1$ for boys and $j = 2$ for girls and $k = 1, \ldots, 12$. Suppose that the Y_{jk}'s are all independent and are Normally distributed with means $\mu_{jk} = E(Y_{jk})$, which may differ among babies, and variance σ^2, which is the same for all of them.

A fairly general model relating birthweight to gestational age is

$$E(Y_{jk}) = \mu_{jk} = \alpha_j + \beta_j x_{jk},$$

where x_{jk} is the gestational age of the kth baby in group j. The intercept parameters α_1 and α_2 are likely to differ because, on average, the boys were heavier than the girls. The slope parameters β_1 and β_2 represent the average increases in birthweight for each additional week of gestational age. The question of interest can be formulated in terms of testing the null hypothesis $H_0 : \beta_1 = \beta_2 = \beta$ (that is, the growth rates are equal and so the lines are parallel), against the alternative hypothesis $H_1 : \beta_1 \neq \beta_2$.

We can test H_0 against H_1 by fitting two models

$$E(Y_{jk}) = \mu_{jk} = \alpha_j + \beta x_{jk}; \quad Y_{jk} \sim N(\mu_{jk}, \sigma^2), \tag{2.6}$$

$$E(Y_{jk}) = \mu_{jk} = \alpha_j + \beta_j x_{jk}; \quad Y_{jk} \sim N(\mu_{jk}, \sigma^2). \tag{2.7}$$

The probability density function for Y_{jk} is

$$f(y_{jk}; \mu_{jk}) = \frac{1}{\sqrt{2\pi\sigma^2}} \exp[-\frac{1}{2\sigma^2}(y_{jk} - \mu_{jk})^2].$$

We begin by fitting the more general Model (2.7). The log-likelihood function is

$$l_1(\alpha_1, \alpha_2, \beta_1, \beta_2; \mathbf{y}) = \sum_{j=1}^{J} \sum_{k=1}^{K} [-\frac{1}{2} \log(2\pi\sigma^2) - \frac{1}{2\sigma^2}(y_{jk} - \mu_{jk})^2]$$

$$= -\frac{1}{2} JK \log(2\pi\sigma^2) - \frac{1}{2\sigma^2} \sum_{j=1}^{J} \sum_{k=1}^{K} (y_{jk} - \alpha_j - \beta_j x_{jk})^2,$$

where $J = 2$ and $K = 12$ in this case. When obtaining maximum likelihood estimates of $\alpha_1, \alpha_2, \beta_1$ and β_2, the parameter σ^2 is treated as a known constant, or **nuisance parameter**, and is not estimated.

The maximum likelihood estimates are the solutions of the simultaneous equations

$$\frac{\partial l_1}{\partial \alpha_j} = \frac{1}{\sigma^2} \sum_k (y_{jk} - \alpha_j - \beta_j x_{jk}) = 0,$$

$$\frac{\partial l_1}{\partial \beta_j} = \frac{1}{\sigma^2} \sum_k x_{jk}(y_{jk} - \alpha_j - \beta_j x_{jk}) = 0, \tag{2.8}$$

where $j = 1$ or 2.

An alternative to maximum likelihood estimation is least squares estimation. For Model (2.7), this involves minimizing the expression

$$S_1 = \sum_{j=1}^{J} \sum_{k=1}^{K} (y_{jk} - \mu_{jk})^2 = \sum_{j=1}^{J} \sum_{k=1}^{K} (y_{jk} - \alpha_j - \beta_j x_{jk})^2. \qquad (2.9)$$

The least squares estimates are the solutions of the equations

$$\frac{\partial S_1}{\partial \alpha_j} = -2 \sum_{k=1}^{K} (y_{jk} - \alpha_j - \beta_j x_{jk}) = 0,$$

$$\frac{\partial S_1}{\partial \beta_j} = -2 \sum_{k=1}^{K} x_{jk} (y_{jk} - \alpha_j - \beta_j x_{jk}) = 0. \qquad (2.10)$$

The equations to be solved in (2.8) and (2.10) are the same and so maximizing l_1 is equivalent to minimizing S_1. For the remainder of this example, we will use the least squares approach.

The estimating Equations (2.10) can be simplified to

$$\sum_{k=1}^{K} y_{jk} - K\alpha_j - \beta_j \sum_{k=1}^{K} x_{jk} = 0,$$

$$\sum_{k=1}^{K} x_{jk} y_{jk} - K\alpha_j \sum_{k=1}^{K} x_{jk} - \beta_j \sum_{k=1}^{K} x_{jk}^2 = 0,$$

for $j = 1$ or 2. These are called the **normal equations**. The solution is

$$b_j = \frac{K \sum_k x_{jk} y_{jk} - (\sum_k x_{jk})(\sum_k y_{jk})}{K \sum_k x_{jk}^2 - (\sum_k x_{jk})^2},$$

$$a_j = \bar{y}_j - b_j \bar{x}_j,$$

where a_j is the estimate of α_j and b_j is the estimate of β_j, for $j = 1$ or 2. By considering the second derivatives of (2.9), it can be verified that the solution of Equations (2.10) does correspond to the minimum of S_1. The numerical value for the minimum value for S_1 for a particular data set can be obtained by substituting the estimates for α_j and β_j and the data values for y_{jk} and x_{jk} into (2.9).

To test $H_0: \beta_1 = \beta_2 = \beta$ against the more general alternative hypothesis H_1, the estimation procedure described above for Model (2.7) is repeated but with the expression in (2.6) used for μ_{jk}. In this case, there are three

Table 2.4 *Summary of data on birthweight and gestational age in Table 2.3 (summation is over k=1,...,K where K=12).*

	Boys ($j = 1$)	Girls ($j = 2$)
$\sum x$	460	465
$\sum y$	36288	34936
$\sum x^2$	17672	18055
$\sum y^2$	110623496	102575468
$\sum xy$	1395370	1358497

parameters, α_1, α_2 and β, instead of four to be estimated. The least squares expression to be minimized is

$$S_0 = \sum_{j=1}^{J} \sum_{k=1}^{K} (y_{jk} - \alpha_j - \beta x_{jk})^2. \tag{2.11}$$

From (2.11) the least squares estimates are given by the solution of the simultaneous equations

$$\frac{\partial S_0}{\partial \alpha_j} = -2 \sum_{k=1}^{K} (y_{jk} - \alpha_j - \beta x_{jk}) = 0,$$

$$\frac{\partial S_0}{\partial \beta} = -2 \sum_{j=1}^{J} \sum_{k=1}^{K} x_{jk}(y_{jk} - \alpha_j - \beta x_{jk}) = 0, \tag{2.12}$$

for $j = 1$ and 2. The solution is

$$b = \frac{K \sum_j \sum_k x_{jk} y_{jk} - \sum_j (\sum_k x_{jk} \sum_k y_{jk})}{K \sum_j \sum_k x_{jk}^2 - \sum_j (\sum_k x_{jk})^2},$$

$$a_j = \bar{y}_j - b\bar{x}_j.$$

These estimates and the minimum value for S_0 can be calculated from the data.

For the example on birthweight and gestational age, the data are summarized in Table 2.4 and the least squares estimates and minimum values for S_0 and S_1 are given in Table 2.5. The fitted values $\hat{y}_{jk}$ are shown in Table 2.6. For Model (2.6), $\hat{y}_{jk} = a_j + b x_{jk}$ is calculated from the estimates in the top part of Table 2.5. For Model (2.7), $\hat{y}_{jk} = a_j + b_j x_{jk}$ is calculated using estimates in the bottom part of Table 2.5. The residual for each observation is $y_{jk} - \hat{y}_{jk}$. The standard deviation s of the residuals can be calculated and used to obtain approximate standardized residuals $(y_{jk} - \hat{y}_{jk})/s$. Figures 2.3 and 2.4 show

Table 2.5 *Analysis of data on birthweight and gestational age in Table 2.3.*

Model	Slopes	Intercepts	Minimum sum of squares
(2.6)	$b = 120.894$	$a_1 = -1610.283$	$\widehat{S}_0 = 658770.8$
		$a_2 = -1773.322$	
(2.7)	$b_1 = 111.983$	$a_1 = -1268.672$	$\widehat{S}_1 = 652424.5$
	$b_2 = 130.400$	$a_2 = -2141.667$	

for Models (2.6) and (2.7), respectively: the standardized residuals plotted against the fitted values, the standardized residuals plotted against gestational age, and Normal probability plots. These types of plots are discussed in Section 2.3.4. The figures show that

1. Standardized residuals show no systematic patterns in relation to either the fitted values or the explanatory variable, gestational age.

2. Standardized residuals are approximately Normally distributed (as the points are near the dotted line in the bottom graph).

3. Very little difference exists between the two models.

The apparent lack of difference between the models can be examined by testing the null hypothesis H_0 (corresponding to Model (2.6)) against the alternative hypothesis H_1 (corresponding to Model (2.7)). If H_0 is correct, then the minimum values $\widehat{S}_1$ and $\widehat{S}_0$ should be nearly equal. If the data support this hypothesis, we would feel justified in using the simpler Model (2.6) to describe the data. On the other hand, if the more general hypothesis H_1 is true then $\widehat{S}_0$ should be much larger than $\widehat{S}_1$ and Model (2.7) would be preferable.

To assess the relative magnitude of the values $\widehat{S}_1$ and $\widehat{S}_0$, we need to use the sampling distributions of the corresponding random variables

$$\widehat{S}_1 = \sum_{j=1}^{J}\sum_{k=1}^{K}(Y_{jk} - a_j - b_j x_{jk})^2$$

and

$$\widehat{S}_0 = \sum_{j=1}^{J}\sum_{k=1}^{K}(Y_{jk} - a_j - b x_{jk})^2.$$

It can be shown (see Exercise 2.3) that

$$\widehat{S}_1 = \sum_{j=1}^{J}\sum_{k=1}^{K}[Y_{jk} - (\alpha_j + \beta_j x_{jk})]^2 - K\sum_{j=1}^{J}(\overline{Y}_j - \alpha_j - \beta_j \overline{x}_j)^2$$

$$- \sum_{j=1}^{J}(b_j - \beta_j)^2(\sum_{k=1}^{K}x_{jk}^2 - K\overline{x}_j^2)$$

Table 2.6 *Observed values and fitted values under Model (2.6) and Model (2.7) for data in Table 2.3.*

Sex	Gestational age	Birthweight	Fitted value under (2.6)	Fitted value under (2.7)
Boys	40	2968	3225.5	3210.6
	38	2795	2983.7	2986.7
	40	3163	3225.5	3210.6
	35	2925	2621.0	2650.7
	36	2625	2741.9	2762.7
	37	2847	2862.8	2874.7
	41	3292	3346.4	3322.6
	40	3473	3225.5	3210.6
	37	2628	2862.8	2874.7
	38	3176	2983.7	2986.7
	40	3421	3225.5	3210.6
	38	2975	2983.7	2986.7
Girls	40	3317	3062.5	3074.3
	36	2729	2578.9	2552.7
	40	2935	3062.5	3074.3
	38	2754	2820.7	2813.5
	42	3210	3304.2	3335.1
	39	2817	2941.6	2943.9
	40	3126	3062.5	3074.3
	37	2539	2699.8	2683.1
	36	2412	2578.9	2552.7
	38	2991	2820.7	2813.5
	39	2875	2941.6	2943.9
	40	3231	3062.5	3074.3

and that the random variables Y_{jk}, $\overline{Y}_j$ and b_j are all independent and have the following distributions:

$$Y_{jk} \sim \mathrm{N}(\alpha_j + \beta_j x_{jk}, \sigma^2),$$
$$\overline{Y}_j \sim \mathrm{N}(\alpha_j + \beta_j \overline{x}_j, \sigma^2/K),$$
$$b_j \sim \mathrm{N}(\beta_j, \sigma^2/(\sum_{k=1}^{K} x_{jk}^2 - K\overline{x}_j^2)).$$

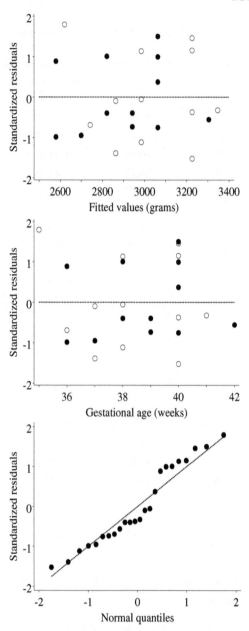

Figure 2.3 *Plots of standardized residuals for Model (2.6) for the data on birth-weight and gestational age (Table 2.3); for the top and middle plots, open circles correspond to data from boys and solid circles correspond to data from girls.*

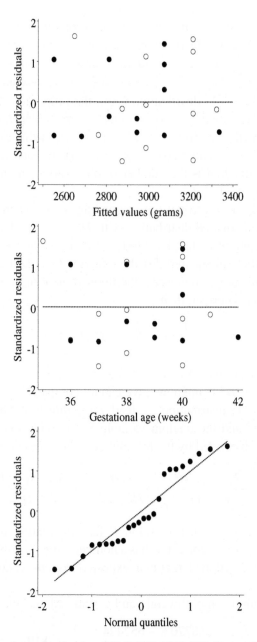

Figure 2.4 *Plots of standardized residuals for Model (2.7) for the data on birth-weight and gestational age (Table 2.3); for the top and middle plots, open circles correspond to data from boys and solid circles correspond to data from girls.*

Therefore, $\widehat{S}_1/\sigma^2$ is a linear combination of sums of squares of random variables with Normal distributions. In general, there are JK random variables $(Y_{jk} - \alpha_j - \beta_j x_{jk})^2/\sigma^2$, J random variables $(\overline{Y}_j - \alpha_j - \beta_j \overline{x}_j)^2 K/\sigma^2$, and J random variables $(b_j - \beta_j)^2(\sum_k x_{jk}^2 - K\overline{x}_j^2)/\sigma^2$. They are all independent and each has the $\chi^2(1)$ distribution. From the properties of the chi-squared distribution in Section 1.5, it follows that $\widehat{S}_1/\sigma^2 \sim \chi^2(JK - 2J)$. Similarly, if H_0 is correct, then $\widehat{S}_0/\sigma^2 \sim \chi^2[JK - (J+1)]$. In this example $J = 2$, so $\widehat{S}_1/\sigma^2 \sim \chi^2(2K - 4)$ and $\widehat{S}_0/\sigma^2 \sim \chi^2(2K - 3)$. In each case the value for the degrees of freedom is the number of observations minus the number of parameters estimated.

If β_1 and β_2 are not equal (corresponding to H_1), then $\widehat{S}_0/\sigma^2$ will have a non-central chi-squared distribution with $JK - (J+1)$ degrees of freedom. On the other hand, provided that Model (2.7) describes the data well, $\widehat{S}_1/\sigma^2$ will have a central chi-squared distribution with $JK - 2J$ degrees of freedom.

The statistic $\widehat{S}_0 - \widehat{S}_1$ represents the improvement in fit of (2.7) compared with (2.6). If H_0 is correct, then

$$\frac{1}{\sigma^2}(\widehat{S}_0 - \widehat{S}_1) \sim \chi^2(J - 1).$$

If H_0 is not correct, then $(\widehat{S}_0 - \widehat{S}_1)/\sigma^2$ has a non-central chi-squared distribution. However, as σ^2 is unknown, we cannot compare $(\widehat{S}_0 - \widehat{S}_1)/\sigma^2$ directly with the $\chi^2(J - 1)$ distribution. Instead we eliminate σ^2 by using the ratio of $(\widehat{S}_0 - \widehat{S}_1)/\sigma^2$ and the random variable $\widehat{S}_1/\sigma^2$ with a central chi-squared distribution, each divided by the relevant degrees of freedom,

$$F = \frac{(\widehat{S}_0 - \widehat{S}_1)/\sigma^2}{(J - 1)} \bigg/ \frac{\widehat{S}_1/\sigma^2}{(JK - 2J)} = \frac{(\widehat{S}_0 - \widehat{S}_1)/(J - 1)}{\widehat{S}_1/(JK - 2J)}.$$

If H_0 is correct, from Section 1.4.4, F has the central distribution $F(J - 1, JK - 2J)$. If H_0 is not correct, F has a non-central F-distribution and the calculated value of F will be larger than expected from the central F-distribution (see Figure 2.5).

For the example on birthweight and gestational age, the value of F is

$$\frac{(658770.8 - 652424.5)/1}{652424.5/20} = 0.19.$$

This value is certainly not statistically significant when compared with the $F(1, 20)$ distribution. Thus, the data do not provide evidence against the hypothesis $H_0 : \beta_0 = \beta_1$, and on the grounds of simplicity, Model (2.6), which specifies the same slopes but different intercepts, is preferable.

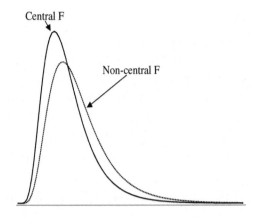

Central F

Non-central F

Figure 2.5 *Central and non-central F-distributions.*

These two examples illustrate the main ideas and methods of statistical modelling which are now discussed more generally.

2.3 Some principles of statistical modelling

Some important principles of statistical modelling are discussed below and also in the Postface in which good statistical practice is discussed.

2.3.1 Exploratory data analysis

Any analysis of data should begin with a consideration of each variable separately, both to check on data quality (for example, are the values plausible?) and to help with model formulation.

1. What is the scale of measurement? Is it continuous or categorical? If it is categorical, how many categories does it have and are they nominal or ordinal?
2. What is the shape of the distribution? This can be examined using frequency tables, dot plots, histograms and other graphical methods.
3. How is it associated with other variables? Cross tabulations for categorical variables, scatter plots for continuous variables, side-by-side box plots for continuous scale measurements grouped according to the factor levels of a categorical variable, and other such summaries can help to identify patterns of association. For example, do the points on a scatter plot suggest linear or non-linear associations? Do the group means increase or decrease consistently with an ordinal variable defining the groups?

2.3.2   Model formulation

The models described in this book involve a single response variable Y (that is, they are *univariate*) and usually several explanatory variables. Knowledge of the context in which the data were obtained, including the substantive questions of interest, theoretical relationships among the variables, the study design, and results of the exploratory data analysis can all be used to help formulate a model. The model has two components:

1. Probability distribution of Y, for example, $Y \sim N(\mu, \sigma^2)$.
2. Equation linking the expected value of Y with a linear combination of the explanatory variables, for example, $E(Y) = \alpha + \beta x$ or $\ln[E(Y)] = \beta_0 + \beta_1 \sin(\alpha x)$.

For generalized linear models the probability distributions all belong to the exponential family of distributions, which includes the Normal, Binomial, Poisson and many other distributions. This family of distributions is discussed in Chapter 3. The equation in the second part of the model has the general form

$$g[E(Y)] = \beta_0 + \beta_1 x_1 + \ldots + \beta_m x_m,$$

where the part $\beta_0 + \beta_1 x_1 + \ldots + \beta_m x_m$ is called the **linear component**. Notation for the linear component is discussed in Section 2.4.

2.3.3   Parameter estimation

The most commonly used estimation methods for classical or frequentist statistical inference are maximum likelihood and least squares. These are described in Section 1.6.1. The alternative approach using Bayesian analysis is introduced in Chapter 12. In this book numerical and graphical methods are used, where appropriate, to complement calculus and algebraic methods of optimization.

2.3.4   Residuals and model checking

Firstly, consider residuals for a model involving the Normal distribution. Suppose that the response variable Y_i is modelled by

$$E(Y_i) = \mu_i; \quad Y_i \sim N(\mu_i, \sigma^2).$$

The fitted values are the estimates $\widehat{\mu}_i$. Residuals can be defined as $y_i - \widehat{\mu}_i$ and the approximate standardized residuals as

$$r_i = (y_i - \widehat{\mu}_i)/\widehat{\sigma},$$

where $\widehat{\sigma}$ is an estimate of the unknown parameter σ. These standardized residuals are slightly correlated because they all depend on the estimates $\widehat{\mu}_i$ and $\widehat{\sigma}$ that were calculated from the observations. Also, they are not exactly Normally distributed because σ has been estimated by $\widehat{\sigma}$. Nevertheless, they are approximately Normally distributed and the adequacy of the approximation can be checked using appropriate graphical methods (see below).

The parameters μ_i are functions of the explanatory variables. If the model is a good description of the association between the response and the explanatory variables, this should be well "captured" or "explained" by the $\widehat{\mu}_i$'s. Therefore, there should be little remaining information in the residuals $y_i - \widehat{\mu}_i$. This too can be checked graphically (see below). Additionally, the sum of squared residuals $\sum(y_i - \widehat{\mu}_i)^2$ provides an overall statistic for assessing the adequacy of the model; in fact, it is the component of the log-likelihood function or least squares expression which is optimized in the estimation process.

Secondly, consider residuals from a Poisson model. Recall the model for chronic medical conditions

$$E(Y_i) = \theta_i; \quad Y_i \sim \text{Po}(\theta_i).$$

In this case approximate standardized residuals are of the form

$$r_i = \frac{y_i - \widehat{\theta}_i}{\sqrt{\widehat{\theta}_i}}.$$

These can be regarded as signed square roots of contributions to the Pearson goodness-of-fit statistic

$$\sum_i \frac{(o_i - e_i)^2}{e_i},$$

where o_i is the observed value y_i and e_i is the fitted value $\widehat{\theta}_i$ "expected" from the model.

For other distributions a variety of definitions of standardized residuals are used. Some of these are transformations of the terms $(y_i - \widehat{\mu}_i)$ designed to improve their Normality or independence (for example, see Kutner et al. 2005, Chapter 9). Others are based on signed square roots of contributions to statistics, such as the log-likelihood function or the sum of squares, which are used as overall measures of the adequacy of the model (for example, see Cox and Snell, 1968, Pregibon, 1981, and Pierce and Schafer, 1986). Many of these residuals are discussed in more detail in McCullagh and Nelder (1989) or Krzanowski (1998).

Residuals are important tools for checking the assumptions made in formulating a model because they should usually be independent and have a distribution which is approximately Normal with a mean of zero and constant variance. They should also be unrelated to the explanatory variables. Therefore, the standardized residuals can be compared with the Normal distribution to assess the adequacy of the distributional assumptions and to identify any unusual values. This can be done by inspecting their frequency distribution and looking for values beyond the likely range; for example, no more than 5% should be less than -1.96 or greater than $+1.96$ and no more than 1% should be beyond ± 2.58.

A more sensitive method for assessing Normality, however, is to use a **Normal probability plot**. This involves plotting the residuals against their expected values, defined according to their rank order, if they were Normally distributed. These values are called the **Normal order statistics** and they depend on the number of observations. Normal probability plots are available in all good statistical software (and analogous probability plots for other distributions are also commonly available). In the plot the points should lie on or near a straight line representing Normality and systematic deviations or outlying observations indicate a departure from this distribution.

The standardized residuals should also be plotted against each of the explanatory variables that are included in the model. If the model adequately describes the effect of the variable, there should be no apparent pattern in the plot. If it is inadequate, the points may display curvature or some other systematic pattern which would suggest that additional or alternative terms may need to be included in the model. The residuals should also be plotted against other potential explanatory variables that are not in the model. If there is any systematic pattern, this suggests that additional variables should be included. Several different residual plots for detecting non-linearity in generalized linear models have been compared by Cai and Tsai (1999).

In addition, the standardized residuals should be plotted against the fitted values $\hat{y}_i$, especially to detect changes in variance. For example, an increase in the spread of the residuals toward the end of the range of fitted values would indicate a departure from the assumption of constant variance (sometimes termed **homoscedasticity**).

Finally, a sequence plot of the residuals should be made using the order in which the values y_i were measured. This might be in time order, spatial order or any other sequential effect that might cause lack of independence among the observations. If the residuals are independent the points should fluctuate randomly without any systematic pattern, such as alternating up and down or steadily increasing or decreasing. If there is evidence of associations among the residuals, this can be checked by calculating serial correlation coefficients among them. If the residuals are correlated, special modelling methods are needed—these are outlined in Chapter 11.

2.3.5 Inference and interpretation

It is sometimes useful to think of scientific data as measurements composed of a message, or **signal**, that is distorted by **noise**. For instance, in the example about birthweight the "signal" is the usual growth rate of babies and the "noise" comes from all the genetic and environmental factors that lead to individual variation. A goal of statistical modelling is to extract as much information as possible about the signal. In practice, this has to be balanced against other criteria such as simplicity. The *Oxford English Dictionary* describes the **law of parsimony** (otherwise known as **Occam's Razor**) as the principle that no more causes should be assumed than will account for the effect. Accordingly a simpler or more parsimonious model that describes the data adequately is preferable to a more complicated one which leaves little of the variability "unexplained." To determine a parsimonious model consistent with the data, we test hypotheses about the parameters. A parsimonious model can be found by balancing good fit and model complexity which we discuss in Section 6.3.3.

Hypothesis testing is performed in the context of model fitting by defining a series of nested models corresponding to different hypotheses. Then the question about whether the data support a particular hypothesis can be formulated in terms of the adequacy of fit of the corresponding model relative to other more complicated models. This logic is illustrated in the examples earlier in this chapter. Chapter 5 provides a more detailed explanation of the concepts and methods used, including the sampling distributions for the statistics used to describe "goodness of fit."

While hypothesis testing is useful for identifying a good model, it is much less useful for interpreting it. Wherever possible, the parameters in a model should have some natural interpretation; for example, the rate of growth of babies, the relative risk of acquiring a disease or the mean difference in profit from two marketing strategies. The estimated magnitude of the parameter and the reliability of the estimate as indicated by its standard error or a confidence interval are far more informative than significance levels or p-values. They make it possible to answer questions such as: Is the effect estimated with sufficient precision to be useful, or is the effect large enough to be of practical, social or biological significance? In many scientific fields, there is increasing emphasis on reporting point estimates and confidence intervals instead of p-values.

2.3.6 Further reading

An excellent discussion of the principles of statistical modelling is in the introductory part of Cox and Snell (1981). The importance of adopting a systematic approach is stressed by Kleinbaum et al. (2007). The various steps of model choice, criticism and validation are outlined by Krzanowski (1998). The use of residuals is described in Kutner et al. (2005), Draper and Smith (1998), Belsley et al. (2004) and Cook and Weisberg (1999).

2.4 Notation and coding for explanatory variables

For the models in this book, the equation linking each response variable Y and a set of explanatory variables $x_1, x_2, \ldots, x_m$ has the form

$$g[E(Y)] = \beta_0 + \beta_1 x_1 + \ldots + \beta_m x_m.$$

For responses $Y_1, \ldots, Y_N$, this can be written in matrix notation as

$$g[E(\mathbf{y})] = \mathbf{X}\boldsymbol{\beta}, \tag{2.13}$$

where

$$\mathbf{y} = \begin{bmatrix} Y_1 \\ \cdot \\ \cdot \\ \cdot \\ Y_N \end{bmatrix} \quad \text{is a vector of responses,}$$

$$g[E(\mathbf{y})] = \begin{bmatrix} g[E(Y_1)] \\ \cdot \\ \cdot \\ \cdot \\ g[E(Y_N)] \end{bmatrix}$$

denotes a vector of functions of the terms $E(Y_i)$ (with the same g for every element),

$$\boldsymbol{\beta} = \begin{bmatrix} \beta_1 \\ \cdot \\ \cdot \\ \cdot \\ \beta_p \end{bmatrix} \quad \text{is a vector of parameters,}$$

and $\mathbf{X}$ is a matrix whose elements are constants representing levels of categorical explanatory variables or measured values of continuous explanatory variables.

For a continuous explanatory variable x (such as gestational age in the example on birthweight), the model contains a term βx where the parameter β represents the change in the response corresponding to a change of one unit in x.

For categorical explanatory variables, there are parameters for the different levels of a factor. The corresponding elements of $\mathbf{X}$ are chosen to exclude or include the appropriate parameters for each observation; they are called **dummy variables**. If they are only zeros and ones, the term **indicator variable** is used.

If there are p parameters in the model and N observations, then $\mathbf{y}$ is an $N \times 1$ random vector, $\boldsymbol{\beta}$ is a $p \times 1$ vector of parameters and $\mathbf{X}$ is an $N \times p$ matrix of known constants. $\mathbf{X}$ is often called the **design matrix** and $\mathbf{X}\boldsymbol{\beta}$ is the **linear component** of the model. Various ways of defining the elements of $\mathbf{X}$ are illustrated in the following examples.

2.4.1   Example: Means for two groups

For the data on chronic medical conditions, the equation in the model

$$E(Y_{jk}) = \theta_j; \quad Y_{jk} \sim \mathrm{Po}(\theta_j), \ j = 1,2$$

can be written in the form of (2.13) with g as the identity function (i.e., $g(\theta_j) = \theta_j$),

$$
\mathbf{y} = \begin{bmatrix} Y_{1,1} \\ Y_{1,2} \\ \vdots \\ Y_{1,26} \\ Y_{2,1} \\ \vdots \\ Y_{2,23} \end{bmatrix}, \quad
\boldsymbol{\beta} = \begin{bmatrix} \theta_1 \\ \theta_2 \end{bmatrix} \quad \text{and} \quad
\mathbf{X} = \begin{bmatrix} 1 & 0 \\ 1 & 0 \\ \vdots & \vdots \\ 1 & 0 \\ 0 & 1 \\ \vdots & \vdots \\ 0 & 1 \end{bmatrix},
$$

The top part of $\mathbf{X}$ picks out the terms θ_1 corresponding to $E(Y_{1k})$ and the bottom part picks out θ_2 for $E(Y_{2k})$. With this model the group means θ_1 and θ_2 can be estimated and compared.

2.4.2 Example: Simple linear regression for two groups

The more general model for the data on birthweight and gestational age is

$$E(Y_{jk}) = \mu_{jk} = \alpha_j + \beta_j x_{jk}; \quad Y_{jk} \sim N(\mu_{jk}, \sigma^2).$$

This can be written in the form of (2.13) if g is the identity function,

$$
\mathbf{y} = \begin{bmatrix} Y_{11} \\ Y_{12} \\ \vdots \\ Y_{1K} \\ Y_{21} \\ \vdots \\ Y_{2K} \end{bmatrix}, \quad
\boldsymbol{\beta} = \begin{bmatrix} \alpha_1 \\ \alpha_2 \\ \beta_1 \\ \beta_2 \end{bmatrix} \quad \text{and} \quad
\mathbf{X} = \begin{bmatrix}
1 & 0 & x_{11} & 0 \\
1 & 0 & x_{12} & 0 \\
\vdots & \vdots & \vdots & \vdots \\
1 & 0 & x_{1K} & 0 \\
0 & 1 & 0 & x_{21} \\
\vdots & \vdots & \vdots & \vdots \\
0 & 1 & 0 & x_{2K}
\end{bmatrix}.
$$

2.4.3 Example: Alternative formulations for comparing the means of two groups

There are several alternative ways of formulating the linear components for comparing means of two groups: $Y_{11}, \ldots, Y_{1K_1}$ and $Y_{21}, \ldots, Y_{2K_2}$.

(a) $E(Y_{1k}) = \beta_1$, and $E(Y_{2k}) = \beta_2$.

This is the version used in Example 2.4.1 above. In this case $\boldsymbol{\beta} = \begin{bmatrix} \beta_1 \\ \beta_2 \end{bmatrix}$ and the rows of $\mathbf{X}$ are as follows

$$\textit{Group } 1 : \begin{bmatrix} 1 & 0 \end{bmatrix}$$
$$\textit{Group } 2 : \begin{bmatrix} 0 & 1 \end{bmatrix}.$$

(b) $E(Y_{1k}) = \mu + \alpha_1$, and $E(Y_{2k}) = \mu + \alpha_2$.

In this version μ represents the overall mean and α_1 and α_2 are the group differences from μ. In this case $\boldsymbol{\beta} = \begin{bmatrix} \mu \\ \alpha_1 \\ \alpha_2 \end{bmatrix}$ and the rows of $\mathbf{X}$ are

$$\textit{Group } 1 : \begin{bmatrix} 1 & 1 & 0 \end{bmatrix}$$
$$\textit{Group } 2 : \begin{bmatrix} 1 & 0 & 1 \end{bmatrix}.$$

This formulation, however, has too many parameters as only two parameters can be estimated from the two sets of observations. Therefore, some modification or constraint is needed.

(c) $E(Y_{1k}) = \mu$ and $E(Y_{2k}) = \mu + \alpha$.

Here Group 1 is treated as the reference group and α represents the additional effect of Group 2. For this version $\boldsymbol{\beta} = \begin{bmatrix} \mu \\ \alpha \end{bmatrix}$ and the rows of $\mathbf{X}$ are

$$Group\ 1 : \begin{bmatrix} 1 & 0 \end{bmatrix}$$
$$Group\ 2 : \begin{bmatrix} 1 & 1 \end{bmatrix}.$$

This is an example of **corner point parameterization** in which group effects are defined as differences from a reference category called the "corner point."

(d) $E(Y_{1k}) = \mu + \alpha$, and $E(Y_{2k}) = \mu - \alpha$.

This version treats the two groups symmetrically; μ is the overall average effect and α represents the group differences. This is an example of a **sum-to-zero constraint** because

$$[E(Y_{1k}) - \mu] + [E(Y_{2k}) - \mu] = \alpha + (-\alpha) = 0.$$

In this case $\boldsymbol{\beta} = \begin{bmatrix} \mu \\ \alpha \end{bmatrix}$ and the rows of $\mathbf{X}$ are

$$Group\ 1 : \begin{bmatrix} 1 & 1 \end{bmatrix}$$
$$Group\ 2 : \begin{bmatrix} 1 & -1 \end{bmatrix}.$$

2.4.4 Example: Ordinal explanatory variables

Let Y_{jk} denote a continuous measurement of quality of life. Data are collected for three groups of patients with mild, moderate or severe disease. The groups can be described by levels of an ordinal variable. This can be specified by defining the model using

$$E(Y_{1k}) = \mu$$
$$E(Y_{2k}) = \mu + \alpha_1$$
$$E(Y_{3k}) = \mu + \alpha_1 + \alpha_2,$$

and hence, $\boldsymbol{\beta} = \begin{bmatrix} \mu \\ \alpha_1 \\ \alpha_2 \end{bmatrix}$ and the rows of $\mathbf{X}$ are

$$Group\ 1 : \begin{bmatrix} 1 & 0 & 0 \end{bmatrix}$$
$$Group\ 2 : \begin{bmatrix} 1 & 1 & 0 \end{bmatrix}$$
$$Group\ 3 : \begin{bmatrix} 1 & 1 & 1 \end{bmatrix}.$$

Thus α_1 represents the effect of Group 2 relative to Group 1 and α_2 represents the effect of Group 3 relative to Group 2.

2.5 Exercises

2.1 Genetically similar seeds are randomly assigned to be raised in either a nutritionally enriched environment (treatment group) or standard conditions (control group) using a completely randomized experimental design. After a predetermined time all plants are harvested, dried and weighed. The results, expressed in grams, for 20 plants in each group are shown in Table 2.7.

Table 2.7 *Dried weight of plants grown under two conditions.*

Treatment group		Control group	
4.81	5.36	4.17	4.66
4.17	3.48	3.05	5.58
4.41	4.69	5.18	3.66
3.59	4.44	4.01	4.50
5.87	4.89	6.11	3.90
3.83	4.71	4.10	4.61
6.03	5.48	5.17	5.62
4.98	4.32	3.57	4.53
4.90	5.15	5.33	6.05
5.75	6.34	5.59	5.14

We want to test whether there is any difference in yield between the two groups. Let Y_{jk} denote the kth observation in the jth group where $j = 1$ for the treatment group, $j = 2$ for the control group and $k = 1, \ldots, 20$ for both groups. Assume that the Y_{jk}'s are independent random variables with $Y_{jk} \sim N(\mu_j, \sigma^2)$. The null hypothesis $H_0 : \mu_1 = \mu_2 = \mu$, that there is no difference, is to be compared with the alternative hypothesis $H_1 : \mu_1 \neq \mu_2$.

a. Conduct an exploratory analysis of the data looking at the distributions for each group (e.g., using dot plots, stem and leaf plots or Normal probability plots) and calculate summary statistics (e.g., means, medians, standard derivations, maxima and minima). What can you infer from these investigations?

b. Perform an unpaired t-test on these data and calculate a 95% confidence interval for the difference between the group means. Interpret these results.

c. The following models can be used to test the null hypothesis H_0 against the alternative hypothesis H_1, where

$$
\begin{aligned}
H_0 : E(Y_{jk}) &= \mu; \quad Y_{jk} \sim N(\mu, \sigma^2), \\
H_1 : E(Y_{jk}) &= \mu_j; \quad Y_{jk} \sim N(\mu_j, \sigma^2),
\end{aligned}
$$

for $j = 1,2$ and $k = 1,\ldots,20$. Find the maximum likelihood and least squares estimates of the parameters μ, μ_1 and μ_2, assuming σ^2 is a known constant.

d. Show that the minimum values of the least squares criteria are

$$
\text{for } H_0, \; \widehat{S_0} = \Sigma\Sigma(Y_{jk} - \overline{Y})^2, \text{ where } \overline{Y} = \sum_{j=1}^{2}\sum_{k=1}^{K} Y_{jk}/40;
$$

$$
\text{for } H_1, \; \widehat{S_1} = \Sigma\Sigma(Y_{jk} - \overline{Y}_j)^2, \text{ where } \overline{Y}_j = \sum_{k=1}^{K} Y_{jk}/20
$$

for $j = 1,2$.

e. Using the results of Exercise 1.4, show that

$$
\frac{1}{\sigma^2}\widehat{S_1} = \frac{1}{\sigma^2}\sum_{j=1}^{2}\sum_{k=1}^{20}(Y_{jk} - \mu_j)^2 - \frac{20}{\sigma^2}\sum_{k=1}^{20}(\overline{Y}_j - \mu_j)^2,
$$

and deduce that if H_1 is true

$$
\frac{1}{\sigma^2}\widehat{S_1} \sim \chi^2(38).
$$

Similarly show that

$$
\frac{1}{\sigma^2}\widehat{S_0} = \frac{1}{\sigma^2}\sum_{j=1}^{2}\sum_{k=1}^{20}(Y_{jk} - \mu)^2 - \frac{40}{\sigma^2}\sum_{j=1}^{2}(\overline{Y} - \mu)^2
$$

and if H_0 is true, then

$$
\frac{1}{\sigma^2}\widehat{S_0} \sim \chi^2(39).
$$

f. Use an argument similar to the one in Example 2.2.2 and the results from (e) to deduce that the statistic

$$F = \frac{\widehat{S}_0 - \widehat{S}_1}{\widehat{S}_1/38}$$

has the central F-distribution $F(1, 38)$ if H_0 is true and a non-central distribution if H_0 is not true.

g. Calculate the F-statistic from (f) and use it to test H_0 against H_1. What do you conclude?

h. Compare the value of F-statistic from (g) with the t-statistic from (b), recalling the relationship between the t-distribution and the F-distribution (see Section 1.4.4) Also compare the conclusions from (b) and (g).

i. Calculate residuals from the model for H_0 and use them to explore the distributional assumptions.

2.2 The weights, in kilograms, of twenty men before and after participation in a "waist loss" program are shown in Table 2.8 (Egger et al. 1999). We want to know if, on average, they retain a weight loss twelve months after the program.

Table 2.8 *Weights of twenty men before and after participation in a "waist loss" program.*

Man	Before	After	Man	Before	After
1	100.8	97.0	11	105.0	105.0
2	102.0	107.5	12	85.0	82.4
3	105.9	97.0	13	107.2	98.2
4	108.0	108.0	14	80.0	83.6
5	92.0	84.0	15	115.1	115.0
6	116.7	111.5	16	103.5	103.0
7	110.2	102.5	17	82.0	80.0
8	135.0	127.5	18	101.5	101.5
9	123.5	118.5	19	103.5	102.6
10	95.0	94.2	20	93.0	93.0

Let Y_{jk} denote the weight of the kth man at the jth time, where $j = 1$ before the program and $j = 2$ twelve months later. Assume the Y_{jk}'s are independent random variables with $Y_{jk} \sim N(\mu_j, \sigma^2)$ for $j = 1, 2$ and $k = 1, \dots, 20$.

a. Use an unpaired t-test to test the hypothesis

$$H_0 : \mu_1 = \mu_2 \quad \text{versus} \quad H_1 : \mu_1 \neq \mu_2.$$

b. Let $D_k = Y_{1k} - Y_{2k}$, for $k = 1, \ldots, 20$. Formulate models for testing H_0 against H_1 using the D_k's. Using analogous methods to Exercise 2.1 above, assuming σ^2 is a known constant, test H_0 against H_1.

c. The analysis in (b) is a paired t-test which uses the natural relationship between weights of the *same* person before and after the program. Are the conclusions the same from (a) and (b)?

d. List the assumptions made for (a) and (b). Which analysis is more appropriate for these data?

2.3 For Model (2.7) for the data on birthweight and gestational age, using methods similar to those for Exercise 1.4, show

$$\widehat{S}_1 = \sum_{j=1}^{J} \sum_{k=1}^{K} (Y_{jk} - a_j - b_j x_{jk})^2$$

$$= \sum_{j=1}^{J} \sum_{k=1}^{K} \left[(Y_{jk} - (\alpha_j + \beta_j x_{jk}) \right]^2 - K \sum_{j=1}^{J} (\overline{Y}_j - \alpha_j - \beta_j \overline{x}_j)^2$$

$$- \sum_{j=1}^{J} (b_j - \beta_j)^2 \left(\sum_{k=1}^{K} x_{jk}^2 - K\overline{x}_j^2 \right)$$

and that the random variables Y_{jk}, $\overline{Y}_j$ and b_j are all independent and have the following distributions

$$Y_{jk} \sim N(\alpha_j + \beta_j x_{jk}, \sigma^2),$$
$$\overline{Y}_j \sim N(\alpha_j + \beta_j \overline{x}_j, \sigma^2/K),$$
$$b_j \sim N(\beta_j, \sigma^2/(\sum_{k=1}^{K} x_{jk}^2 - K\overline{x}_j^2)).$$

2.4 Suppose you have the following data

x:	1.0	1.2	1.4	1.6	1.8	2.0
y:	3.15	4.85	6.50	7.20	8.25	16.50

and you want to fit a model with

$$E(Y) = \ln(\beta_0 + \beta_1 x + \beta_2 x^2).$$

Write this model in the form of (2.13) specifying the vectors **y** and **β** and the matrix **X**.

2.5 The model for two-factor analysis of variance with two levels of one factor, three levels of the other and no replication is

$$E(Y_{jk}) = \mu_{jk} = \mu + \alpha_j + \beta_k; \qquad Y_{jk} \sim N(\mu_{jk}, \sigma^2),$$

where $j = 1, 2$; $k = 1, 2, 3$ and, using the sum-to-zero constraints, $\alpha_1 + \alpha_2 = 0, \beta_1 + \beta_2 + \beta_3 = 0$. Also the Y_{jk}'s are assumed to be independent. Write the equation for $E(Y_{jk})$ in matrix notation. (Hint: Let $\alpha_2 = -\alpha_1$, and $\beta_3 = -\beta_1 - \beta_2$).

Chapter 3

Exponential Family and Generalized Linear Models

3.1 Introduction

Linear models of the form

$$E(Y_i) = \mu_i = \mathbf{x}_i^T \boldsymbol{\beta}; \qquad Y_i \sim N(\mu_i, \sigma^2), \qquad (3.1)$$

where the random variables Y_i are independent are the basis of most analyses of continuous data. Note that the random variables Y_i for different subjects, indexed by the subscript i, may have different expected values μ_i. Sometimes there may only be one observation y_i for each Y_i, but on other occasions there may be several observations y_{ij} $(j = 1, \ldots, n_i)$ for each Y_i. The transposed vector $\mathbf{x}_i^T$ represents the ith row of the design matrix $\mathbf{X}$. The example about the association between birthweight and gestational age is of this form, see Section 2.2.2. So is the exercise on plant growth where Y_i is the dry weight of plants and $\mathbf{X}$ has elements to identify the treatment and control groups (Exercise 2.1). Generalizations of these examples to the association between a continuous response and several explanatory variables (multiple regression) and comparisons of more than two means (analysis of variance) are also of this form.

Advances in statistical theory and computer software allow us to use methods analogous to those developed for linear models in the following more general situations:

1. Response variables have distributions other than the Normal distribution— they may even be categorical rather than continuous.

2. Association between the response and explanatory variables need not be of the simple linear form in (3.1).

One of these advances has been the recognition that many of the "nice" properties of the Normal distribution are shared by a wider class of distributions called the **exponential family of distributions**. These distributions and their properties are discussed in the next section.

A second advance is the extension of the numerical methods to estimate the parameters $\boldsymbol{\beta}$ from the linear model described in (3.1) to the situation where there is some non-linear function relating $E(Y_i) = \mu_i$ to the linear component $\mathbf{x}_i^T \boldsymbol{\beta}$, that is,

$$g(\mu_i) = \mathbf{x}_i^T \boldsymbol{\beta}$$

(see Section 2.4). The function g is called the **link function**. In the initial formulation of generalized linear models by Nelder and Wedderburn (1972) and in most of the examples considered in this book, g is a simple mathematical function. These models have now been further generalized to situations where functions may be estimated numerically; such models are called **generalized additive models** (see Hastie and Tibshirani, 1990). In theory, the estimation is straightforward. In practice, it may require a considerable amount of computation involving numerical optimization of non-linear functions. Procedures to do these calculations are now included in many statistical programs.

This chapter introduces the exponential family of distributions and defines generalized linear models. Methods for parameter estimation and hypothesis testing are developed in Chapters 4 and 5, respectively.

3.2   Exponential family of distributions

Consider a single random variable Y whose probability distribution depends on a single parameter θ. The distribution belongs to the exponential family if it can be written in the form

$$f(y; \theta) = s(y)t(\theta)e^{a(y)b(\theta)}, \tag{3.2}$$

where a, b, s and t are known functions. Notice the symmetry between y and θ. This is emphasized if Equation (3.2) is rewritten as

$$f(y; \theta) = \exp[a(y)b(\theta) + c(\theta) + d(y)], \tag{3.3}$$

where $s(y) = \exp d(y)$ and $t(\theta) = \exp c(\theta)$.

If $a(y) = y$, the distribution is said to be in **canonical** (that is, standard) form and $b(\theta)$ is sometimes called the **natural parameter** of the distribution.

If there are other parameters, in addition to the parameter of interest θ, they are regarded as **nuisance parameters** forming parts of the functions a, b, c and d, and they are treated as though they are known.

Many well-known distributions belong to the exponential family. For example, the Poisson, Normal and Binomial distributions can all be written in the canonical form—see Table 3.1.

Table 3.1 *Poisson, Normal and Binomial distributions as members of the exponential family.*

Distribution	Natural parameter	c	d
Poisson	$\log \theta$	$-\theta$	$-\log y!$
Normal	$\dfrac{\mu}{\sigma^2}$	$-\dfrac{\mu^2}{2\sigma^2} - \dfrac{1}{2}\log\left(2\pi\sigma^2\right)$	$-\dfrac{y^2}{2\sigma^2}$
Binomial	$\log\left(\dfrac{\pi}{1-\pi}\right)$	$n\log\left(1-\pi\right)$	$\log\binom{n}{y}$

3.2.1 Poisson distribution

The probability function for the discrete random variable Y is

$$f(y, \theta) = \frac{\theta^y e^{-\theta}}{y!},$$

where y takes the values $0, 1, 2, \ldots$. This can be rewritten as

$$f(y, \theta) = \exp(y\log\theta - \theta - \log y!),$$

which is in the canonical form because $a(y) = y$. Also the natural parameter is $\log \theta$.

The Poisson distribution, denoted by $Y \sim \text{Po}(\theta)$, is used to model count data. Typically these are the number of occurrences of some event in a defined time period or space, when the probability of an event occurring in a very small time (or space) is low and the events occur independently. Examples include the number of medical conditions reported by a person (Example 2.2.1), the number of tropical cyclones during a season (Example 1.6.5), the number of spelling mistakes on the page of a newspaper, or the number of faulty components in a computer or in a batch of manufactured items. If a random variable has the Poisson distribution, its expected value and variance are equal. Real data that might be plausibly modelled by the Poisson distribution often have a larger variance and are said to be **overdispersed**, and the model may have to be adapted to reflect this feature. Chapter 9 describes various models based on the Poisson distribution.

3.2.2   Normal distribution

The probability density function is

$$f(y; \mu) = \frac{1}{(2\pi\sigma^2)^{1/2}} \exp\left[-\frac{1}{2\sigma^2}(y-\mu)^2\right],$$

where μ is the parameter of interest and σ^2 is regarded as a nuisance parameter. This can be rewritten as

$$f(y; \mu) = \exp\left[-\frac{y^2}{2\sigma^2} + \frac{y\mu}{\sigma^2} - \frac{\mu^2}{2\sigma^2} - \frac{1}{2}\log(2\pi\sigma^2)\right].$$

This is in the canonical form. The natural parameter is $b(\mu) = \mu/\sigma^2$ and the other terms in (3.3) are

$$c(\mu) = -\frac{\mu^2}{2\sigma^2} - \frac{1}{2}\log(2\pi\sigma^2) \text{ and } d(y) = -\frac{y^2}{2\sigma^2}$$

(alternatively, the term $-\frac{1}{2}\log(2\pi\sigma^2)$ could be included in $d(y)$).

The Normal distribution is used to model continuous data that have a symmetric distribution. It is widely used for three main reasons. First, many naturally occurring phenomena are well described by the Normal distribution; for example, height or blood pressure of people. Second, even if data are not Normally distributed (e.g., if their distribution is skewed) the average or total of a random sample of values will be approximately Normally distributed; this result is proved in the Central Limit Theorem. Third, there is a great deal of statistical theory developed for the Normal distribution, including sampling distributions derived from it and approximations to other distributions. For these reasons, if continuous data **y** are not Normally distributed it is often worthwhile trying to identify a transformation, such as $y' = \log y$ or $y' = \sqrt{y}$, which produces data **y'** that are approximately Normal.

3.2.3   Binomial distribution

Consider a series of binary events, called "trials," each with only two possible outcomes: "success" or "failure." Let the random variable Y be the number of "successes" in n independent trials in which the probability of success, π, is the same in all trials. Then Y has the Binomial distribution with probability density function

$$f(y; \pi) = \binom{n}{y} \pi^y (1-\pi)^{n-y},$$

where y takes the values $0, 1, 2, \ldots, n$ and

$$\binom{n}{y} = \frac{n!}{y!(n-y)!}.$$

This is denoted by $Y \sim \text{Bin}(n, \pi)$. Here π is the parameter of interest and n is assumed to be known. The probability function can be rewritten as

$$f(y; \pi) = \exp \left[y \log \pi - y \log(1 - \pi) + n \log(1 - \pi) + \log \binom{n}{y} \right],$$

which is of the form (3.3) with $b(\pi) = \log \pi - \log(1 - \pi) = \log[\pi/(1 - \pi)]$.

The Binomial distribution is usually the model of first choice for observations of a process with binary outcomes. Examples include the number of candidates who pass a test (the possible outcomes for each candidate being to pass or to fail) or the number of patients with some disease who are alive at a specified time since diagnosis (the possible outcomes being survival or death).

Other examples of distributions belonging to the exponential family are given in the exercises at the end of the chapter; not all of them are of the canonical form.

3.3 Properties of distributions in the exponential family

Expressions are needed for the expected value and variance of $a(Y)$. To find these the following results are used that apply for any probability density function provided that the order of integration and differentiation can be interchanged. From the definition of a probability density function, the area under the curve is unity so

$$\int f(y; \theta) \, dy = 1, \tag{3.4}$$

where integration is over all possible values of y. (If the random variable Y is discrete, then integration is replaced by summation.)

If both sides of (3.4) are differentiated with respect to θ, this gives

$$\frac{d}{d\theta} \int f(y; \theta) \, dy = \frac{d}{d\theta} . 1 = 0. \tag{3.5}$$

If the order of integration and differentiation in the first term is reversed, then (3.5) becomes

$$\int \frac{df(y; \theta)}{d\theta} \, dy = 0. \tag{3.6}$$

Similarly if (3.4) is differentiated twice with respect to θ and the order of integration and differentiation is reversed, this gives

$$\int \frac{d^2 f(y;\theta)}{d\theta^2} dy = 0. \tag{3.7}$$

These results can now be used for distributions in the exponential family. From (3.3)

$$f(y;\theta) = \exp\left[a(y)b(\theta) + c(\theta) + d(y)\right],$$

so

$$\frac{df(y;\theta)}{d\theta} = \left[a(y)b'(\theta) + c'(\theta)\right] f(y;\theta).$$

By (3.6)

$$\int \left[a(y)b'(\theta) + c'(\theta)\right] f(y;\theta)dy = 0.$$

This can be simplified to

$$b'(\theta)\mathrm{E}[a(y)] + c'(\theta) = 0 \tag{3.8}$$

because $\int a(y)f(y;\theta)dy = \mathrm{E}[a(y)]$ by the definition of the expected value and $\int c'(\theta)f(y;\theta)dy = c'(\theta)$ by (3.4). Rearranging (3.8) gives

$$\mathrm{E}[a(Y)] = -c'(\theta)/b'(\theta). \tag{3.9}$$

A similar argument can be used to obtain $\mathrm{var}[a(Y)]$:

$$\frac{d^2 f(y;\theta)}{d\theta^2} = \left[a(y)b''(\theta) + c''(\theta)\right] f(y;\theta) + \left[a(y)b'(\theta) + c'(\theta)\right]^2 f(y;\theta) \tag{3.10}$$

The second term on the right-hand side of (3.10) can be rewritten as

$$[b'(\theta)]^2 \{a(y) - E[a(Y)]\}^2 f(y;\theta)$$

using (3.9). Then by (3.7)

$$\int \frac{d^2 f(y;\theta)}{d\theta^2} dy = b''(\theta)\mathrm{E}[a(Y)] + c''(\theta) + [b'(\theta)]^2 \mathrm{var}[a(Y)] = 0 \tag{3.11}$$

because $\int \{a(y) - E[a(Y)]\}^2 f(y;\theta)dy = \mathrm{var}[a(Y)]$ by definition.

Rearranging (3.11) and substituting (3.9) gives

$$\mathrm{var}[a(Y)] = \frac{b''(\theta)c'(\theta) - c''(\theta)b'(\theta)}{[b'(\theta)]^3}. \tag{3.12}$$

Equations (3.9) and (3.12) can readily be verified for the Poisson, Normal and Binomial distributions (see Exercise 3.4) and used to obtain the expected value and variance for other distributions in the exponential family.

Expressions are also needed for the expected value and variance of the derivatives of the log-likelihood function. From (3.3), the log-likelihood function for a distribution in the exponential family is

$$l(\theta;y) = a(y)b(\theta) + c(\theta) + d(y).$$

The derivative of $l(\theta;y)$ with respect to θ is

$$U(\theta;y) = \frac{dl(\theta;y)}{d\theta} = a(y)b'(\theta) + c'(\theta).$$

The function U is called the **score statistic**, and as it depends on y, it can be regarded as a random variable, that is,

$$U = a(Y)b'(\theta) + c'(\theta). \tag{3.13}$$

Its expected value is

$$E(U) = b'(\theta)E[a(Y)] + c'(\theta).$$

From (3.9)

$$E(U) = b'(\theta)\left[-\frac{c'(\theta)}{b'(\theta)}\right] + c'(\theta) = 0. \tag{3.14}$$

The variance of U is called the **information** and will be denoted by $\mathfrak{I}$. Using the formula for the variance of a linear transformation of random variables (see (1.3) and (3.13)),

$$\mathfrak{I} = \text{var}(U) = \left[b'(\theta)^2\right]\text{var}[a(Y)].$$

Substituting (3.12) gives

$$\text{var}(U) = \frac{b''(\theta)c'(\theta)}{b'(\theta)} - c''(\theta). \tag{3.15}$$

The score statistic U is used for inference about parameter values in generalized linear models (see Chapter 5).

Another property of U which will be used later is

$$\text{var}(U) = E(U^2) = -E(U'). \tag{3.16}$$

The first equality follows from the general result

$$\text{var}(X) = \text{E}(X^2) - [\text{E}(X)]^2$$

for any random variable, and the fact that $\text{E}(U) = 0$ from (3.14). To obtain the second equality, U is differentiated with respect to θ; from (3.13)

$$U' = \frac{dU}{d\theta} = a(Y)b''(\theta) + c''(\theta).$$

Therefore, the expected value of U' is

$$\begin{aligned} \text{E}(U') &= b''(\theta)\text{E}[a(Y)] + c''(\theta) \\ &= b''(\theta)\left[-\frac{c'(\theta)}{b'(\theta)}\right] + c''(\theta) \\ &= -\text{var}(U) = -\mathfrak{J} \end{aligned} \tag{3.17}$$

by substituting (3.9) and then using (3.15).

3.4   Generalized linear models

The unity of many statistical methods was demonstrated by Nelder and Wedderburn (1972) using the idea of a generalized linear model. This model is defined in terms of a set of independent random variables $Y_1, \ldots, Y_N$, each with a distribution from the exponential family and the following properties:

1. The distribution of each Y_i has the canonical form and depends on a single parameter θ_i (the θ_i's do not all have to be the same); thus,

$$f(y_i; \theta_i) = \exp[y_i b_i(\theta_i) + c_i(\theta_i) + d_i(y_i)].$$

2. The distributions of all the Y_i's are of the same form (e.g., all Normal or all Binomial) so that the subscripts on b, c and d are not needed.

Thus, the joint probability density function of $Y_1, \ldots, Y_N$ is

$$f(y_1, \ldots, y_N; \theta_1, \ldots, \theta_N) = \prod_{i=1}^{N} \exp[y_i b(\theta_i) + c(\theta_i) + d(y_i)] \tag{3.18}$$

$$= \exp\left[\sum_{i=1}^{N} y_i b(\theta_i) + \sum_{i=1}^{N} c(\theta_i) + \sum_{i=1}^{N} d(y_i)\right]. \tag{3.19}$$

The parameters θ_i are typically not of direct interest (since there may be one for each observation). For model specification we are usually interested in a smaller set of parameters $\beta_1, \ldots, \beta_p$ (where $p < N$). Suppose that $E(Y_i) = \mu_i$, where μ_i is some function of θ_i. For a generalized linear model there is a transformation of μ_i such that

$$g(\mu_i) = \mathbf{x}_i^T \boldsymbol{\beta}.$$

In this equation:

a. g is a monotone, differentiable function called the **link function**; that is, it is flat, or increasing or decreasing with μ_i, but it cannot be increasing for some values of μ_i and decreasing for other values.

b. The vector $\mathbf{x}_i$ is a $p \times 1$ vector of explanatory variables (covariates and dummy variables for levels of factors),

$$\mathbf{x}_i = \begin{bmatrix} x_{i1} \\ \vdots \\ x_{ip} \end{bmatrix} \quad \text{so } \mathbf{x}_i^T = \begin{bmatrix} x_{i1} & \cdots & x_{ip} \end{bmatrix}$$

and

c. $\boldsymbol{\beta}$ is the $p \times 1$ vector of parameters $\boldsymbol{\beta} = \begin{bmatrix} \beta_1 \\ \vdots \\ \beta_p \end{bmatrix}$.

The vector $\mathbf{x}_i^T$ is the ith row of the design matrix $\mathbf{X}$.

Thus, a generalized linear model has three components:

1. Response variables $Y_1, \ldots, Y_N$, which are assumed to share the same distribution from the exponential family;

2. A set of parameters $\boldsymbol{\beta}$ and explanatory variables

$$\mathbf{X} = \begin{bmatrix} \mathbf{x}_1^T \\ \vdots \\ \mathbf{x}_N^T \end{bmatrix} = \begin{bmatrix} x_{11} & \cdots & x_{1p} \\ \vdots & & \vdots \\ x_{N1} & & x_{Np} \end{bmatrix};$$

3. A monotone link function g such that

$$g(\mu_i) = \mathbf{x}_i^T \boldsymbol{\beta},$$

where

$$\mu_i = E(Y_i).$$

This chapter concludes with three examples of generalized linear models.

3.5   Examples

3.5.1   Normal linear model

The best known special case of a generalized linear model is the model

$$E(Y_i) = \mu_i = \mathbf{x}_i^T \boldsymbol{\beta}; \qquad Y_i \sim N(\mu_i, \sigma^2),$$

where $Y_1, \ldots, Y_N$ are independent. Here the link function is the identity function, $g(\mu_i) = \mu_i$. This model is usually written in the form

$$\mathbf{y} = \mathbf{X}\boldsymbol{\beta} + \mathbf{e},$$

where $\mathbf{e} = \begin{bmatrix} e_1 \\ \vdots \\ e_N \end{bmatrix}$ and the e_i's are independent, identically distributed random variables with $e_i \sim N(0, \sigma^2)$ for $i = 1, \ldots, N$.

In this form, the linear component $\boldsymbol{\mu} = \mathbf{X}\boldsymbol{\beta}$ represents the "signal" and $\mathbf{e}$ represents the "noise," random variation or "error." Multiple regression, analysis of variance and analysis of covariance are all of this form. These models are considered in Chapter 6.

3.5.2   Historical linguistics

Consider a language which is the descendant of another language; for example, modern Greek is a descendant of ancient Greek, and the Romance languages are descendants of Latin. A simple model for the change in vocabulary is that if the languages are separated by time t, then the probability that they have cognate words for a particular meaning is $e^{-\theta t}$, where θ is a parameter (see Figure 3.1). It is believed that θ is approximately the same for many commonly used meanings. For a test list of N different commonly used meanings suppose that a linguist judges, for each meaning, whether the corresponding words in two languages are cognate or not cognate. We can develop a generalized linear model to describe this situation.

Define random variables $Y_1, \ldots, Y_N$ as follows:

$$Y_i = \begin{cases} 1 & \text{if the languages have cognate words for meaning } i, \\ 0 & \text{if the words are not cognate.} \end{cases}$$

Then

$$P(Y_i = 1) = e^{-\theta t} = \pi$$

and

$$P(Y_i = 0) = 1 - e^{-\theta t} = 1 - \pi.$$

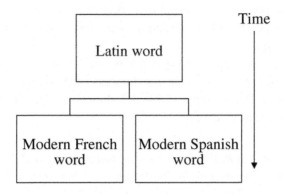

Figure 3.1 *Schematic diagram for the example on historical linguistics.*

This is an example of the Bernoulli distribution $B(\pi)$, which is used to describe the probability of a binary random variable. It is a special case of the Binomial distribution $\text{Bin}(n, \pi)$ with $n = 1$ and $\text{E}(Y_i) = \pi$. In this case we are interested in the parameter θ rather than π, so the link function g is taken as logarithmic

$$g(\pi) = \log \pi = -\theta t,$$

and $g[\text{E}(Y)]$ is linear in the parameter θ. In the notation used above, $\mathbf{x}_i = [-t]$ (the same for all i) and $\boldsymbol{\beta} = [\theta]$.

3.5.3 Mortality rates

For a large population the probability of a randomly chosen individual dying at a particular time is small. If we assume that deaths from a non-infectious disease are independent events, then the number of deaths Y in a population can be modelled by a Poisson distribution

$$f(y; \mu) = \frac{\mu^y e^{-\mu}}{y!},$$

where y can take the values $0, 1, 2, \ldots$ and $\mu = \text{E}(Y)$ is the expected number of deaths in a specified time period, such as a year.

The parameter μ will depend on the population size, the period of observation and various characteristics of the population (e.g., age, sex and medical history). It can be modelled, for example, by

$$\text{E}(Y) = \mu = n\lambda(\mathbf{x}^T \boldsymbol{\beta}),$$

where n is the population size and $\lambda(\mathbf{x}^T\boldsymbol{\beta})$ is the rate per 100,000 people per year (which depends on the population characteristics described by the linear component $\mathbf{x}^T\boldsymbol{\beta}$).

Changes in mortality with age can be modelled by taking independent random variables $Y_1, \ldots, Y_N$ to be the numbers of deaths occurring in successive age groups. For example, Table 3.2 shows age-specific data for deaths from coronary heart disease.

Table 3.2 *Numbers of deaths from coronary heart disease and population sizes by 5-year age groups for men in the Hunter region of New South Wales, Australia in 1991.*

Age group (years)	Number of deaths, y_i	Population size, n_i	Rate per 100,000 men per year, $y_i/n_i \times 100,000$
30–34	1	17,742	5.6
35–39	5	16,554	30.2
40–44	5	16,059	31.1
45–49	12	13,083	91.7
50–54	25	10,784	231.8
55–59	38	9,645	394.0
60–64	54	10,706	504.4
65–69	65	9,933	654.4

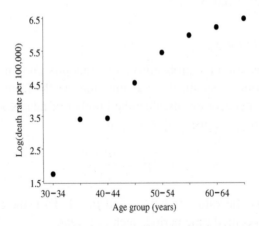

Figure 3.2 *Death rate per 100,000 men (on a logarithmic scale) plotted against age.*

Figure 3.2 shows how the mortality rate $y_i/n_i \times 100,000$ increases with age. Note that a logarithmic scale has been used on the vertical axis. On this scale the scatter plot is approximately linear, suggesting that the association between y_i/n_i and age group i is approximately exponential. Therefore, a possible model is

$$E(Y_i) = \mu_i = n_i e^{\theta i} \quad ; \quad Y_i \sim \text{Po}(\mu_i),$$

where $i = 1$ for the age group 30–34 years, $i = 2$ for 35–39 years, $\ldots, i = 8$ for 65–69 years.

This can be written as a generalized linear model using the logarithmic link function

$$g(\mu_i) = \log \mu_i = \log n_i + \theta i,$$

which has the linear component $\mathbf{x}_i^T \boldsymbol{\beta}$ with $\mathbf{x}_i^T = [\log n_i \ i]$ and $\boldsymbol{\beta} = \begin{bmatrix} 1 \\ \theta \end{bmatrix}$.

3.6 Exercises

3.1 The following associations can be described by generalized linear models. For each one, identify the response variable and the explanatory variables, select a probability distribution for the response (justifying your choice) and write down the linear component.

    a. The effect of age, sex, height, mean daily food intake and mean daily energy expenditure on a person's weight.

    b. The proportions of laboratory mice that became infected after exposure to bacteria when five different exposure levels are used and 20 mice are exposed at each level.

    c. The association between the number of trips per week to the supermarket for a household and the number of people in the household, the household income and the distance to the supermarket.

3.2 If the random variable Y has the **Gamma distribution** with a scale parameter β, which is the parameter of interest, and a known shape parameter α, then its probability density function is

$$f(y;\beta) = \frac{\beta^\alpha}{\Gamma(\alpha)} y^{\alpha-1} e^{-y\beta}.$$

Show that this distribution belongs to the exponential family and find the natural parameter. Also using results in this chapter, find $E(Y)$ and $\text{var}(Y)$.

3.3 Show that the following probability density functions belong to the exponential family:

    a. Pareto distribution $f(y;\theta) = \theta y^{-\theta-1}$.

    b. Exponential distribution $f(y;\theta) = \theta e^{-y\theta}$.

c. Negative Binomial distribution

$$f(y; \theta) = \binom{y+r-1}{r-1} \theta^r (1-\theta)^y,$$

where r is known.

3.4 Use results (3.9) and (3.12) to verify the following results:

a. For $Y \sim \text{Po}(\theta)$, $E(Y) = \text{var}(Y) = \theta$.
b. For $Y \sim \text{N}(\mu, \sigma^2)$, $E(Y) = \mu$ and $\text{var}(Y) = \sigma^2$.
c. For $Y \sim \text{Bin}(n, \pi)$, $E(Y) = n\pi$ and $\text{var}(Y) = n\pi(1-\pi)$.

3.5 a. For a Negative Binomial distribution $Y \sim \text{NBin}(r, \theta)$, find $E(Y)$ and $\text{var}(Y)$.

b. Notice that for the Poisson distribution $E(Y) = \text{var}(Y)$, for the Binomial distribution $E(Y) > \text{var}(Y)$ and for the Negative Binomial distribution $E(Y) < \text{var}(Y)$. How might these results affect your choice of a model?

3.6 Do you consider the model suggested in Example 3.5.3 to be adequate for the data shown in Figure 3.2? Justify your answer. Use simple linear regression (with suitable transformations of the variables) to obtain a model for the change of death rates with age. How well does the model fit the data? (Hint: Compare observed and expected numbers of deaths in each group.)

3.7 Consider N independent binary random variables $Y_1, \ldots, Y_N$ with

$$P(Y_i = 1) = \pi_i \text{ and } P(Y_i = 0) = 1 - \pi_i.$$

The probability function of Y_i, the Bernoulli distribution $B(\pi)$, can be written as

$$\pi_i^{y_i} (1-\pi_i)^{1-y_i},$$

where $y_i = 0$ or 1.

a. Show that this probability function belongs to the exponential family of distributions.

b. Show that the natural parameter is

$$\log \left(\frac{\pi_i}{1-\pi_i} \right).$$

This function, the logarithm of the **odds** $\pi_i/(1-\pi_i)$, is called the **logit** function.

c. Show that $E(Y_i) = \pi_i$.

d. If the link function is

$$g(\pi) = \log\left(\frac{\pi}{1-\pi}\right) = \mathbf{x}^T\boldsymbol{\beta},$$

show that this is equivalent to modelling the probability π as

$$\pi = \frac{e^{\mathbf{x}^T\boldsymbol{\beta}}}{1+e^{\mathbf{x}^T\boldsymbol{\beta}}}.$$

e. In the particular case where $\mathbf{x}^T\boldsymbol{\beta} = \beta_1 + \beta_2 x$, this gives

$$\pi = \frac{e^{\beta_1+\beta_2 x}}{1+e^{\beta_1+\beta_2 x}},$$

which is the **logistic function**.

Sketch the graph of π against x in this case, taking β_1 and β_2 as constants. How would you interpret this graph if x is the dose of an insecticide and π is the probability of an insect dying?

3.8 Is the **extreme value (Gumbel) distribution**, with probability density function

$$f(y;\theta) = \frac{1}{\phi}\exp\left\{\frac{(y-\theta)}{\phi} - \exp\left[\frac{(y-\theta)}{\phi}\right]\right\}$$

(where $\phi > 0$ is regarded as a nuisance parameter) a member of the exponential family?

3.9 Suppose $Y_1,\ldots,Y_N$ are independent random variables each with the Pareto distribution and

$$E(Y_i) = (\beta_0 + \beta_1 x_i)^2.$$

Is this a generalized linear model? Give reasons for your answer.

3.10 Let $Y_1,\ldots,Y_N$ be independent random variables with

$$E(Y_i) = \mu_i = \beta_0 + \log(\beta_1 + \beta_2 x_i); \quad Y_i \sim N(\mu, \sigma^2)$$

for all $i = 1,\ldots,N$. Is this a generalized linear model? Give reasons for your answer.

3.11 For the Pareto distribution, find the score statistics U and the information $\mathfrak{I} = \text{var}(U)$. Verify that $E(U) = 0$.

3.12 See some more relationships between distributions in Figure 3.3.

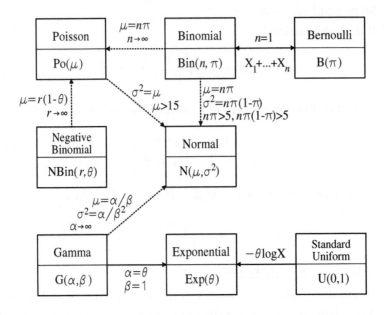

Figure 3.3 *Some relationships among distributions in the exponential family. Dotted lines indicate an asymptotic relationship and solid lines a transformation.*

a. Show that the Exponential distribution $\text{Exp}(\theta)$ is a special case of the Gamma distribution $G(\alpha, \beta)$.

b. If X has the Uniform distribution $U[0, 1]$, that is, $f(x) = 1$ for $0 < x < 1$, show that $Y = -\theta \log X$ has the distribution $\text{Exp}(\theta)$.

c. Use the moment generating functions (or other methods) to show

   i. $\text{Bin}(n, \pi) \to \text{Po}(\lambda)$ as $n \to \infty$.

   ii. $\text{NBin}(r, \theta) \to \text{Po}(r(1 - \theta))$ as $r \to \infty$.

d. Use the Central Limit Theorem to show

   i. $\text{Po}(\lambda) \to N(\mu, \mu)$ for large μ.

   ii. $\text{Bin}(n, \pi) \to N(n\pi, n\pi(1 - \pi))$ for large n, provided neither $n\pi$ nor $n\pi(1 - \pi)$ is too small.

   iii. $G(\alpha, \beta) \to N(\alpha/\beta, \alpha/\beta^2)$ for large α.

Chapter 4

Estimation

4.1 Introduction

This chapter is about obtaining point and interval estimates of parameters
for generalized linear models using methods based on maximum likelihood.
Although explicit mathematical expressions can be found for estimators in
some special cases, numerical methods are usually needed. Typically these
methods are iterative and are based on the Newton–Raphson algorithm. To
illustrate this principle, the chapter begins with a numerical example. Then
the theory of estimation for generalized linear models is developed. Finally
there is another numerical example to demonstrate the methods in detail.

4.2 Example: Failure times for pressure vessels

The data in Table 4.1 are the lifetimes (times to failure in hours) of Kevlar
epoxy strand pressure vessels at 70% stress level. They are given in Table 29.1
of the book of data sets by Andrews and Herzberg (1985).

Figure 4.1 shows the shape of their distribution.

Table 4.1 *Lifetimes of N = 49 pressure vessels.*

1051	4921	7886	10861	13520
1337	5445	8108	11026	13670
1389	5620	8546	11214	14110
1921	5817	8666	11362	14496
1942	5905	8831	11604	15395
2322	5956	9106	11608	16179
3629	6068	9711	11745	17092
4006	6121	9806	11762	17568
4012	6473	10205	11895	17568
4063	7501	10396	12044	

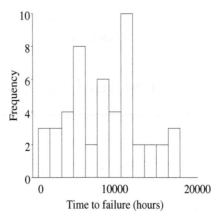

Figure 4.1 *Distribution of lifetimes of pressure vessels.*

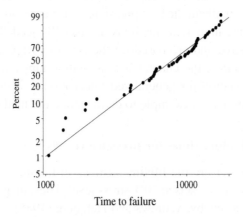

Figure 4.2 *Probability plot of the data on lifetimes of pressure vessels compared with the Weibull distribution with shape parameter = 2.*

A commonly used model for times to failure (or survival times) is the **Weibull distribution** which has the probability density function

$$f(y; \lambda, \theta) = \frac{\lambda y^{\lambda-1}}{\theta^\lambda} \exp\left[-\left(\frac{y}{\theta}\right)^\lambda\right], \tag{4.1}$$

where $y > 0$ is the time to failure, λ is a parameter that determines the shape of the distribution and θ is a parameter that determines the scale. Figure 4.2 is a probability plot of the data in Table 4.1 compared with the Weibull distribution with $\lambda = 2$. Although there are discrepancies between the distribution and the data for some of the shorter times, for most of the observations the distribution appears to provide a good model for the data. Therefore, we will use a Weibull distribution with $\lambda = 2$ and estimate θ.

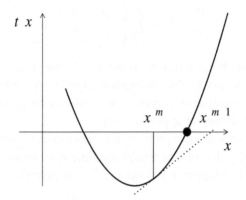

Figure 4.3 *One step of the Newton–Raphson method for finding a solution (•) of the equation t(x)=0.*

The distribution in (4.1) can be written as

$$f(y; \theta) = \exp\left[\log\lambda + (\lambda - 1)\log y - \lambda\log\theta - (y/\theta)^\lambda\right].$$

This belongs to the exponential family (3.2) with

$$a(y) = y^\lambda, b(\theta) = -\theta^{-\lambda}, c(\theta) = \log\lambda - \lambda\log\theta, \text{ and } d(y) = (\lambda - 1)\log y,$$
$$(4.2)$$

where λ is a nuisance parameter. This is not in the canonical form (unless $\lambda = 1$, corresponding to the exponential distribution) and so it cannot be used directly in the specification of a generalized linear model. However, it is suitable for illustrating the estimation of parameters for distributions in the exponential family.

Let $Y_1, \ldots, Y_N$ denote the data, with $N = 49$. If the data are from a random sample of pressure vessels, we assume the Y_i's are independent random variables. If they all have the Weibull distribution with the same parameters, their joint probability distribution is

$$f(y_1, \ldots, y_N; \theta, \lambda) = \prod_{i=1}^{N} \frac{\lambda y_i^{\lambda-1}}{\theta^\lambda} \exp\left[-\left(\frac{y_i}{\theta}\right)^\lambda\right].$$

The log-likelihood function is

$$l(\theta; y_1, \ldots, y_N, \lambda) = \sum_{i=1}^{N}\left[[(\lambda - 1)\log y_i + \log\lambda - \lambda\log\theta] - \left(\frac{y_i}{\theta}\right)^\lambda\right]. \quad (4.3)$$

To maximize this function requires the derivative with respect to θ. This is the score function

$$\frac{dl}{d\theta} = U = \sum_{i=1}^{N} \left[\frac{-\lambda}{\theta} + \frac{\lambda y_i^{\lambda}}{\theta^{\lambda+1}} \right]. \qquad (4.4)$$

The maximum likelihood estimator $\widehat{\theta}$ is the solution of the equation $U(\theta) = 0$. In this case it is easy to find an explicit expression for $\widehat{\theta}$ if λ is a known constant, but for illustrative purposes, we will obtain a numerical solution using the Newton–Raphson approximation.

Figure 4.3 shows the principle of the Newton–Raphson algorithm. The aims is to find the value of x at which the function t crosses the x-axis, i.e., where $t(x) = 0$. The slope of t at a value $x^{(m-1)}$ is given by

$$\left[\frac{dt}{dx} \right]_{x=x^{(m-1)}} = t'(x^{(m-1)}) = \frac{t(x^{(m)}) - t(x^{(m-1)})}{x^{(m)} - x^{(m-1)}}, \qquad (4.5)$$

where the distance $x^{(m)} - x^{(m-1)}$ is small. If $x^{(m)}$ is the required solution so that $t(x^m) = 0$, then (4.5) can be re-arranged to give

$$x^{(m)} = x^{(m-1)} - \frac{t(x^{(m-1)})}{t'(x^{(m-1)})}. \qquad (4.6)$$

This is the Newton–Raphson formula for solving $t(x) = 0$. Starting with an initial guess $x^{(1)}$ successive approximations are obtained using (4.6) until the iterative process converges.

For maximum likelihood estimation using the score function, the estimating equation equivalent to (4.6) is

$$\theta^{(m)} = \theta^{(m-1)} - \frac{U^{(m-1)}}{U'^{(m-1)}}. \qquad (4.7)$$

From (4.4), for the Weibull distribution with $\lambda = 2$,

$$U = -\frac{2 \times N}{\theta} + \frac{2 \times \sum y_i^2}{\theta^3}, \qquad (4.8)$$

which is evaluated at successive estimates $\theta^{(m)}$. The derivative of U, obtained by differentiating (4.4), is

$$\frac{dU}{d\theta} = U' = \sum_{i=1}^{N} \left[\frac{\lambda}{\theta^2} - \frac{\lambda(\lambda+1)y_i^{\lambda}}{\theta^{\lambda+2}} \right]$$

$$= \frac{2 \times N}{\theta^2} - \frac{2 \times 3 \times \sum y_i^2}{\theta^4}. \qquad (4.9)$$

Table 4.2 *Details of Newton–Raphson iterations to obtain a maximum likelihood estimate for the scale parameter for the Weibull distribution to model the data in Table 4.1.*

Iteration	1	2	3	4
θ	8805.7	9633.9	9876.4	9892.1
$U \times 10^6$	2915.10	552.80	31.78	0.21
$U' \times 10^6$	-3.52	-2.28	-2.02	-2.00
$E(U') \times 10^6$	-2.53	-2.11	-2.01	-2.00
U/U'	-827.98	-242.46	-15.73	-0.105
$U/E(U')$	-1152.21	-261.99	-15.81	-0.105

For maximum likelihood estimation, it is common to approximate U' by its expected value $E(U')$. For distributions in the exponential family, this is readily obtained using expression (3.17). The information $\mathfrak{I}$ is

$$\mathfrak{I} = E(-U') = E\left[-\sum_{i=1}^{N} U'_i\right] = \sum_{i=1}^{N} \left[E(-U'_i)\right]$$

$$= \sum_{i=1}^{N} \left[\frac{b''(\theta)c'(\theta)}{b'(\theta)} - c''(\theta)\right]$$

$$= \frac{\lambda^2 N}{\theta^2}, \tag{4.10}$$

where U_i is the score for Y_i and expressions for b and c are given in (4.2). Thus an alternative estimating equation is

$$\theta^{(m)} = \theta^{(m-1)} + \frac{U^{(m-1)}}{\mathfrak{I}^{(m-1)}}. \tag{4.11}$$

This is called the **method of scoring**.

Table 4.2 shows the results of using equation (4.7) iteratively taking the mean of the data in Table 4.1, $\bar{y} = 8805.7$, as the initial value $\theta^{(1)}$; this and subsequent approximations are shown in the top row of Table 4.2. Numbers in the second row were obtained by evaluating (4.8) at $\theta^{(m)}$ and the data values; they approach zero rapidly. The third and fourth rows, U' and $E(U') = -\mathfrak{I}$, have similar values illustrating that either could be used; this is further shown by the similarity of the numbers in the fifth and sixth rows. The final estimate is $\theta^{(5)} = 9892.1 - (-0.105) = 9892.2$—this is the maximum likelihood estimate $\hat{\theta}$ for these data. At this value the log-likelihood function, calculated from (4.3), is $l = -480.850$.

Figure 4.4 shows the log-likelihood function for these data and the Weibull distribution with $\lambda = 2$. The maximum value is at $\widehat{\theta} = 9892.2$. The curvature of the function in the vicinity of the maximum determines the reliability of $\widehat{\theta}$. The curvature of l is defined by the rate of change of U, that is, by U'. If U', or $E(U')$, is small then l is flat so that U is approximately zero for a wide interval of θ values. In this case $\widehat{\theta}$ is not well determined and its standard error is large. In fact, it is shown in Chapter 5 that the variance of $\widehat{\theta}$ is inversely related to $\mathfrak{I} = E(-U')$ and the standard error of $\widehat{\theta}$ is approximately

$$s.e.(\widehat{\theta}) = \sqrt{1/\mathfrak{I}}. \tag{4.12}$$

For this example, at $\widehat{\theta} = 9892.2, \mathfrak{I} = -E(U') = 2.00 \times 10^{-6}$, so $s.e.(\widehat{\theta}) = 1/\sqrt{0.000002} = 707$. If the sampling distribution of $\widehat{\theta}$ is approximately Normal, a 95% confidence interval for θ is given approximately by

$$9892 \pm 1.96 \times 707, \text{ or } (8506, 11278).$$

The methods illustrated in this example are now developed for generalized linear models.

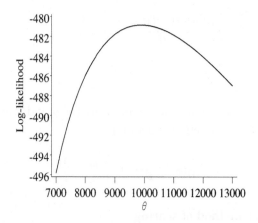

Figure 4.4 *Log-likelihood function for the pressure vessel data in Table 4.1.*

4.3   Maximum likelihood estimation

Consider independent random variables $Y_1, \ldots, Y_N$ satisfying the properties of a generalized linear model. We wish to estimate parameters $\boldsymbol{\beta}$ which are related to the Y_i's through $E(Y_i) = \mu_i$ and $g(\mu_i) = \mathbf{x}_i^T \boldsymbol{\beta}$.

For each Y_i, the log-likelihood function is

$$l_i = y_i b(\theta_i) + c(\theta_i) + d(y_i), \tag{4.13}$$

where the functions b, c and d are defined in (3.3). Also,

$$E(Y_i) = \mu_i = -c'(\theta_i)/b'(\theta_i) \tag{4.14}$$

$$\text{var}(Y_i) = \left[b''(\theta_i)c'(\theta_i) - c''(\theta_i)b'(\theta_i) \right] / \left[b'(\theta_i) \right]^3 \tag{4.15}$$

$$\text{and } g(\mu_i) = \mathbf{x}_i^T \boldsymbol{\beta} = \eta_i, \tag{4.16}$$

where $\mathbf{x}_i$ is a vector with elements $x_{ij}, j = 1, \ldots, p$.

The log-likelihood function for all the Y_i's is

$$l = \sum_{i=1}^{N} l_i = \sum y_i b(\theta_i) + \sum c(\theta_i) + \sum d(y_i).$$

To obtain the maximum likelihood estimator for the parameter β_j we need

$$\frac{\partial l}{\partial \beta_j} = U_j = \sum_{i=1}^{N} \left[\frac{\partial l_i}{\partial \beta_j} \right] = \sum_{i=1}^{N} \left[\frac{\partial l_i}{\partial \theta_i} \cdot \frac{\partial \theta_i}{\partial \mu_i} \cdot \frac{\partial \mu_i}{\partial \beta_j} \right], \tag{4.17}$$

using the chain rule for differentiation. We will consider each term on the right-hand side of (4.17) separately. First

$$\frac{\partial l_i}{\partial \theta_i} = y_i b'(\theta_i) + c'(\theta_i) = b'(\theta_i)(y_i - \mu_i)$$

by differentiating (4.13) and substituting (4.14). Next

$$\frac{\partial \theta_i}{\partial \mu_i} = 1 \left/ \left(\frac{\partial \mu_i}{\partial \theta_i} \right) \right. .$$

Differentiation of (4.14) gives

$$\frac{\partial \mu_i}{\partial \theta_i} = \frac{-c''(\theta_i)}{b'(\theta_i)} + \frac{c'(\theta_i)b''(\theta_i)}{[b'(\theta_i)]^2}$$

$$= b'(\theta_i)\text{var}(Y_i)$$

from (4.15). Finally, from (4.16)

$$\frac{\partial \mu_i}{\partial \beta_j} = \frac{\partial \mu_i}{\partial \eta_i} \cdot \frac{\partial \eta_i}{\partial \beta_j} = \frac{\partial \mu_i}{\partial \eta_i} x_{ij}.$$

Hence the score, given in (4.17), is

$$U_j = \sum_{i=1}^{N} \left[\frac{(y_i - \mu_i)}{\text{var}(Y_i)} x_{ij} \left(\frac{\partial \mu_i}{\partial \eta_i} \right) \right]. \tag{4.18}$$

The variance–covariance matrix of the U_j's has terms

$$\Im_{jk} = E\left[U_j U_k\right],$$

which form the **information matrix** $\Im$. From (4.18)

$$
\begin{aligned}
\Im_{jk} &= E\left\{\sum_{i=1}^{N}\left[\frac{(Y_i - \mu_i)}{\text{var}(Y_i)}x_{ij}\left(\frac{\partial \mu_i}{\partial \eta_i}\right)\right]\sum_{l=1}^{N}\left[\frac{(Y_l - \mu_l)}{\text{var}(Y_l)}x_{lk}\left(\frac{\partial \mu_l}{\partial \eta_l}\right)\right]\right\}\\
&= \sum_{i=1}^{N}\frac{E\left[(Y_i - \mu_i)^2\right]x_{ij}x_{ik}}{[\text{var}(Y_i)]^2}\left(\frac{\partial \mu_i}{\partial \eta_i}\right)^2 \qquad\qquad (4.19)
\end{aligned}
$$

because $E[(Y_i - \mu_i)(Y_l - \mu_l)] = 0$ for $i \neq l$ as the Y_i's are independent. Using $E\left[(Y_i - \mu_i)^2\right] = \text{var}(Y_i)$, (4.19) can be simplified to

$$\Im_{jk} = \sum_{i=1}^{N}\frac{x_{ij}x_{ik}}{\text{var}(Y_i)}\left(\frac{\partial \mu_i}{\partial \eta_i}\right)^2. \qquad\qquad (4.20)$$

The estimating Equation (4.11) for the method of scoring generalizes to

$$\mathbf{b}^{(m)} = \mathbf{b}^{(m-1)} + \left[\Im^{(m-1)}\right]^{-1}\mathbf{U}^{(m-1)}, \qquad\qquad (4.21)$$

where $\mathbf{b}^{(m)}$ is the vector of estimates of the parameters $\beta_1, \ldots, \beta_p$ at the mth iteration. In Equation (4.21), $\left[\Im^{(m-1)}\right]^{-1}$ is the inverse of the information matrix with elements $\Im_{jk}$ given by (4.20), and $\mathbf{U}^{(m-1)}$ is the vector of elements given by (4.18), all evaluated at $\mathbf{b}^{(m-1)}$. Multiplying both sides of Equation (4.21) by $\Im^{(m-1)}$ gives

$$\Im^{(m-1)}\mathbf{b}^{(m)} = \Im^{(m-1)}\mathbf{b}^{(m-1)} + \mathbf{U}^{(m-1)}. \qquad\qquad (4.22)$$

From (4.20) $\Im$ can be written as

$$\Im = \mathbf{X}^T\mathbf{W}\mathbf{X},$$

where $\mathbf{W}$ is the $N \times N$ diagonal matrix with elements

$$w_{ii} = \frac{1}{\text{var}(Y_i)}\left(\frac{\partial \mu_i}{\partial \eta_i}\right)^2. \qquad\qquad (4.23)$$

The expression on the right-hand side of (4.22) is the vector with elements

$$\sum_{k=1}^{p}\sum_{i=1}^{N}\frac{x_{ij}x_{ik}}{\text{var}(Y_i)}\left(\frac{\partial \mu_i}{\partial \eta_i}\right)^2 b_k^{(m-1)} + \sum_{i=1}^{N}\frac{(y_i - \mu_i)x_{ij}}{\text{var}(Y_i)}\left(\frac{\partial \mu_i}{\partial \eta_i}\right)$$

Table 4.3 *Data for Poisson regression example.*

y_i	2	3	6	7	8	9	10	12	15
x_i	−1	−1	0	0	0	0	1	1	1

evaluated at $\mathbf{b}^{(m-1)}$; this follows from equations (4.20) and (4.18). Thus the right-hand side of Equation (4.22) can be written as

$$\mathbf{X}^T \mathbf{W} \mathbf{z},$$

where $\mathbf{z}$ has elements

$$z_i = \sum_{k=1}^{p} x_{ik} b_k^{(m-1)} + (y_i - \mu_i) \left(\frac{\partial \eta_i}{\partial \mu_i} \right) \tag{4.24}$$

with μ_i and $\partial \eta_i / \partial \mu_i$ evaluated at $\mathbf{b}^{(m-1)}$.

Hence the iterative Equation (4.22), can be written as

$$\mathbf{X}^T \mathbf{W} \mathbf{X} \mathbf{b}^{(m)} = \mathbf{X}^T \mathbf{W} \mathbf{z}. \tag{4.25}$$

This is the same form as the normal equations for a linear model obtained by weighted least squares, except that it has to be solved iteratively because, in general, $\mathbf{z}$ and $\mathbf{W}$ depend on $\mathbf{b}$. Thus for generalized linear models, maximum likelihood estimators are obtained by an **iterative weighted least squares** procedure (Charnes et al. 1976).

Most statistical packages that include procedures for fitting generalized linear models have an efficient algorithm based on (4.25). They begin by using some initial approximation $\mathbf{b}^{(0)}$ to evaluate $\mathbf{z}$ and $\mathbf{W}$, then (4.25) is solved to give $\mathbf{b}^{(1)}$, which in turn is used to obtain better approximations for $\mathbf{z}$ and $\mathbf{W}$, and so on until adequate convergence is achieved. When the difference between successive approximations $\mathbf{b}^{(m-1)}$ and $\mathbf{b}^{(m)}$ is sufficiently small, $\mathbf{b}^{(m)}$ is taken as the maximum likelihood estimate.

The example below illustrates the use of this estimation procedure.

4.4 Poisson regression example

The artificial data in Table 4.3 are counts y observed at various values of a covariate x. They are plotted in Figure 4.5.

Let us assume that the responses Y_i are Poisson random variables. In practice, such an assumption would be made either on substantive grounds or from

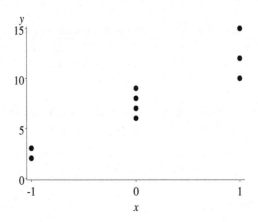

Figure 4.5 *Poisson regression example (data in Table 4.3).*

noticing that in Figure 4.5 the variability increases with Y. This observation supports the use of the Poisson distribution which has the property that the expected value and variance of Y_i are equal

$$E(Y_i) = \text{var}(Y_i). \tag{4.26}$$

Let us model the association between Y_i and x_i by the straight line

$$E(Y_i) = \mu_i = \beta_1 + \beta_2 x_i$$
$$= \mathbf{x}_i^T \boldsymbol{\beta},$$

where

$$\boldsymbol{\beta} = \left[\begin{array}{c} \beta_1 \\ \beta_2 \end{array} \right] \text{ and } \mathbf{x}_i = \left[\begin{array}{c} 1 \\ x_i \end{array} \right]$$

for $i = 1, \ldots, N$. Thus we take the link function $g(\mu_i)$ to be the identity function

$$g(\mu_i) = \mu_i = \mathbf{x}_i^T \boldsymbol{\beta} = \eta_i.$$

Therefore $\partial \mu_i / \partial \eta_i = 1$, which simplifies Equations (4.23) and (4.24). From (4.23) and (4.26)

$$w_{ii} = \frac{1}{\text{var}(Y_i)} = \frac{1}{\beta_1 + \beta_2 x_i}.$$

Using the estimate $\mathbf{b} = \left[\begin{array}{c} b_1 \\ b_2 \end{array} \right]$ for $\boldsymbol{\beta}$, Equation (4.24) becomes

$$z_i = b_1 + b_2 x_i + (y_i - b_1 - b_2 x_i) = y_i.$$

Also

$$J = \mathbf{X}^T\mathbf{W}\mathbf{X} = \begin{bmatrix} \sum_{i=1}^N \dfrac{1}{b_1 + b_2 x_i} & \sum_{i=1}^N \dfrac{x_i}{b_1 + b_2 x_i} \\[2ex] \sum_{i=1}^N \dfrac{x_i}{b_1 + b_2 x_i} & \sum_{i=1}^N \dfrac{x_i^2}{b_1 + b_2 x_i} \end{bmatrix}$$

and

$$\mathbf{X}^T\mathbf{W}\mathbf{z} = \begin{bmatrix} \sum_{i=1}^N \dfrac{y_i}{b_1 + b_2 x_i} \\[2ex] \sum_{i=1}^N \dfrac{x_i y_i}{b_1 + b_2 x_i} \end{bmatrix}.$$

The maximum likelihood estimates are obtained iteratively from the equations

$$(\mathbf{X}^T\mathbf{W}\mathbf{X})^{(m-1)}\mathbf{b}^{(m)} = \mathbf{X}^T\mathbf{W}\mathbf{z}^{(m-1)},$$

where the superscript $^{(m-1)}$ denotes evaluation at $\mathbf{b}^{(m-1)}$.

For these data, $N = 9$,

$$\mathbf{y} = \mathbf{z} = \begin{bmatrix} 2 \\ 3 \\ \vdots \\ 15 \end{bmatrix} \quad \text{and} \quad \mathbf{X} = \begin{bmatrix} \mathbf{x}_1^T \\ \mathbf{x}_2^T \\ \vdots \\ \mathbf{x}_9^T \end{bmatrix} = \begin{bmatrix} 1 & -1 \\ 1 & -1 \\ \vdots & \vdots \\ 1 & 1 \end{bmatrix}.$$

From Figure 4.5 the initial estimates $b_1^{(1)} = 7$ and $b_2^{(1)} = 5$ are obtained. Therefore,

$$(\mathbf{X}^T\mathbf{W}\mathbf{X})^{(1)} = \begin{bmatrix} 1.821429 & -0.75 \\ -0.75 & 1.25 \end{bmatrix}, \qquad (\mathbf{X}^T\mathbf{W}\mathbf{z})^{(1)} = \begin{bmatrix} 9.869048 \\ 0.583333 \end{bmatrix},$$

so
$$\mathbf{b}^{(2)} = \left[(\mathbf{X}^T\mathbf{W}\mathbf{X})^{(1)}\right]^{-1}(\mathbf{X}^T\mathbf{W}\mathbf{z})^{(1)}$$

$$= \begin{bmatrix} 0.729167 & 0.4375 \\ 0.4375 & 1.0625 \end{bmatrix}\begin{bmatrix} 9.869048 \\ 0.583333 \end{bmatrix}$$

$$= \begin{bmatrix} 7.4514 \\ 4.9375 \end{bmatrix}.$$

This iterative process is continued until it converges. The results are shown in Table 4.4.

The maximum likelihood estimates are $\widehat{\beta}_1 = 7.45163$ and $\widehat{\beta}_2 = 4.93530$. At these values the inverse of the information matrix $J = \mathbf{X}^T\mathbf{W}\mathbf{X}$ is

$$J^{-1} = \begin{bmatrix} 0.7817 & 0.4166 \\ 0.4166 & 1.1863 \end{bmatrix}$$

(this is the variance–covariance matrix for $\widehat{\boldsymbol{\beta}}$—see Section 5.4). Thus the estimated standard error for b_1 is $\sqrt{0.7817} = 0.8841$ and the estimated standard error for b_2 is $\sqrt{1.1863} = 1.0892$. So, for example, an approximate 95% confidence interval for the slope β_2 is

$$4.9353 \pm 1.96 \times 1.0892 \text{ or } (2.80, 7.07).$$

The Stata command that produces the same results is as follows:

```
──────────── Stata code (Poisson regression) ────────────
.glm y x, family(poisson) link(identity) vce(eim)
```

where the last term indicates that the variance–covariance matrix is obtained from expected information matrix $\mathfrak{I}$ (other options for estimating this matrix are also possible).

The R commands are below and start by calling the 'dobson' library which contains all the data sets from this book

```
──────────── R code (Poisson regression) ────────────
>library(dobson)
>data(poisson)
>res.p=glm(y~x,family=poisson(link="identity"),data=poisson)
>summary(res.p)
```

In both cases the parameter estimates and their standard errors are given as

$$b_1(s.e) = 7.4516(0.8841) \text{ and } b_2(s.e) = 4.9353(1.0892).$$

4.5 Exercises

4.1 The data in Table 4.5 show the numbers of cases of AIDS in Australia by date of diagnosis for successive 3-month periods from 1984 to 1988. (Data from National Centre for HIV Epidemiology and Clinical Research, 1994.)

Table 4.4 *Successive approximations for regression coefficients in the Poisson regression example.*

m	1	2	3	4
$b_1^{(m)}$	7	7.45139	7.45163	7.45163
$b_2^{(m)}$	5	4.93750	4.93531	4.93530

Table 4.5 *Numbers of cases of AIDS in Australia for successive quarter from 1984 to 1988.*

	Quarter			
Year	1	2	3	4
1984	1	6	16	23
1985	27	39	31	30
1986	43	51	63	70
1987	88	97	91	104
1988	110	113	149	159

In this early phase of the epidemic, the numbers of cases seemed to be increasing exponentially.

a. Plot the number of cases y_i against time period i ($i = 1, \ldots, 20$).

b. A possible model is the Poisson distribution with parameter $\lambda_i = i^\theta$, or equivalently

$$\log \lambda_i = \theta \log i.$$

Plot $\log y_i$ against $\log i$ to examine this model.

c. Fit a generalized linear model to these data using the Poisson distribution, the log-link function and the equation

$$g(\lambda_i) = \log \lambda_i = \beta_1 + \beta_2 x_i,$$

where $x_i = \log i$. Firstly, do this from first principles, working out expressions for the weight matrix $\mathbf{W}$ and other terms needed for the iterative equation

$$\mathbf{X}^T \mathbf{W} \mathbf{X} \mathbf{b}^{(m)} = \mathbf{X}^T \mathbf{W} \mathbf{z}$$

and using software which can perform matrix operations to carry out the calculations.

d. Fit the model described in (c) using statistical software which can perform Poisson regression. Compare the results with those obtained in (c).

4.2 The data in Table 4.6 are times to death, y_i, in weeks from diagnosis and $\log_{10}$ (initial white blood cell count), x_i, for seventeen patients suffering from leukemia. (This is Example U from Cox and Snell, 1981.)

a. Plot y_i against x_i. Do the data show any trend?

Table 4.6 *Survival time, y_i, in weeks and $\log_{10}$ (initial white blood cell count), x_i, for seventeen leukemia patients.*

y_i	65	156	100	134	16	108	121	4	39
x_i	3.36	2.88	3.63	3.41	3.78	4.02	4.00	4.23	3.73

y_i	143	56	26	22	1	1	5	65
x_i	3.85	3.97	4.51	4.54	5.00	5.00	4.72	5.00

b. A possible specification for $E(Y)$ is

$$E(Y_i) = \exp(\beta_1 + \beta_2 x_i),$$

which will ensure that $E(Y)$ is non-negative for all values of the parameters and all values of x. Which link function is appropriate in this case?

c. The Exponential distribution is often used to describe survival times. The probability distribution is $f(y; \theta) = \theta e^{-y\theta}$. This is a special case of the Gamma distribution with shape parameter $\phi = 1$ (see Exercise 3.12(a)). Show that $E(Y) = 1/\theta$ and $\text{var}(Y) = 1/\theta^2$.

d. Fit a model with the equation for $E(Y_i)$ given in (b) and the Exponential distribution using appropriate statistical software.

e. For the model fitted in (d), compare the observed values y_i and fitted values $\hat{y}_i = \exp(\hat{\beta}_1 + \hat{\beta}_2 x_i)$, and use the standardized residuals $r_i = (y_i - \hat{y}_i)/\hat{y}_i$ to investigate the adequacy of the model. (Note: $\hat{y}_i$ is used as the denominator of r_i because it is an estimate of the standard deviation of Y_i—see (c) above.)

4.3 Let $Y_1, \ldots, Y_N$ be a random sample from the Normal distribution $Y_i \sim N(\log \beta, \sigma^2)$ where σ^2 is known. Find the maximum likelihood estimator of β from first principles. Also verify Equations (4.18) and (4.25) in this case.

Chapter 5

Inference

5.1 Introduction

The two main tools of statistical inference are confidence intervals and hypothesis tests. Their derivation and use for generalized linear models are covered in this chapter from a classical or frequentist perspective. In Chapter 12 they are discussed from a Bayesian perspective.

Confidence intervals, also known as **interval estimates**, are increasingly regarded as more useful than hypothesis tests because the width of a confidence interval provides a measure of the precision with which inferences can be made. It does so in a way which is conceptually simpler than the power of a statistical test (Altman et al. 2000).

Hypothesis tests in a statistical modelling framework are performed by comparing how well two related models fit the data (see the examples in Chapter 2). For generalized linear models, the two models should have the same probability distribution and the same link function, but the linear component of one model has more parameters than the other. The simpler model, corresponding to the null hypothesis H_0, must be a special case of the other more general model. If the simpler model fits the data as well as the more general model does, then it is preferred on the grounds of parsimony and H_0 is retained. If the more general model fits significantly better, then H_0 is rejected in favor of an alternative hypothesis H_1, which corresponds to the more general model. Summary statistics are used to compare how well the models fit the data. These **goodness of fit statistics** may be based on the maximum value of the likelihood function, the maximum value of the log-likelihood function, the minimum value of the sum of squares criterion or a composite statistic based on the residuals. The process and logic can be summarized as follows:

1. Specify a model M_0 corresponding to H_0. Specify a more general model M_1 (with M_0 as a special case of M_1).

2. Fit M_0 and calculate the goodness of fit statistic G_0. Fit M_1 and calculate the goodness of fit statistic G_1.

3. Calculate the improvement in fit, usually $G_1 - G_0$ but G_1/G_0 is another possibility.

4. Use the sampling distribution of $G_1 - G_0$ (or some related statistic) to test the null hypothesis that $G_1 = G_0$ against the alternative hypothesis $G_1 \neq G_0$.

5. If the hypothesis that $G_1 = G_0$ is not rejected, then H_0 is not rejected and M_0 is the preferred model. If the hypothesis $G_1 = G_0$ is rejected, then H_0 is rejected and M_1 is regarded as the better model.

For both forms of inference, sampling distributions are required. To calculate a confidence interval, the sampling distribution of the estimator is required. To test a hypothesis, the sampling distribution of the goodness of fit statistic is required. This chapter is about the relevant sampling distributions for generalized linear models.

If the response variables are Normally distributed, the sampling distributions used for inference can often be determined exactly. For other distributions we need to rely on large-sample asymptotic results based on the Central Limit Theorem. The rigorous development of these results requires careful attention to various regularity conditions. For independent observations from distributions which belong to the exponential family and in particular for generalized linear models, the necessary conditions are indeed satisfied. In this book we consider only the major steps and not the finer points involved in deriving the sampling distributions. Details of the distribution theory for generalized linear models were developed by Fahrmeir and Kaufmann (1985).

The basic idea is that under appropriate conditions, if S is a statistic of interest, then approximately

$$\frac{S - E(S)}{\sqrt{\mathrm{var}(S)}} \sim N(0, 1)$$

or equivalently

$$\frac{[S - E(S)]^2}{\mathrm{var}(S)} \sim \chi^2(1),$$

where $E(S)$ and $\mathrm{var}(S)$ are the expectation and variance of S, respectively.

If there is a vector of statistics of interest $\mathbf{s} = \begin{bmatrix} S_1 \\ \vdots \\ S_p \end{bmatrix}$ with asymptotic expectation $E(\mathbf{s})$ and asymptotic variance–covariance matrix $\mathbf{V}$, then approximately

$$[\mathbf{s} - E(\mathbf{s})]^T \mathbf{V}^{-1} [\mathbf{s} - E(\mathbf{s})] \sim \chi^2(p), \tag{5.1}$$

provided $\mathbf{V}$ is non-singular so a unique inverse matrix $\mathbf{V}^{-1}$ exists.

5.2 Sampling distribution for score statistics

Suppose $Y_1, \ldots, Y_N$ are independent random variables in a generalized linear model with parameters $\boldsymbol{\beta}$, where $E(Y_i) = \mu_i$ and $g(\mu_i) = \mathbf{x}_i^T \boldsymbol{\beta} = \eta_i$. From Equation (4.18) the score statistics are

$$U_j = \frac{\partial l}{\partial \beta_j} = \sum_{i=1}^{N} \left[\frac{(Y_i - \mu_i)}{\operatorname{var}(Y_i)} x_{ij} \left(\frac{\partial \mu_i}{\partial \eta_i} \right) \right] \qquad \text{for } j = 1, \ldots, p.$$

As $E(Y_i) = \mu_i$ for all i,

$$E(U_j) = 0 \qquad \text{for } j = 1, \ldots, p \tag{5.2}$$

consistent with the general result (3.14). The variance–covariance matrix of the score statistics is the information matrix $\mathfrak{I}$ with elements

$$\mathfrak{I}_{jk} = E[U_j U_k]$$

given by Equation (4.20).

If there is only one parameter β, the score statistic has the asymptotic sampling distribution

$$\frac{U}{\sqrt{\mathfrak{I}}} \sim N(0, 1), \text{ or equivalently } \frac{U^2}{\mathfrak{I}} \sim \chi^2(1)$$

because $E(U) = 0$ and $\operatorname{var}(U) = \mathfrak{I}$.

If there is a vector of parameters

$$\boldsymbol{\beta} = \begin{bmatrix} \beta_1 \\ \vdots \\ \beta_p \end{bmatrix}, \text{ then the score vector } \mathbf{U} = \begin{bmatrix} U_1 \\ \vdots \\ U_p \end{bmatrix}$$

has the multivariate Normal distribution $\mathbf{U} \sim MVN(\mathbf{0}, \mathfrak{I})$, at least asymptotically, and so

$$\mathbf{U}^T \mathfrak{I}^{-1} \mathbf{U} \sim \chi^2(p) \tag{5.3}$$

for large samples.

5.2.1   Example: Score statistic for the Normal distribution

Let $Y_1, \ldots, Y_N$ be independent, identically distributed random variables with $Y_i \sim N(\mu, \sigma^2)$, where μ is the parameter of interest and σ^2 is a known constant. The log-likelihood function is

$$l = -\frac{1}{2\sigma^2} \sum_{i=1}^{N} (y_i - \mu)^2 - N \log(\sigma\sqrt{2\pi}).$$

The score statistic is

$$U = \frac{dl}{d\mu} = \frac{1}{\sigma^2} \sum (Y_i - \mu) = \frac{N}{\sigma^2} (\bar{Y} - \mu),$$

so the maximum likelihood estimator, obtained by solving the equation $U = 0$, is $\hat{\mu} = \bar{Y}$. The expected value of the statistic U is

$$E(U) = \frac{1}{\sigma^2} \sum [E(Y_i) - \mu]$$

from Equation (1.2). As $E(Y_i) = \mu$, it follows that $E(U) = 0$ as expected. The variance of U is

$$\mathfrak{I} = \text{var}(U) = \frac{1}{\sigma^4} \sum \text{var}(Y_i) = \frac{N}{\sigma^2}$$

from Equation (1.3) and $\text{var}(Y_i) = \sigma^2$. Therefore,

$$\frac{U}{\sqrt{\mathfrak{I}}} = \frac{(\bar{Y} - \mu)}{\sigma/\sqrt{N}}.$$

According to result (5.1) this has the asymptotic distribution $N(0, 1)$. In fact, the result is exact because $\bar{Y} \sim N(\mu, \sigma^2/N)$ (see Exercise 1.4(a)). Similarly

$$U^T \mathfrak{I}^{-1} U = \frac{U^2}{\mathfrak{I}} = \frac{(\bar{Y} - \mu)^2}{\sigma^2/N} \sim \chi^2(1)$$

is an exact result.

The sampling distribution of U can be used to make inferences about μ. For example, a 95% confidence interval for μ is $\bar{y} \pm 1.96\sigma/\sqrt{N}$, where σ is assumed to be known.

5.2.2   Example: Score statistic for the Binomial distribution

If $Y \sim \text{Bin}(n, \pi)$, the log-likelihood function is

$$l(\pi; y) = y \log \pi + (n - y) \log(1 - \pi) + \log \binom{n}{y},$$

so the score statistic is

$$U = \frac{dl}{d\pi} = \frac{Y}{\pi} - \frac{n-Y}{1-\pi} = \frac{Y - n\pi}{\pi(1-\pi)}.$$

But $E(Y) = n\pi$ and so $E(U) = 0$ as expected. Also $\text{var}(Y) = n\pi(1-\pi)$, so

$$\mathfrak{I} = \text{var}(U) = \frac{1}{\pi^2(1-\pi)^2}\text{var}(Y) = \frac{n}{\pi(1-\pi)},$$

and hence,

$$\frac{U}{\sqrt{\mathfrak{I}}} = \frac{Y - n\pi}{\sqrt{n\pi(1-\pi)}} \sim N(0,1)$$

approximately. This is the Normal approximation to Binomial distribution (without any continuity correction). It is used to find confidence intervals for, and test hypotheses about, π.

5.3 Taylor series approximations

To obtain the asymptotic sampling distributions for various other statistics, it is useful to use Taylor series approximations. The Taylor series approximation for a function $f(x)$ of a single variable x about a value t is

$$f(x) = f(t) + (x-t)\left[\frac{df}{dx}\right]_{x=t} + \frac{1}{2}(x-t)^2\left[\frac{d^2f}{dx^2}\right]_{x=t} + \cdots$$

provided that x is near t.

For a log-likelihood function of a single parameter β, the first three terms of the Taylor series approximation near an estimate b are

$$l(\beta) = l(b) + (\beta - b)U(b) + \frac{1}{2}(\beta - b)^2 U'(b),$$

where $U(b) = dl/d\beta$ is the score function evaluated at $\beta = b$. If $U' = d^2l/d\beta^2$ is approximated by its expected value $E(U') = -\mathfrak{I}$, the approximation becomes

$$l(\beta) = l(b) + (\beta - b)U(b) - \frac{1}{2}(\beta - b)^2\mathfrak{I}(b),$$

where $\mathfrak{I}(b)$ is the information evaluated at $\beta = b$. The corresponding approximation for the log-likelihood function for a vector parameter $\boldsymbol{\beta}$ is

$$l(\boldsymbol{\beta}) = l(\mathbf{b}) + (\boldsymbol{\beta} - \mathbf{b})^T\mathbf{U}(\mathbf{b}) - \frac{1}{2}(\boldsymbol{\beta} - \mathbf{b})^T\mathfrak{I}(\mathbf{b})(\boldsymbol{\beta} - \mathbf{b}), \qquad (5.4)$$

where $\mathbf{U}$ is the vector of scores and $\mathfrak{I}$ is the information matrix.

For the score function of a single parameter β, the first two terms of the Taylor series approximation near an estimate b give

$$U(\beta) = U(b) + (\beta - b)U'(b).$$

If U' is approximated by $E(U') = -\mathfrak{I}$ then

$$U(\beta) = U(b) - (\beta - b)\mathfrak{I}(b).$$

The corresponding expression for a vector parameter $\boldsymbol{\beta}$ is

$$\mathbf{U}(\boldsymbol{\beta}) = \mathbf{U}(\mathbf{b}) - \mathfrak{I}(\mathbf{b})(\boldsymbol{\beta} - \mathbf{b}). \tag{5.5}$$

5.4 Sampling distribution for maximum likelihood estimators

Equation (5.5) can be used to obtain the sampling distribution of the maximum likelihood estimator $\mathbf{b} = \widehat{\boldsymbol{\beta}}$. By definition, $\mathbf{b}$ is the estimator which maximizes $l(\mathbf{b})$ and so $\mathbf{U}(\mathbf{b}) = \mathbf{0}$. Therefore,

$$\mathbf{U}(\boldsymbol{\beta}) = -\mathfrak{I}(\mathbf{b})(\boldsymbol{\beta} - \mathbf{b}),$$

or equivalently,

$$(\mathbf{b} - \boldsymbol{\beta}) = \mathfrak{I}^{-1}\mathbf{U}$$

provided that $\mathfrak{I}$ is non-singular. If $\mathfrak{I}$ is regarded as constant, then $E(\mathbf{b} - \boldsymbol{\beta}) = 0$ because $E(\mathbf{U}) = \mathbf{0}$ by Equation (5.2). Therefore, $E(\mathbf{b}) = \boldsymbol{\beta}$, at least asymptotically, so $\mathbf{b}$ is a consistent estimator of $\boldsymbol{\beta}$. The variance–covariance matrix for $\mathbf{b}$ is

$$E\left[(\mathbf{b} - \boldsymbol{\beta})(\mathbf{b} - \boldsymbol{\beta})^T\right] = \mathfrak{I}^{-1}E(\mathbf{U}\mathbf{U}^T)\mathfrak{I}^{-1} = \mathfrak{I}^{-1} \tag{5.6}$$

because $\mathfrak{I} = E(\mathbf{U}\mathbf{U}^T)$ and $(\mathfrak{I}^{-1})^T = \mathfrak{I}^{-1}$ as $\mathfrak{I}$ is symmetric.

The asymptotic sampling distribution for $\mathbf{b}$, by (5.1), is

$$(\mathbf{b} - \boldsymbol{\beta})^T\mathfrak{I}(\mathbf{b})(\mathbf{b} - \boldsymbol{\beta}) \sim \chi^2(p). \tag{5.7}$$

This is the **Wald statistic**. For the one-parameter case, the more commonly used form is

$$b \sim N(\beta, \mathfrak{I}^{-1}). \tag{5.8}$$

If the response variables in the generalized linear model are Normally distributed then (5.7) and (5.8) are exact results (see Example 5.4.1 below).

5.4.1 Example: Maximum likelihood estimators for the Normal linear model

Consider the model

$$E(Y_i) = \mu_i = \mathbf{x}_i^T \boldsymbol{\beta}; \quad Y_i \sim N(\mu_i, \sigma^2), \tag{5.9}$$

where the Y_i's are N independent random variables and $\boldsymbol{\beta}$ is a vector of p parameters $(p < N)$. This is a generalized linear model with the identity function as the link function. This model is discussed in more detail in Chapter 6.

As the link function is the identity, in equation (4.16) $\mu_i = \eta_i$ and so $\partial \mu_i / \partial \eta_i = 1$. The elements of the information matrix, given in Equation (4.20), have the simpler form

$$\mathfrak{J}_{jk} = \sum_{i=1}^{N} \frac{x_{ij} x_{ik}}{\sigma^2}$$

because $\text{var}(Y_i) = \sigma^2$. Therefore, the information matrix can be written as

$$\mathfrak{J} = \frac{1}{\sigma^2} \mathbf{X}^T \mathbf{X}. \tag{5.10}$$

Similarly the expression in (4.24) has the simpler form

$$z_i = \sum_{k=1}^{p} x_{ik} b_k^{(m-1)} + (y_i - \mu_i).$$

But μ_i evaluated at $\mathbf{b}^{(m-1)}$ is $\mathbf{x}_i^T \mathbf{b}^{(m-1)} = \sum_{k=1}^{p} x_{ik} b_k^{(m-1)}$. Therefore, $z_i = y_i$ in this case.

The estimating Equation (4.25) is

$$\frac{1}{\sigma^2} \mathbf{X}^T \mathbf{X} \mathbf{b} = \frac{1}{\sigma^2} \mathbf{X}^T \mathbf{y},$$

and hence the maximum likelihood estimator is

$$\mathbf{b} = (\mathbf{X}^T \mathbf{X})^{-1} \mathbf{X}^T \mathbf{y}. \tag{5.11}$$

The model (5.9) can be written in vector notation as $\mathbf{y} \sim \text{MVN}(\mathbf{X}\boldsymbol{\beta}, \sigma^2 \mathbf{I})$, where $\mathbf{I}$ is the $N \times N$ unit matrix with ones on the diagonal and zeros elsewhere. From (5.11)

$$E(\mathbf{b}) = (\mathbf{X}^T \mathbf{X})^{-1} (\mathbf{X}^T \mathbf{X} \boldsymbol{\beta}) = \boldsymbol{\beta},$$

so **b** is an unbiased estimator of $\boldsymbol{\beta}$.

To obtain the variance–covariance matrix for **b** we use

$$\begin{aligned}
\mathbf{b} - \boldsymbol{\beta} &= (\mathbf{X}^T\mathbf{X})^{-1}\mathbf{X}^T\mathbf{y} - \boldsymbol{\beta} \\
&= (\mathbf{X}^T\mathbf{X})^{-1}\mathbf{X}^T(\mathbf{y} - \mathbf{X}\boldsymbol{\beta}).
\end{aligned}$$

Hence,

$$\begin{aligned}
\mathrm{E}\left[(\mathbf{b} - \boldsymbol{\beta})(\mathbf{b} - \boldsymbol{\beta})^T\right] &= (\mathbf{X}^T\mathbf{X})^{-1}\mathbf{X}^T\mathrm{E}\left[(\mathbf{y} - \mathbf{X}\boldsymbol{\beta})(\mathbf{y} - \mathbf{X}\boldsymbol{\beta})^T\right]\mathbf{X}(\mathbf{X}^T\mathbf{X})^{-1} \\
&= (\mathbf{X}^T\mathbf{X})^{-1}\mathbf{X}^T\left[\mathrm{var}(\mathbf{y})\right]\mathbf{X}(\mathbf{X}^T\mathbf{X})^{-1} \\
&= \sigma^2(\mathbf{X}^T\mathbf{X})^{-1}.
\end{aligned}$$

But $\sigma^2(\mathbf{X}^T\mathbf{X})^{-1} = \mathfrak{I}^{-1}$ from (5.10), so the variance–covariance matrix for **b** is $\mathfrak{I}^{-1}$ as in (5.6).

The maximum likelihood estimator **b** is a linear combination of the elements Y_i of **y**, from (5.11). As the Y_is are Normally distributed, from the results in Section 1.4.1, the elements of **b** are also Normally distributed. Hence, the exact sampling distribution of **b**, in this case, is

$$\mathbf{b} \sim \mathrm{N}(\boldsymbol{\beta}, \mathfrak{I}^{-1})$$

or

$$(\mathbf{b} - \boldsymbol{\beta})^T\mathfrak{I}(\mathbf{b} - \boldsymbol{\beta}) \sim \chi^2(p).$$

5.5   Log-likelihood ratio statistic

One way of assessing the adequacy of a model is to compare it with a more general model with the maximum number of parameters that can be estimated. This is called a **saturated model**. It is a generalized linear model with the same distribution and same link function as the model of interest.

If there are N observations $Y_i, i = 1, \ldots, N$, all with potentially different values for the linear component $\mathbf{x}_i^T\boldsymbol{\beta}$, then a saturated model can be specified with N parameters. This is also called a **maximal** or **full model**.

If some of the observations have the same linear component or covariate pattern, that is, they correspond to the same combination of factor levels and have the same values of any continuous explanatory variables, they are called **replicates**. In this case, the maximum number of parameters that can be estimated for the saturated model is equal to the number of potentially different linear components, which may be less than N.

In general, let m denote the maximum number of parameters that can be estimated. Let $\boldsymbol{\beta}_{\max}$ denote the parameter vector for the saturated model and $\mathbf{b}_{max}$ denote the maximum likelihood estimator of $\boldsymbol{\beta}_{max}$. The likelihood function for the saturated model evaluated at $\mathbf{b}_{max}$, $L(\mathbf{b}_{max}; \mathbf{y})$, will be larger than any other likelihood function for these observations, with the same assumed distribution and link function, because it provides the most complete description of the data. Let $L(\mathbf{b}; \mathbf{y})$ denote the maximum value of the likelihood function for the model of interest. Then the likelihood ratio

$$\lambda = \frac{L(\mathbf{b}_{max}; \mathbf{y})}{L(\mathbf{b}; \mathbf{y})}$$

provides a way of assessing the goodness of fit for the model. In practice, the logarithm of the likelihood ratio, which is the difference between the log-likelihood functions,

$$\log \lambda = l(\mathbf{b}_{max}; \mathbf{y}) - l(\mathbf{b}; \mathbf{y})$$

is used. Large values of $\log \lambda$ suggest that the model of interest is a poor description of the data relative to the saturated model. To determine the critical region for $\log \lambda$, we need its sampling distribution.

In the next section we see that $2 \log \lambda$ has a chi-squared distribution. Therefore, $2 \log \lambda$ rather than $\log \lambda$ is the more commonly used statistic. It was called the **deviance** by Nelder and Wedderburn (1972).

5.6   Sampling distribution for the deviance

The deviance, also called the **log-likelihood (ratio) statistic**, is

$$D = 2[l(\mathbf{b}_{max}; \mathbf{y}) - l(\mathbf{b}; \mathbf{y})].$$

From Equation (5.4), if $\mathbf{b}$ is the maximum likelihood estimator of the parameter $\boldsymbol{\beta}$ (so that $\mathbf{U}(\mathbf{b}) = \mathbf{0}$),

$$l(\boldsymbol{\beta}) - l(\mathbf{b}) = -\frac{1}{2}(\boldsymbol{\beta} - \mathbf{b})^T \mathfrak{I}(\mathbf{b})(\boldsymbol{\beta} - \mathbf{b})$$

approximately. Therefore, the statistic

$$2[l(\mathbf{b}; \mathbf{y}) - l(\boldsymbol{\beta}; \mathbf{y})] = (\boldsymbol{\beta} - \mathbf{b})^T \mathfrak{I}(\mathbf{b})(\boldsymbol{\beta} - \mathbf{b}),$$

which has the chi-squared distribution $\chi^2(p)$, where p is the number of parameters, from (5.7).

From this result the sampling distribution for the deviance can be derived

$$
\begin{aligned}
D &= 2[l(\mathbf{b}_{max}; \mathbf{y}) - l(\mathbf{b}; \mathbf{y})] \\
&= 2[l(\mathbf{b}_{max}; \mathbf{y}) - l(\boldsymbol{\beta}_{max}; \mathbf{y})] \\
&\quad - 2[l(\mathbf{b}; \mathbf{y}) - l(\boldsymbol{\beta}; \mathbf{y})] + 2[l(\boldsymbol{\beta}_{max}; \mathbf{y}) - l(\boldsymbol{\beta}; \mathbf{y})]. \quad (5.12)
\end{aligned}
$$

The first term in square brackets in (5.12) has the distribution $\chi^2(m)$ where m is the number of parameters in the saturated model. The second term has the distribution $\chi^2(p)$, where p is the number of parameters in the model of interest. The third term, $\upsilon = 2[l(\boldsymbol{\beta}_{\max};\mathbf{y}) - l(\boldsymbol{\beta};\mathbf{y})]$, is a positive constant which will be near zero if the model of interest fits the data almost as well as the saturated model fits. Therefore, the sampling distribution of the deviance is approximately

$$D \sim \chi^2(m-p, \upsilon),$$

where υ is the non-centrality parameter, by the results in Section 1.5. The deviance forms the basis for most hypothesis testing for generalized linear models. This is described in Section 5.7.

If the response variables Y_i are Normally distributed, then D has a chi-squared distribution exactly. In this case, however, D depends on $\text{var}(Y_i) = \sigma^2$, which, in practice, is usually unknown. This means that D cannot be used directly as a goodness of fit statistic (see Example 5.6.2).

For Y_i's with other distributions, the sampling distribution of D may be only approximately chi-squared. However, for the Binomial and Poisson distributions, for example, D can be calculated and used directly as a goodness of fit statistic (see Examples 5.6.1 and 5.6.3).

5.6.1   Example: Deviance for a Binomial model

If the response variables $Y_1, \ldots, Y_N$ are independent and $Y_i \sim \text{Bin}(n_i, \pi_i)$, then the log-likelihood function is

$$l(\boldsymbol{\beta};\mathbf{y}) = \sum_{i=1}^{N} \left[y_i \log \pi_i - y_i \log(1 - \pi_i) + n_i \log(1 - \pi_i) + \log \binom{n_i}{y_i} \right].$$

For a saturated model, the π_i's are all different, so $\boldsymbol{\beta} = [\pi_1, \ldots, \pi_N]^T$. The maximum likelihood estimates are $\widehat{\pi}_i = y_i/n_i$, so the maximum value of the log-likelihood function is

$$
\begin{aligned}
l(\mathbf{b}_{\max};\mathbf{y}) \;=\; &\sum \left[y_i \log\left(\frac{y_i}{n_i}\right) - y_i \log\left(\frac{n_i - y_i}{n_i}\right) \right. \\
&\left. + n_i \log\left(\frac{n_i - y_i}{n_i}\right) + \log \binom{n_i}{y_i} \right].
\end{aligned}
$$

For any other model with $p < N$ parameters, let $\widehat{\pi}_i$ denote the maximum likelihood estimates for the probabilities and let $\widehat{y}_i = n_i \widehat{\pi}_i$ denote the fitted values. Then the log-likelihood function evaluated at these values is

$$l(\mathbf{b};\mathbf{y}) = \sum \left[y_i \log\left(\frac{\widehat{y}_i}{n_i}\right) - y_i \log\left(\frac{n_i - \widehat{y}_i}{n_i}\right) \right. \\ \left. + n_i \log\left(\frac{n_i - \widehat{y}_i}{n_i}\right) + \log\left(\frac{n_i}{y_i}\right) \right].$$

Therefore, the deviance is

$$D = 2\left[l(\mathbf{b}_{\max};\mathbf{y}) - l(\mathbf{b};\mathbf{y})\right]$$

$$= 2\sum_{i=1}^{N} \left[y_i \log\left(\frac{y_i}{\widehat{y}_i}\right) + (n_i - y_i)\log\left(\frac{n_i - y_i}{n_i - \widehat{y}_i}\right) \right].$$

5.6.2 Example: Deviance for a Normal linear model

Consider the model

$$E(Y_i) = \mu_i = \mathbf{x}_i^T \boldsymbol{\beta}; \quad Y_i \sim N(\mu_i, \sigma^2), i = 1, \ldots, N$$

where the Y_i's are independent. The log-likelihood function is

$$l(\boldsymbol{\beta};\mathbf{y}) = -\frac{1}{2\sigma^2} \sum_{i=1}^{N} (y_i - \mu_i)^2 - \frac{1}{2} N \log(2\pi\sigma^2).$$

For a saturated model all the μ_i's can be different, so $\boldsymbol{\beta}$ has N elements $\mu_1, \ldots, \mu_N$. Differentiating the log-likelihood function with respect to each μ_i and solving the estimating equations gives $\widehat{\mu}_i = y_i$. Therefore, the maximum value of the log-likelihood function for the saturated model is

$$l(\mathbf{b}_{\max};\mathbf{y}) = -\frac{1}{2} N \log(2\pi\sigma^2).$$

For any other model with $p < N$ parameters, let

$$\mathbf{b} = (\mathbf{X}^T\mathbf{X})^{-1}\mathbf{X}^T\mathbf{y}$$

be the maximum likelihood estimator (from Equation (5.11)). The corresponding maximum value for the log-likelihood function is

$$l(\mathbf{b};\mathbf{y}) = -\frac{1}{2\sigma^2} \sum (y_i - \mathbf{x}_i^T\mathbf{b})^2 - \frac{1}{2} N \log(2\pi\sigma^2).$$

Therefore, the deviance is

$$
\begin{aligned}
D &= 2[l(\mathbf{b}_{\max};\mathbf{y}) - l(\mathbf{b};\mathbf{y})] \\
&= \frac{1}{\sigma^2}\sum_{i=1}^{N}(y_i - \mathbf{x}_i^T\mathbf{b})^2 \qquad (5.13) \\
&= \frac{1}{\sigma^2}\sum_{i=1}^{N}(y_i - \widehat{\mu}_i)^2, \qquad (5.14)
\end{aligned}
$$

where $\widehat{\mu}_i$ denotes the fitted value $\mathbf{x}_i^T\mathbf{b}$.

In the particular case where there is only one parameter, for example, when $E(Y_i) = \mu$ for all i, $\mathbf{X}$ is a vector of N ones and so $b = \widehat{\mu} = \sum_{i=1}^{N} y_i/N = \bar{y}$ and $\widehat{\mu}_i = \bar{y}$ for all i. Therefore,

$$
D = \frac{1}{\sigma^2}\sum_{i=1}^{N}(y_i - \bar{y})^2.
$$

But this statistic is related to the sample variance S^2

$$
S^2 = \frac{1}{N-1}\sum_{i=1}^{N}(y_i - \bar{y})^2 = \frac{\sigma^2 D}{N-1}.
$$

From Exercise 1.4(d) $(N-1)S^2/\sigma^2 \sim \chi^2(N-1)$, so $D \sim \chi^2(N-1)$ exactly. More generally, from (5.13)

$$
\begin{aligned}
D &= \frac{1}{\sigma^2}\sum(y_i - \mathbf{x}_i^T\mathbf{b})^2 \\
&= \frac{1}{\sigma^2}(\mathbf{y} - \mathbf{Xb})^T(\mathbf{y} - \mathbf{Xb}),
\end{aligned}
$$

where the design matrix $\mathbf{X}$ has rows $\mathbf{x}_i$. The term $(\mathbf{y} - \mathbf{Xb})$ can be written as

$$
\begin{aligned}
\mathbf{y} - \mathbf{Xb} &= \mathbf{y} - \mathbf{X}(\mathbf{X}^T\mathbf{X})^{-1}\mathbf{X}^T\mathbf{y} \\
&= [\mathbf{I} - \mathbf{X}(\mathbf{X}^T\mathbf{X})^{-1}\mathbf{X}^T]\mathbf{y} = [\mathbf{I} - \mathbf{H}]\mathbf{y},
\end{aligned}
$$

where $\mathbf{H} = \mathbf{X}(\mathbf{X}^T\mathbf{X})^{-1}\mathbf{X}^T$, which is called the **"hat" matrix**. Therefore, the quadratic form in D can be written as

$$
(\mathbf{y} - \mathbf{Xb})^T(\mathbf{y} - \mathbf{Xb}) = \{[\mathbf{I} - \mathbf{H}]\mathbf{y}\}^T[\mathbf{I} - \mathbf{H}]\mathbf{y} = \mathbf{y}^T[\mathbf{I} - \mathbf{H}]\mathbf{y}
$$

because $\mathbf{H}$ is idempotent (i.e., $\mathbf{H} = \mathbf{H}^T$ and $\mathbf{HH} = \mathbf{H}$). The rank of $\mathbf{I}$ is n and the rank of $\mathbf{H}$ is p so the rank of $\mathbf{I} - \mathbf{H}$ is $n - p$ so, from Section 1.4.2, part

8, D has a chi-squared distribution with $n - p$ degrees of freedom and non-centrality parameter $\lambda = (\mathbf{X}\boldsymbol{\beta})^T(\mathbf{I} - \mathbf{H})(\mathbf{X}\boldsymbol{\beta})/\sigma^2$. But $(\mathbf{I} - \mathbf{H})\mathbf{X} = 0$, so D has the central distribution $\chi^2(N - p)$ exactly (for more details, see Forbes, Evans, Hastings, and Peacock, 2010).

The term **scaled deviance** is sometimes used for

$$\sigma^2 D = \sum(y_i - \widehat{\mu}_i)^2.$$

If the model fits the data well, then $D \sim \chi^2(N - p)$. The expected value for a random variable with the distribution $\chi^2(N - p)$ is $N - p$ (from Section 1.4.2, part 2), so the expected value of D is $N - p$.

This provides an estimate of σ^2 as

$$\widetilde{\sigma}^2 = \frac{\sum(y_i - \widehat{\mu}_i)^2}{N - p}.$$

Some statistical programs output the scaled deviance for a Normal linear model and call $\widetilde{\sigma}^2$ the scale parameter.

The deviance is also related to the sum of squares of the standardized residuals (see Section 2.3.4)

$$\sum_{i=1}^{N} r_i^2 = \frac{1}{\widehat{\sigma}^2} \sum_{i=1}^{N} (y_i - \widehat{\mu}_i)^2,$$

where $\widehat{\sigma}^2$ is an estimate of σ^2. This provides a rough rule of thumb for the overall magnitude of the standardized residuals. If the model fits well so that $D \sim \chi^2(N - p)$, you could expect $\sum r_i^2 = N - p$, approximately.

5.6.3   Example: Deviance for a Poisson model

If the response variables $Y_1, \ldots, Y_N$ are independent and $Y_i \sim \text{Po}(\lambda_i)$, the log-likelihood function is

$$l(\boldsymbol{\beta}; \mathbf{y}) = \sum y_i \log \lambda_i - \sum \lambda_i - \sum \log y_i!.$$

For the saturated model, the λ_i's are all different, so $\boldsymbol{\beta} = [\lambda_1, \ldots, \lambda_N]^T$. The maximum likelihood estimates are $\widehat{\lambda}_i = y_i$, so the maximum value of the log-likelihood function is

$$l(\mathbf{b}_{\max}; \mathbf{y}) = \sum y_i \log y_i - \sum y_i - \sum \log y_i!.$$

Suppose the model of interest has $p < N$ parameters. The maximum likelihood estimator $\mathbf{b}$ can be used to calculate estimates $\widehat{\lambda}_i$ and, hence, fitted values $\widehat{y}_i = \widehat{\lambda}_i$ because $E(Y_i) = \lambda_i$. The maximum value of the log-likelihood in this case is

$$l(\mathbf{b};\mathbf{y}) = \sum y_i \log \widehat{y}_i - \widehat{y}_i - \sum \log y_i!.$$

Therefore, the deviance is

$$
\begin{aligned}
D &= 2[l(\mathbf{b}_{\max};\mathbf{y}) - l(\mathbf{b};\mathbf{y})] \\
&= 2\left[\sum y_i \log(y_i/\widehat{y}_i) - \sum (y_i - \widehat{y}_i)\right].
\end{aligned}
$$

For most models it can be shown that $\sum y_i = \sum \widehat{y}_i$ (see Exercise 9.1). Therefore, D can be written in the form

$$D = 2\sum o_i \log(o_i/e_i)$$

if o_i is used to denote the observed value y_i and e_i is used to denote the estimated expected value $\widehat{y}_i$.

The value of D can be calculated from the data in this case (unlike the case for the Normal distribution, where D depends on the unknown constant σ^2). This value can be compared with the distribution $\chi^2(N-p)$. The following example illustrates the idea.

The data in Table 5.1 relate to Example 4.4 where a straight line was fitted to Poisson responses. The fitted values are

$$\widehat{y}_i = b_1 + b_2 x_i,$$

where $b_1 = 7.45163$ and $b_2 = 4.93530$ (from Table 4.4). The value of D is $D = 2 \times (0.94735 - 0) = 1.8947$, which is small relative to the degrees of freedom, $N - p = 9 - 2 = 7$. This value of D is given in the output from the model fitting using Stata or R in Section 4.4. In fact, D is below the lower 5% tail of the distribution $\chi^2(7)$ indicating that the model fits the data well— perhaps not surprisingly for such a small set of artificial data!

5.7 Hypothesis testing

Hypotheses about a parameter vector $\boldsymbol{\beta}$ of length p can be tested using the sampling distribution of the Wald statistic $(\widehat{\boldsymbol{\beta}} - \boldsymbol{\beta})^T \mathfrak{I}(\widehat{\boldsymbol{\beta}} - \boldsymbol{\beta}) \sim \chi^2(p)$ from (5.7). Occasionally the score statistic is used: $\mathbf{U}^T \mathfrak{I}^{-1} \mathbf{U} \sim \chi^2(p)$ from (5.3).

An alternative approach, outlined in Section 5.1 and used in Chapter 2, is to compare the goodness of fit of two models. The models need to be **nested** or **hierarchical**, that is, they have the same probability distribution and the

Table 5.1 *Results from the Poisson regression Example 4.4.*

x_i	y_i	$\widehat{y}_i$	$y_i \log(y_i/\widehat{y}_i)$
-1	2	2.51633	-0.45931
-1	3	2.51633	0.52743
0	6	7.45163	-1.30004
0	7	7.45163	-0.43766
0	8	7.45163	0.56807
0	9	7.45163	1.69913
1	10	12.38693	-2.14057
1	12	12.38693	-0.38082
1	15	12.38693	2.87112
Sum	72	72	0.94735

same link function, but the linear component of the simpler model M_0 is a special case of the linear component of the more general model M_1.

Consider the null hypothesis

$$H_0 : \boldsymbol{\beta} = \boldsymbol{\beta}_0 = \begin{bmatrix} \beta_1 \\ \vdots \\ \beta_q \end{bmatrix}$$

corresponding to model M_0 and a more general hypothesis

$$H_1 : \boldsymbol{\beta} = \boldsymbol{\beta}_1 = \begin{bmatrix} \beta_1 \\ \vdots \\ \beta_p \end{bmatrix}$$

corresponding to M_1, with $q < p < N$.

We can test H_0 against H_1 using the difference of the deviance statistics

$$\begin{aligned} \triangle D &= D_0 - D_1 = 2[l(\mathbf{b}_{\max};\mathbf{y}) - l(\mathbf{b}_0;\mathbf{y})] - 2[l(\mathbf{b}_{\max};\mathbf{y}) - l(\mathbf{b}_1;\mathbf{y})] \\ &= 2[l(\mathbf{b}_1;\mathbf{y}) - l(\mathbf{b}_0;\mathbf{y})]. \end{aligned}$$

If both models describe the data well, then $D_0 \sim \chi^2(N - q)$ and $D_1 \sim \chi^2(N - p)$ so that $\triangle D \sim \chi^2(p - q)$, provided that certain independence conditions hold (see Section 1.5). If the value of $\triangle D$ is consistent with the $\chi^2(p - q)$ distribution we would generally choose the model M_0 corresponding to H_0 because it is simpler. Thus we would use Occam's Razor although this principle needs to be tempered with judgment about the practical importance of the extra parameters, as opposed to their statistical significance.

If model M_0 does not describe the data well, then D_0 will be bigger than would be expected for a value from $\chi^2(N-q)$. In fact the sampling distribution of D_0 might be better described by a non-central χ^2 distribution which has a larger expected value than the corresponding central χ^2 distribution. If model M_1 does describe the data set well so that $D_1 \sim \chi^2(N-p)$ but M_0 does not describe the data well, then $\triangle D$ will be bigger than expected from $\chi^2(p-q)$.

This result is used to test the hypothesis H_1 as follows: if the value of $\triangle D$ is in the critical region (i.e., greater than the upper tail $100 \times \alpha\%$ point of the $\chi^2(p-q)$ distribution), then we would reject H_0 in favour of H_1 on the grounds that model M_1 provides a significantly better description of the data (even though it too may not fit the data particularly well).

Provided that the deviance can be calculated from the data, $\triangle D$ provides a good method for hypothesis testing. The sampling distribution of $\triangle D$ is usually better approximated by the chi-squared distribution than is the sampling distribution of a single deviance.

For models based on the Normal distribution, or other distributions with nuisance parameters that are not estimated, the deviance may not be fully determined from the data. The following example shows how this problem may be overcome.

5.7.1   Example: Hypothesis testing for a Normal linear model

For the Normal linear model

$$E(Y_i) = \mu_i = \mathbf{x}_i^T \boldsymbol{\beta}; \quad Y_i \sim N(\mu_i, \sigma^2)$$

for independent random variables $Y_1, \ldots, Y_N$, the deviance is

$$D = \frac{1}{\sigma^2} \sum_{i=1}^{N} (y_i - \widehat{\mu}_i)^2,$$

from Equation (5.13).

Let $\widehat{\mu}_i(0)$ and $\widehat{\mu}_i(1)$ denote the fitted values for model M_0 (corresponding to null hypothesis H_0) and model M_1 (corresponding to the alternative hypothesis H_1), respectively. Then

$$D_0 = \frac{1}{\sigma^2} \sum_{i=1}^{N} [y_i - \widehat{\mu}_i(0)]^2$$

and

$$D_1 = \frac{1}{\sigma^2} \sum_{i=1}^{N} [y_i - \widehat{\mu}_i(1)]^2.$$

It is usual to assume that M_1 fits the data well (and so H_1 is correct) so that $D_1 \sim \chi^2(N-p)$. If M_0 also fits well, then $D_0 \sim \chi^2(N-q)$ and so $\triangle D = D_0 - D_1 \sim \chi^2(p-q)$. If M_0 does not fit well (i.e., H_0 is not correct), then $\triangle D$ will have a non-central χ^2 distribution. To eliminate the term σ^2 the following ratio is used

$$
\begin{aligned}
F &= \frac{D_0 - D_1}{p-q} \bigg/ \frac{D_1}{N-p} \\
&= \frac{\left\{ \sum [y_i - \widehat{\mu}_i(0)]^2 - \sum [y_i - \widehat{\mu}_i(1)]^2 \right\} / (p-q)}{\sum [y_i - \widehat{\mu}_i(1)]^2 / (N-p)}.
\end{aligned}
$$

Thus F can be calculated directly from the fitted values. If H_0 is correct, F will have the central $F(p-q, N-p)$ distribution (at least approximately). If H_0 is not correct, the value of F will be larger than expected from the distribution $F(p-q, N-p)$.

A numerical illustration is provided by the example on birthweights and gestational age in Section 2.2.2. The models are given in (2.6) and (2.7). The minimum values of the sums of squares are related to the deviances by $\widehat{S}_0 = \sigma^2 D_0$ and $\widehat{S}_1 = \sigma^2 D_1$. There are $N = 24$ observations. The simpler model (2.6) has $q = 3$ parameters to be estimated, and the more general model (2.7) has $p = 4$ parameters to be estimated. From Table 2.5

$$
\begin{aligned}
D_0 &= 658770.8/\sigma^2 \quad \text{with } N-q = 21 \text{ degrees of freedom} \\
\text{and } D_1 &= 652424.5/\sigma^2 \quad \text{with } N-p = 20 \text{ degrees of freedom.}
\end{aligned}
$$

Therefore,

$$
F = \frac{(658770.8 - 652424.5)/1}{652424.5/20} = 0.19,
$$

which is certainly not significant compared with the $F(1,20)$ distribution. So the data are consistent with model (2.6) in which birthweight increases with gestational age at the same rate for boys and girls.

5.8   Exercises

5.1 Consider the single response variable Y with $Y \sim \text{Bin}(n, \pi)$.

   a. Find the Wald statistic $(\widehat{\pi} - \pi)^T \mathfrak{J}(\widehat{\pi} - \pi)$, where $\widehat{\pi}$ is the maximum likelihood estimator of π and $\mathfrak{J}$ is the information.

   b. Verify that the Wald statistic is the same as the score statistic $U^T \mathfrak{J}^{-1} U$ in this case (see Example 5.2.2).

c. Find the deviance

$$2[l(\widehat{\pi};y) - l(\pi;y)].$$

d. For large samples, both the Wald/score statistic and the deviance approximately have the $\chi^2(1)$ distribution. For $n = 10$ and $y = 3$, use both statistics to assess the adequacy of the models:

(i) $\pi = 0.1$;   (ii) $\pi = 0.3$;   (iii) $\pi = 0.5$.

Do the two statistics lead to the same conclusions?

5.2 Consider a random sample $Y_1, \ldots, Y_N$ with the exponential distribution

$$f(y_i; \theta_i) = \theta_i \exp(-y_i \theta_i).$$

Derive the deviance by comparing the maximal model with different values of θ_i for each Y_i and the model with $\theta_i = \theta$ for all i.

5.3 Suppose $Y_1, \ldots, Y_N$ are independent identically distributed random variables with the Pareto distribution with parameter θ.

a. Find the maximum likelihood estimator $\widehat{\theta}$ of θ.

b. Find the Wald statistic for making inferences about θ (Hint: Use the results from Exercise 3.10).

c. Use the Wald statistic to obtain an expression for an approximate 95% confidence interval for θ.

d. Random variables Y with the Pareto distribution with the parameter θ can be generated from random numbers U, which are uniformly distributed between 0 and 1 using the relationship $Y = (1/U)^{1/\theta}$ (Evans et al. 2000). Use this relationship to generate a sample of 100 values of Y with $\theta = 2$. From these data calculate an estimate $\widehat{\theta}$. Repeat this process 20 times and also calculate 95% confidence intervals for θ. Compare the average of the estimates $\widehat{\theta}$ with $\theta = 2$. How many of the confidence intervals contain θ?

5.4 For the leukemia survival data in Exercise 4.2:

a. Use the Wald statistic to obtain an approximate 95% confidence interval for the parameter β_1.

b. By comparing the deviances for two appropriate models, test the null hypothesis $\beta_2 = 0$ against the alternative hypothesis $\beta_2 \neq 0$. What can you conclude about the use of the initial white blood cell count as a predictor of survival time?

## Chapter 6

Normal Linear Models

6.1 Introduction

This chapter is about models of the form

$$E(Y_i) = \mu_i = \mathbf{x}_i^T \boldsymbol{\beta}; \quad Y_i \sim N(\mu_i, \sigma^2), \tag{6.1}$$

where $Y_1, \ldots, Y_N$ are independent random variables. The link function is the identity function, that is, $g(\mu_i) = \mu_i$. This model is usually written as

$$\mathbf{y} = \mathbf{X}\boldsymbol{\beta} + \mathbf{e}, \tag{6.2}$$

where

$$\mathbf{y} = \begin{bmatrix} Y_1 \\ \vdots \\ Y_N \end{bmatrix}, \ \mathbf{X} = \begin{bmatrix} \mathbf{x}_1^T \\ \vdots \\ \mathbf{x}_N^T \end{bmatrix}, \ \boldsymbol{\beta} = \begin{bmatrix} \beta_1 \\ \vdots \\ \beta_p \end{bmatrix}, \ \mathbf{e} = \begin{bmatrix} e_1 \\ \vdots \\ e_N \end{bmatrix}$$

and the e_i's are independent identically distributed random variables with $e_i \sim N(0, \sigma^2)$ for $i = 1, \ldots, N$. Multiple linear regression, analysis of variance (ANOVA) and analysis of covariance (ANCOVA) are all of this form and together are sometimes called **general linear models**.

The coverage in this book is not detailed, rather the emphasis is on those aspects which are particularly relevant for the model fitting approach to statistical analysis. Many books provide much more detail, for example, Kutner et al. (2005).

The chapter begins with a summary of basic results, mainly derived in previous chapters. Then the main issues are illustrated through five numerical examples.

6.2 Basic results

6.2.1 *Maximum likelihood estimation*

From Section 5.4.1, the maximum likelihood estimator of $\boldsymbol{\beta}$ is given by

$$\mathbf{b} = (\mathbf{X}^T\mathbf{X})^{-1}\mathbf{X}^T\mathbf{y}, \tag{6.3}$$

provided $(\mathbf{X}^T\mathbf{X})$ is non-singular. As $E(\mathbf{b}) = \boldsymbol{\beta}$, the estimator is unbiased. It has variance–covariance matrix $\sigma^2(\mathbf{X}^T\mathbf{X})^{-1} = \mathfrak{J}^{-1}$.

In the context of generalized linear models, σ^2 is treated as a nuisance parameter. However, it can be shown that

$$\widehat{\sigma}^2 = \frac{1}{N-p}(\mathbf{y} - \mathbf{Xb})^T(\mathbf{y} - \mathbf{Xb}) \tag{6.4}$$

is an unbiased estimator of σ^2 and this can be used to estimate $\mathfrak{J}$ and hence make inferences about $\mathbf{b}$.

6.2.2 *Least squares estimation*

If $E(\mathbf{y}) = \mathbf{X}\boldsymbol{\beta}$ and $E[(\mathbf{y} - \mathbf{X}\boldsymbol{\beta})(\mathbf{y} - \mathbf{X}\boldsymbol{\beta})^T] = \mathbf{V}$, where $\mathbf{V}$ is known, we can obtain the least squares estimator $\widetilde{\boldsymbol{\beta}}$ of $\boldsymbol{\beta}$ without making any further assumptions about the distribution of $\mathbf{y}$. We minimize

$$S_w = (\mathbf{y} - \mathbf{X}\boldsymbol{\beta})^T\mathbf{V}^{-1}(\mathbf{y} - \mathbf{X})\boldsymbol{\beta}.$$

The solution of

$$\frac{\partial S_w}{\partial \boldsymbol{\beta}} = -2\mathbf{X}^T\mathbf{V}^{-1}(\mathbf{y} - \mathbf{X}\boldsymbol{\beta}) = 0$$

is

$$\widetilde{\boldsymbol{\beta}} = (\mathbf{X}^T\mathbf{V}^{-1}\mathbf{X})^{-1}\mathbf{X}^T\mathbf{V}^{-1}\mathbf{y},$$

provided the matrix inverses exist. In particular, for Model (6.1), where the elements of $\mathbf{y}$ are independent and have a common variance then

$$\widetilde{\boldsymbol{\beta}} = (\mathbf{X}^T\mathbf{X})^{-1}\mathbf{X}^T\mathbf{y}.$$

So in this case, maximum likelihood estimators and least squares estimators are the same.

6.2.3   Deviance

The residual corresponding to Y_i is $y_i - \widehat{\mu}_i = \mathbf{x}_i^T \mathbf{b}$. Therefore, the vector of residuals is $\mathbf{y} - \mathbf{Xb}$ and the sum of squares of the residuals, or residual sum of squares, is $(\mathbf{y} - \mathbf{Xb})^T (\mathbf{y} - \mathbf{Xb})$. The residual sum of squares can be calculated from the data, whereas the deviance

$$D = \frac{1}{\sigma^2}(\mathbf{y} - \mathbf{Xb})^T (\mathbf{y} - \mathbf{Xb})$$

includes the unknown constant σ^2 and so cannot be calculated directly. The residual sum of squares is sometimes called the scaled deviance because

$$(\mathbf{y} - \mathbf{Xb})^T (\mathbf{y} - \mathbf{Xb}) = \sigma^2 D.$$

From Section 5.6.1 the quadratic form in the deviance can be expanded to give

$$
\begin{aligned}
D &= \frac{1}{\sigma^2}(\mathbf{y}^T\mathbf{y} - 2\mathbf{b}^T\mathbf{X}^T\mathbf{y} + \mathbf{b}^T\mathbf{X}^T\mathbf{Xb}) \\
&= \frac{1}{\sigma^2}(\mathbf{y}^T\mathbf{y} - \mathbf{b}^T\mathbf{X}^T\mathbf{y})
\end{aligned}
\tag{6.5}
$$

because $\mathbf{X}^T\mathbf{Xb} = \mathbf{X}^T\mathbf{y}$ from Equation (6.3). This alternative formulation for D is used in the next section.

6.2.4   Hypothesis testing

Consider a null hypothesis H_0 and a more general hypothesis H_1 specified as

$$H_0 : \boldsymbol{\beta} = \boldsymbol{\beta}_0 = \begin{bmatrix} \beta_1 \\ \vdots \\ \beta_q \end{bmatrix} \quad \text{and} \quad H_1 : \boldsymbol{\beta} = \boldsymbol{\beta}_1 = \begin{bmatrix} \beta_1 \\ \vdots \\ \beta_p \end{bmatrix},$$

where $q < p < N$. Let $\mathbf{X}_0$ and $\mathbf{X}_1$ denote the corresponding design matrices, $\mathbf{b}_0$ and $\mathbf{b}_1$ the maximum likelihood estimators, and D_0 and D_1 the deviances. The hypothesis H_0 is tested against H_1 using

$$
\begin{aligned}
\triangle D &= D_0 - D_1 = \frac{1}{\sigma^2}\left[(\mathbf{y}^T\mathbf{y} - \mathbf{b}_0^T\mathbf{X}_0^T\mathbf{y}) - (\mathbf{y}^T\mathbf{y} - \mathbf{b}_1^T\mathbf{X}_1^T\mathbf{y})\right] \\
&= \frac{1}{\sigma^2}(\mathbf{b}_1^T\mathbf{X}_1^T\mathbf{y} - \mathbf{b}_0^T\mathbf{X}_0^T\mathbf{y})
\end{aligned}
$$

by (6.5). As the model corresponding to H_1 is more general, it is more likely to fit the data well, so D_1 is assumed to have the central distribution $\chi^2(N-p)$.

Table 6.1 *Analysis of Variance table.*

Source of variance	Degrees of freedom	Sum of squares	Mean square
Model with β_0	q	$\mathbf{b}_0^T \mathbf{X}_0^T \mathbf{y}$	
Improvement due to model with β_1	$p - q$	$\mathbf{b}_1^T \mathbf{X}_1^T \mathbf{y} - \mathbf{b}_0^T \mathbf{X}_0^T \mathbf{y}$	$\dfrac{\mathbf{b}_1^T \mathbf{X}_1^T \mathbf{y} - \mathbf{b}_0^T \mathbf{X}_0^T \mathbf{y}}{p - q}$
Residual	$N - p$	$\mathbf{y}^T \mathbf{y} - \mathbf{b}_1^T \mathbf{X}_1^T \mathbf{y}$	$\dfrac{\mathbf{y}^T \mathbf{y} - \mathbf{b}_1^T \mathbf{X}_1^T \mathbf{y}}{N - p}$
Total	N	$\mathbf{y}^T \mathbf{y}$	

On the other hand, D_0 may have a non-central distribution $\chi^2(N - q, v)$ if H_0 is not correct (see Section 5.2). In this case, $\triangle D = D_0 - D_1$ would have the non-central distribution $\chi^2(p - q, v)$ (provided appropriate conditions are satisfied, see Section 1.5). Therefore, the statistic

$$F = \frac{D_0 - D_1}{p - q} \bigg/ \frac{D_1}{N - p} = \frac{(\mathbf{b}_1^T \mathbf{X}_1^T \mathbf{y} - \mathbf{b}_0^T \mathbf{X}_0^T \mathbf{y})}{p - q} \bigg/ \frac{(\mathbf{y}^T \mathbf{y} - \mathbf{b}_1^T \mathbf{X}_1^T \mathbf{y})}{N - p}$$

will have the central distribution $F(p - q, N - p)$ if H_0 is correct; F will otherwise have a non-central distribution. Therefore, values of F that are large relative to the distribution $F(p - q, N - p)$ provide evidence against H_0 (see Figure 2.5). An alternative view of the F statistic is in terms of the residual sums of squares

$$F = \frac{S_0 - S_1}{p - q} \bigg/ \frac{S_1}{N - p},$$

where $S_0 = \mathbf{y}^T \mathbf{y} - \mathbf{b}_0^T \mathbf{X}_0^T \mathbf{y}$ and $S_1 = \mathbf{y}^T \mathbf{y} - \mathbf{b}_1^T \mathbf{X}_1^T \mathbf{y}$. This hypothesis test is often summarized by the ANOVA table shown in Table 6.1.

6.2.5   Orthogonality

Usually inferences about a parameter for one explanatory variable depend on which other explanatory variables are included in the model. An exception is when the design matrix can be partitioned into components $\mathbf{X}_1, \ldots, \mathbf{X}_m$ corresponding to submodels of interest,

$$\mathbf{X} = [\mathbf{X}_1, \ldots, \mathbf{X}_m] \quad \text{for } m \le p,$$

where $\mathbf{X}_j^T \mathbf{X}_k = \mathbf{O}$, a matrix of zeros, for each $j \ne k$. In this case, $\mathbf{X}$ is said to be **orthogonal**. Let $\boldsymbol{\beta}$ have corresponding components $\boldsymbol{\beta}_1, \ldots, \boldsymbol{\beta}_m$ so that

$$\mathrm{E}(\mathbf{y}) = \mathbf{X}\boldsymbol{\beta} = \mathbf{X}_1 \boldsymbol{\beta}_1 + \mathbf{X}_2 \boldsymbol{\beta}_2 + \ldots + \mathbf{X}_m \boldsymbol{\beta}_m.$$

Table 6.2 *Multiple hypothesis tests when the design matrix* **X** *is orthogonal.*

Source of variance	Degrees of freedom	Sum of squares
Model corresponding to H_1	p_1	$\mathbf{b}_1^T \mathbf{X}_1^T \mathbf{y}$
$\vdots$	$\vdots$	$\vdots$
Model corresponding to H_m	p_m	$\mathbf{b}_m^T \mathbf{X}_m^T \mathbf{y}$
Residual	$N - \sum_{j=1}^m p_j$	$\mathbf{y}^T \mathbf{y} - \mathbf{b}^T \mathbf{X}^T \mathbf{y}$
Total	N	$\mathbf{y}^T \mathbf{y}$

Typically, the components correspond to individual covariates or groups of associated explanatory variables such as dummy variables denoting levels of a factor. If **X** can be partitioned in this way, then $\mathbf{X}^T\mathbf{X}$ is a block diagonal matrix

$$\mathbf{X}^T\mathbf{X} = \begin{bmatrix} \mathbf{X}_1^T\mathbf{X}_1 & & \mathbf{O} \\ & \ddots & \\ \mathbf{O} & & \mathbf{X}_m^T\mathbf{X}_m \end{bmatrix}. \quad \text{Also} \quad \mathbf{X}^T\mathbf{y} = \begin{bmatrix} \mathbf{X}_1^T\mathbf{y} \\ \vdots \\ \mathbf{X}_m^T\mathbf{y} \end{bmatrix}.$$

Therefore, the estimates $\mathbf{b}_j = (\mathbf{X}_j^T\mathbf{X}_j)^{-1}\mathbf{X}_j^T\mathbf{y}$ are unaltered by the inclusion of other elements in the model and also

$$\mathbf{b}^T\mathbf{X}^T\mathbf{y} = \mathbf{b}_1^T\mathbf{X}_1^T\mathbf{y} + \ldots + \mathbf{b}_m^T\mathbf{X}_m^T\mathbf{y}.$$

Consequently, the hypotheses

$$H_1 : \boldsymbol{\beta}_1 = \mathbf{0}, \ldots, H_m : \boldsymbol{\beta}_m = \mathbf{0}$$

can be tested independently as shown in Table 6.2.

In practice, except for some well-designed experiments, the design matrix **X** is hardly ever orthogonal. Therefore, inferences about any subset of parameters, $\boldsymbol{\beta}_j$, say, depend on the order in which other terms are included in the model. To overcome this ambiguity, many statistical programs provide tests based on all other terms being included before $\mathbf{X}_j\boldsymbol{\beta}_j$ is added. The resulting sums of squares and hypothesis tests are sometimes called **Type III tests** (if the tests depend on the sequential order of fitting terms they are called Type I).

6.2.6 *Residuals*

Corresponding to the model formulation (6.2), the residuals are defined as

$$\widehat{e}_i = y_i - \mathbf{x}_i^T\mathbf{b} = y_i - \widehat{\mu}_i,$$

where $\widehat{\mu}_i$ is the fitted value. The variance–covariance matrix of the vector of residuals $\widehat{\mathbf{e}}$ is

$$
\begin{aligned}
\mathrm{E}(\widehat{\mathbf{e}}\widehat{\mathbf{e}}^T) &= \mathrm{E}[(\mathbf{y}-\mathbf{Xb})(\mathbf{y}-\mathbf{Xb})^T] \\
&= \mathrm{E}(\mathbf{yy}^T) - \mathbf{X}\mathrm{E}(\mathbf{bb}^T)\mathbf{X}^T \\
&= \sigma^2[\mathbf{I}-\mathbf{X}(\mathbf{X}^T\mathbf{X})^{-1}\mathbf{X}^T],
\end{aligned}
$$

where $\mathbf{I}$ is the unit matrix. So the **standardized residuals** are

$$
r_i = \frac{\widehat{e}_i}{\widehat{\sigma}(1-h_{ii})^{1/2}},
$$

where h_{ii} is the ith element on the diagonal of the **projection** or **hat matrix** $\mathbf{H}=\mathbf{X}(\mathbf{X}^T\mathbf{X})^{-1}\mathbf{X}^T$ and $\widehat{\sigma}^2$ is an estimate of σ^2.

These residuals should be used to check the adequacy of the fitted model using the various plots and other methods discussed in Section 2.3.4. These diagnostic tools include checking linearity of associations between variables, serial independence of observations, Normality of residuals and associations with other potential explanatory variables that are not included in the model.

6.2.7　Other diagnostics

In addition to residuals, there are numerous other methods to assess the adequacy of a model and to identify unusual or influential observations.

An **outlier** is an observation which is not well fitted by the model. An **influential observation** is one which has a relatively large effect on inferences based on the model. Influential observations may or may not be outliers and vice versa.

The value h_{ii}, the ith element on the diagonal of the hat matrix, is called the **leverage** of the ith observation. An observation with high leverage can make a substantial difference to the fit of the model. As a rule of thumb, if h_{ii} is greater than two or three times p/N, it may be a concern (where p is the number of parameters and N the number of observations).

Measures which combine standardized residuals and leverage include

$$
\mathrm{DFITS}_i = r_i\left(\frac{h_{ii}}{1-h_{ii}}\right)^{1/2}
$$

and **Cook's distance**

$$
D_i = \frac{1}{p}\left(\frac{h_{ii}}{1-h_{ii}}\right)r_i^2.
$$

Large values of these statistics indicate that the ith observation is influential;

for example, it is suggested that values of Cook's distance greater than unity may require further investigation. Details of hypothesis tests for these and related statistics are given, for example, by Cook and Weisberg (1999).

Another approach to identifying influential observations is to fit a model with and without each observation and see what difference this makes to the estimates **b** and the overall goodness of fit statistics, such as the deviance or the minimum value of the sum of squares criterion. For example, the statistic **delta-beta** is defined by

$$\triangle_i \widehat{\beta}_j = b_j - b_{j(i)}$$

where $b_{j(i)}$ denotes the estimate of β_j obtained when the ith observation is omitted from the data. These statistics can be standardized by dividing by their standard errors, and then they can be compared with the standard Normal distribution to identify unusually large ones. They can be plotted against the observation numbers i so that the "offending" observations can be easily identified.

The delta-betas can be combined over all parameters using

$$D_i = \frac{1}{p\widehat{\sigma}^2} \left(\mathbf{b} - \mathbf{b}_{(i)}\right)^T \mathbf{X}^T \mathbf{X}(\mathbf{b} - \mathbf{b}_{(i)}),$$

where $\mathbf{b}_{(i)}$ denotes the vector of estimates $b_{j(i)}$. This statistic is, in fact, equal to the Cook's distance (Kutner et al. 2005).

Similarly the influence of the ith observation on the deviance, called **delta-deviance**, can be calculated as the difference between the deviance for the model fitted from all the data and the deviance for the same model with the ith observation omitted.

For Normal linear models there are algebraic simplifications of these statistics which mean that, in fact, the models do not have to be refitted omitting one observation at a time. The statistics can be calculated easily and are provided routinely by most statistical software. More detail of these diagnostic tools is given in many textbooks on regression, such as Weisberg (2014) and Fox and Weisberg (2011) which has a related R package.

Once an influential observation or an outlier is detected, the first step is to determine whether it might be a measurement error, transcription error or some other mistake. It should be removed from the data set only if there is a good substantive reason for doing so. Otherwise a possible solution is to retain it and report the results that are obtained with and without its inclusion in the calculations.

6.3 Multiple linear regression

If the explanatory variables are all continuous, the design matrix has a column of ones, corresponding to an intercept term in the linear component, and all the other elements are observed values of the explanatory variables. Multiple linear regression is the simplest Normal linear model for this situation. The following example provides an illustration.

6.3.1 Example: Carbohydrate diet

The data in Table 6.3 show percentages of total calories obtained from complex carbohydrates, for twenty male insulin-dependent diabetics who had been on a high-carbohydrate diet for six months. Compliance with the regime was thought to be related to age (in years), body weight (relative to "ideal" weight for height) and other components of the diet, such as the percentage of calories as protein. These other variables are treated as explanatory variables.

The first model is

$$E(Y_i) = \mu_i = \beta_0 + \beta_1 x_{i1} + \beta_2 x_{i2} + \beta_3 x_{i3} \quad ; \quad Y_i \sim N(\mu_i, \sigma^2), \qquad (6.6)$$

in which carbohydrate Y is linearly related to age x_1, relative weight x_2 and protein x_3 ($i = 1, \ldots, N = 20$). In this case

$$\mathbf{y} = \begin{bmatrix} Y_1 \\ \vdots \\ Y_N \end{bmatrix}, \quad \mathbf{X} = \begin{bmatrix} 1 & x_{11} & x_{12} & x_{13} \\ \vdots & \vdots & \vdots & \vdots \\ 1 & x_{N1} & x_{N2} & x_{N3} \end{bmatrix} \quad \text{and} \quad \boldsymbol{\beta} = \begin{bmatrix} \beta_0 \\ \vdots \\ \beta_3 \end{bmatrix}.$$

For these data

$$\mathbf{X}^T \mathbf{y} = \begin{bmatrix} 752 \\ 34596 \\ 82270 \\ 12105 \end{bmatrix}$$

and

$$\mathbf{X}^T \mathbf{X} = \begin{bmatrix} 20 & 923 & 2214 & 318 \\ 923 & 45697 & 102003 & 14780 \\ 2214 & 102003 & 250346 & 35306 \\ 318 & 14780 & 35306 & 5150 \end{bmatrix}.$$

Therefore, the solution of $\mathbf{X}^T \mathbf{X} \mathbf{b} = \mathbf{X}^T \mathbf{y}$ is

$$\mathbf{b} = \begin{bmatrix} 36.9601 \\ -0.1137 \\ -0.2280 \\ 1.9577 \end{bmatrix}.$$

Table 6.3 *Carbohydrate, age, relative weight and protein for twenty male insulin-dependent diabetics; for units, see text (data from K. Webb, personal communication, 1982).*

Carbohydrate	Age	Weight	Protein
y	x_1	x_2	x_3
33	33	100	14
40	47	92	15
37	49	135	18
27	35	144	12
30	46	140	15
43	52	101	15
34	62	95	14
48	23	101	17
30	32	98	15
38	42	105	14
50	31	108	17
51	61	85	19
30	63	130	19
36	40	127	20
41	50	109	15
42	64	107	16
46	56	117	18
24	61	100	13
35	48	118	18
37	28	102	14

and

$$(\mathbf{X}^T\mathbf{X})^{-1} = \begin{bmatrix} 4.8158 & -0.0113 & -0.0188 & -0.1362 \\ -0.0113 & 0.0003 & 0.0000 & -0.0004 \\ -0.0188 & 0.0000 & 0.0002 & -0.0002 \\ -0.1362 & -0.0004 & -0.0002 & 0.0114 \end{bmatrix}$$

correct to four decimal places. Also $\mathbf{y}^T\mathbf{y} = 29368$, $N\bar{y}^2 = 28275.2$ and $\mathbf{b}^T\mathbf{X}^T\mathbf{y} = 28800.337$, and so the residual sum of squares is $29368 - 28800.337 = 567.663$ for Model (6.6). Using (6.4) to obtain an unbiased estimator of σ^2, gives $\hat{\sigma}^2 = 35.479$, and hence the standard errors for elements of $\mathbf{b}$ are obtained, which are shown in Table 6.4.

The use of the deviance is illustrated by testing the hypothesis, H_0, that the

Table 6.4 *Estimates for Model (6.6).*

Term	Estimate b_j	Standard error*
Constant	36.960	13.071
Coefficient for age	−0.114	0.109
Coefficient for weight	−0.228	0.083
Coefficient for protein	1.958	0.635

*Values calculated using more significant figures for $(\mathbf{X}^T\mathbf{X})^{-1}$ than shown above.

Table 6.5 *Analysis of Variance table comparing Models (6.6) and (6.7).*

Source variation	Degrees of freedom	Sum of squares	Mean square
Model (6.7)	3	28761.978	
Improvement due to Model (6.6)	1	38.359	38.359
Residual	16	567.663	35.489
Total	20	29368.000	

response does not depend on age, that is, $\beta_1 = 0$. The corresponding model is

$$E(Y_i) = \beta_0 + \beta_2 x_{i2} + \beta_3 x_{i3}. \tag{6.7}$$

The matrix $\mathbf{X}$ for this model is obtained from the previous one by omitting the second column so that

$$\mathbf{X}^T\mathbf{y} = \begin{bmatrix} 752 \\ 82270 \\ 12105 \end{bmatrix}, \quad \mathbf{X}^T\mathbf{X} = \begin{bmatrix} 20 & 2214 & 318 \\ 2214 & 250346 & 35306 \\ 318 & 35306 & 5150 \end{bmatrix},$$

and hence,

$$\mathbf{b} = \begin{bmatrix} 33.130 \\ -0.222 \\ 1.824 \end{bmatrix}.$$

For Model (6.7), $\mathbf{b}^T\mathbf{X}^T\mathbf{y} = 28761.978$ so that the residual sum of squares is $29368 - 28761.978 = 606.022$. Therefore, the difference in the residual sums of squares for Models (6.7) and (6.6) is $606.022 - 567.663 = 38.359$. The significance test for H_0 is summarized in Table 6.5. The value $F = 38.359/35.489 = 1.08$ is not significant compared with the $F(1, 16)$ distribution, so the data provide no evidence against H_0, that is, the response appears to be unrelated to age.

These results can be reproduced by fitting models using software such as

R or Stata. For example, for R the command lm is used for linear regression when the response variable is assumed to be Normally distributed and the link is assumed to be the identity function. Parameter estimates, standard errors and residual sums of squares for Models (6.6) and (6.7) can be obtained using the following R commands.

```
────────────────────── R code (linear model) ──────────────────────
>res.lm=lm(carbohydrate~age+weight+protein,data=carbohydrate)
>summary(res.lm)
>res.lm1=lm(carbohydrate~weight+protein,data=carbohydrate)
>summary(res.lm1)
>anova(res.lm, res.lm1)
```

For Stata, the corresponding commands are

```
────────────────── Stata code (linear model) ──────────────────
.regress carbohydrate age weight protein
.regress carbohydrate weight protein
```

The results can also be obtained using the generalized linear model commands in either R or Stata which requires explicit specification of the distribution of the response variable and the link function. In this case the residual sum of squares in the output is called *residual deviance* in R or the *deviance* in Stata. The command for Model (6.6) using R is

```
──────────── R code (generalized linear model) ────────────
>data(carbohydrate)
>res.glm=glm(carbohydrate~age+weight+protein,
 family=gaussian,data=carbohydrate)
```

and using Stata is

```
──────────── Stata code (generalized linear model) ────────────
.glm carbohydrate age weight protein, family(gaussian)
 link(identity)
```

For Model (6.7) these commands can be repeated without the explanatory variable for age.

Notice also that the parameter estimates for Models (6.6) and (6.7) differ; for example, the coefficient for protein is 1.958 for the model including a term for age but 1.824 when the age term is omitted. This is an example of lack of orthogonality. It is illustrated further in Exercise 6.3(c) as the ANOVA table for testing the hypothesis that the coefficient for age is zero when both weight and protein are in the model, Table 6.5, differs from the ANOVA table when weight is not included.

Diagnostic results from fitting Model (6.6) are shown in Table 6.6. Plots of these results are in Figure 6.1. There is little evidence against the assumptions of Normality and homoscedasticity, or that any of the observations is unduly influential.

None of the standardized residuals is large relative to the standard Normal distribution, and this is shown in the Normal probability plot in Figure 6.1.

Table 6.6 *Diagnostics for Model (6.6) fitted to the data in Table 6.3.*

Obser -vation	Carbo -hydrate	Fitted value	Residual	Std residual	DFIT	Cook's distance
1	33	37.8150	−4.8150	−0.8756	−0.3615	0.03318
2	40	40.0054	−0.0054	−0.0010	−0.0003	0.00000
3	37	35.8464	1.1536	0.2155	0.1019	0.00276
4	27	23.6394	3.3606	0.7936	0.7754	0.15403
5	30	29.1742	0.8258	0.1590	0.0864	0.00199
6	43	37.3848	5.6152	0.9866	0.3043	0.02320
7	34	35.6584	−1.6584	−0.3161	−0.1650	0.00722
8	48	44.5969	3.4031	0.6811	0.4343	0.04884
9	30	40.3424	−10.3424	−1.8828	−0.8663	0.15577
10	38	35.6518	2.3482	0.4139	0.1288	0.00438
11	50	42.0913	7.9087	1.4475	0.6528	0.09875
12	51	47.8409	3.1591	0.6576	0.4731	0.05808
13	30	37.3527	−7.3527	−1.4451	−0.9134	0.19342
14	36	42.6090	−6.6090	−1.3170	−0.8636	0.17731
15	41	35.7880	5.2120	0.9057	0.2404	0.01462
16	42	36.6103	5.3897	0.9852	0.4238	0.04498
17	46	39.1550	6.8450	1.2252	0.4608	0.05131
18	24	32.6743	−8.6743	−1.6879	−1.0563	0.24457
19	35	39.8364	−4.8364	−0.8568	−0.2860	0.02081
20	37	37.9273	−0.9273	−0.1727	−0.0805	0.00172

6.3.2   Coefficient of determination, R^2

A commonly used measure of goodness of fit for multiple linear regression models is based on a comparison with the simplest or **minimal model** using the least squares criterion (in contrast to the maximal model and the log-likelihood function, which are used to define the deviance). For the model specified in (6.2), the least squares criterion is

$$S = \sum_{i=1}^{N} e_i^2 = \mathbf{e}^T \mathbf{e} = (\mathbf{Y} - \mathbf{X}\boldsymbol{\beta})^T (\mathbf{Y} - \mathbf{X}\boldsymbol{\beta})$$

and, from Section 6.2.2, the least squares estimate is $\mathbf{b} = (\mathbf{X}^T\mathbf{X})^{-1}\mathbf{X}^T\mathbf{y}$ so the minimum value of S is

$$\widehat{S} = (\mathbf{y} - \mathbf{X}\mathbf{b})^T (\mathbf{y} - \mathbf{X}\mathbf{b}) = \mathbf{y}^T\mathbf{y} - \mathbf{b}^T\mathbf{X}^T\mathbf{y}.$$

The simplest model is $E(Y_i) = \mu$ for all i. In this case, $\boldsymbol{\beta}$ has the single element μ and $\mathbf{X}$ is a vector of N ones. So $\mathbf{X}^T\mathbf{X} = N$ and $\mathbf{X}^T\mathbf{y} = \sum y_i$ so that $\mathbf{b} = \widehat{\mu} = \bar{y}$. In this case, the value of S is

$$\widehat{S}_0 = \mathbf{y}^T\mathbf{y} - N\bar{y}^2 = \sum (y_i - \bar{y})^2.$$

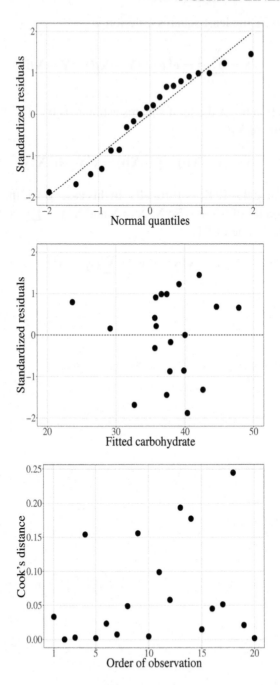

Figure 6.1 *Diagnostic plots for Model (6.6) fitted to the data in Table 6.3—results shown in Table 6.6.*

So $\widehat{S}_0$ is proportional to the variance of the observations and it is the largest or "worst possible" value of S. The relative improvement in fit for any other model is

$$R^2 = \frac{\widehat{S}_0 - \widehat{S}}{\widehat{S}_0} = \frac{\mathbf{b}^T \mathbf{X}^T \mathbf{y} - N\bar{y}^2}{\mathbf{y}^T \mathbf{y} - N\bar{y}^2}.$$

R^2 is called the **coefficient of determination**. It can be interpreted as the proportion of the total variation in the data which is explained by the model. For example, for the carbohydrate data $R^2 = 0.481$ for Model (6.6), so 48.1% of the variation is "explained" by the model. If the term for age is dropped, for Model (6.7) $R^2 = 0.445$, so 44.5% of variation is "explained."

If the model does not fit the data much better than the minimal model, then $\widehat{S}$ will be almost equal to $\widehat{S}_0$ and R^2 will be almost zero. On the other hand if the maximal model is fitted, with one parameter μ_i for each observation Y_i, then $\boldsymbol{\beta}$ has N elements, $\mathbf{X}$ is the $N \times N$ unit matrix $\mathbf{I}$ and $\mathbf{b} = \mathbf{y}$ (i.e., $\widehat{\mu}_i = y_i$). So for the maximal model $\mathbf{b}^T \mathbf{X}^T \mathbf{y} = \mathbf{y}^T \mathbf{y}$ and hence $\widehat{S} = 0$ and $R^2 = 1$, corresponding to a "perfect" fit. In general, $0 < R^2 < 1$. The square root of R^2 is called the **multiple correlation coefficient**.

Despite its popularity and ease of interpretation R^2 has limitations as a measure of goodness of fit. Its sampling distribution is not readily determined. Also it always increases as more parameters are added to the model, so modifications of R^2 have to be used to adjust for the number of parameters.

6.3.3   Model selection

Many applications of multiple linear regression involve numerous explanatory variables and it is often desirable to identify a subset of these variables that provides a good, yet parsimonious, model for the response.

In the previous section R^2 was mentioned as an estimate of goodness of fit, but one that is unsuitable for model selection because it always increases as explanatory variables are added which can lead to over-fitting. This problem can be overcome using **cross-validation** where the data are randomly split into training and test samples. A model is built using the training sample and prediction errors are estimated using the test sample. If the first m observations are used as the training sample, then the stages for multiple linear regression would be

$$\hat{\boldsymbol{\beta}}^* = (\mathbf{X}_{1:m}^T \mathbf{X}_{1:m})^{-1} \mathbf{X}_{1:m}^T \mathbf{y}_{1:m},$$

$$\hat{e}_i^* = y_i - \sum_{j=0}^{p} \hat{\beta}_j^* x_i, \qquad i = m+1, \ldots, n,$$

$$\text{RMSE} = \sqrt{\sum_i (\hat{e}_i^*)^2 / (n-m)}, \qquad i = m+1, \ldots, n,$$

where $\hat{\mathbf{e}}^*$ are the prediction errors and RMSE is the root mean square prediction error.

Rather than using the first m observations for the training sample it is usually better to select m observations at random. To increase robustness multiple test and training samples are usually created. For k-fold cross-validation the sample is randomly split into k non-overlapping samples or "folds" of size $m = n/k$. The k-fold cross-validation can also be replicated multiple times to further increase robustness.

Cross-validation can be used to choose between two alternative models, or where the number of explanatory variables is reasonably small it is feasible to use an exhaustive search over all possible models. This approach for the carbohydrate diet data is illustrated in Figure 6.2. For three explanatory variables there are eight different combinations of variables ranging from none included to all three included. The root mean square prediction errors were based on 5-fold cross-validation replicated 10 times. The model with the lowest errors has the two explanatory variables, protein and weight, and this model generally has lower errors than the full model which also includes age. The three models with the worst prediction error all include age, suggesting that age is not a useful explanatory variable.

The following code estimates the cross-validated prediction error for the carbohydrate diet example in R using the "cvTools" library (Alfons 2012).

────────── R code (cross-validated prediction error) ──────────

```
>library(cvTools)
>full.model <- lm(carbohydrate ~ ., data = carbohydrate)
>cvFit(fit, data = carbohydrate, K = 5, R = 10,
   y = carbohydrate$carbohydrate)
```

An alternative model selection procedure is to add or delete terms sequentially from the model; this is called **stepwise regression** and it might be useful when there are many variables to choose from. Details of the methods are given in standard textbooks on regression such as Kutner et al. (2005).

With stepwise regression there are two possible approaches. Explanatory

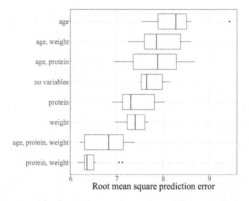

Figure 6.2 *Boxplots of cross-validated root mean square prediction errors for an exhaustive search over all possible models using the three explanatory variables for the carbohydrate diet data. Results are ordered by the mean error.*

variables can be added one at a time, the one with the smallest p-value first, then the next smallest p-value, and so on. This is called forward selection and may be the only feasible approach if there are very large numbers of possible explanatory variables. Alternatively the process can begin with all variables in the model and those with the largest p-values sequentially deleted until all those remaining have p-values below a specified level. This is called backward selection and is generally preferable to forward selection. However, forward and backward selection may result in the inclusion of different variables and sometimes both approaches are used alternately.

The stepwise method is illustrated by a simple example using the carbohydrate diet data with three potential explanatory variables in Table 6.7. Variables are selected based on the p-value obtained from an F-test from a comparison of models as shown in Table 6.5. In the first step, protein is added as the first explanatory variable and in the second step weight is added as a second explanatory variable. The steps stop here, so age is not selected.

Table 6.7 *Example of stepwise selection using the carbohydrate data.*

Step	Variable added	R^2	p-value	Model
1	Protein	0.463	0.0399	Carbohydrate = 12.5 $+1.58\times$Protein
2	Weight	0.667	0.0067	Carbohydrate = 33.1 + $1.82\times$Protein $-0.22\times$Weight

The code to fit a stepwise model in R using the "olsrr" library (Hebbali 2017) is shown below.

──────────── R code (stepwise model selection) ────────

```
>library(olsrr)
>full.model <- lm(carbohydrate ~ ., data = carbohydrate)
>ols_stepwise(full.model, details=TRUE)
```

The Stata code for stepwise selection of explanatory variables reflects the forward and backward approaches and the user needs to specify p-values for removal of variables in backward selection and inclusion in forward selection. The following Stata code produces the model shown at step 2 in Table 6.7.

──────────── Stata code (backward model selection) ────────

```
. stepwise, pr(0.05): regress carbohydrate age weight protein
```

Stepwise selection only involves a relatively small subset of models, and some potentially useful explanatory variables, or combinations of explanatory variables, may never be considered. Also including or excluding variables one at a time can result in widely varying models and the estimated optimal set of variables can change greatly for just a small amount of additional data (Tibshirani 1994).

An alternative variable selection technique is the **least absolute shrinkage and selection operator** (lasso) which includes all explanatory variables but shrinks the parameter estimates towards zero (Hastie et al. 2009). It does this by adding a penalty to the standard least squares regression equation (Section 6.2.2) and selects the optimal β's by minimising:

$$\frac{1}{2}\left(\sum_{i=1}^{N} y_i - \beta_0 - \sum_{j=1}^{p} x_{ij}\beta_j\right)^2 + \lambda \sum_{j=1}^{p} |\beta_j|$$

where λ is a penalty term that punishes higher absolute estimates of β creating a trade-off between a good fit and parsimony. The penalty applies to the p explanatory variables, excluding the intercept β_0 which is an anchor point and not part of optimising the fit. A λ close to zero will mean the β estimates are similar to standard regression. As λ increases the β estimates are shrunk towards zero and each other, hence the apposite acronym "lasso." The lasso will still effectively select some explanatory variables, because for the optimal λ there are often explanatory variables with a $\hat{\beta}$ of zero. The optimal λ is estimated using cross-validation to minimise the overall prediction error using a range of λ's that cover no shrinkage ($\lambda = 0$) to almost complete shrinkage.

An example of using the lasso for the carbohydrate diet data is shown in Figure 6.3. The estimates on the far left, when λ is zero, are the same as least squares regression, and the estimates shrink towards zero for higher values of λ. At the optimal estimate for λ of 1.78 (estimated using cross-validation) the estimate for age is zero and only protein and weight remain in the model, but the absolute values for both estimates are less than half those from using least squares (Table 6.8).

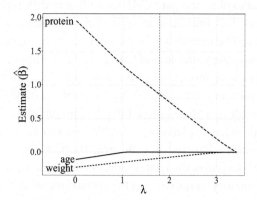

Figure 6.3 *Lasso selection of the three explanatory variables for the carbohydrate diet data. The vertical line indicates the optimal λ penalty.*

Table 6.8 *Table of estimates for the carbohydrate diet data using three alternative model selection approaches. Age was not selected by the stepwise or lasso.*

Variable	Full model (no selection)	Stepwise selection	Lasso
Intercept	36.96	33.13	34.20
Age	−0.114	-	-
Protein	1.958	1.824	0.858
Weight	−0.228	−0.222	−0.093

The code to apply the lasso in R using the "glmnet" library (Friedman et al. 2010) is below.

```
———————————————— R code (lasso) ——————————————
>library(glmnet)
>y = carbohydrate$carbohydrate
>x = as.matrix(carbohydrate[,c('age','weight','protein')])
>fit = glmnet(x, y)
>plot(fit, xvar='lambda')
>cvfit = cv.glmnet(x, y)
>coef(cvfit, s = "lambda.1se")
```

In Stata, the lasso and other regression improvements suggested by (Hastie et al. 2009) can be used from the package 'elasticregress' from http://fmwww.bc.edu/RePEc/bocode/e.

For all model selection techniques it is important to remember that there may not always be one "best" model and instead there may be multiple competing models that explain the data well but with very different interpretations of cause and effect. It is important to spend time considering which explanatory variables should be included in any model selection algorithm rather than including every possible variable and letting the algorithm decide what is important. This is because model selection algorithms are driven by achieving a good fit to the data and including every available variable in the selection procedure may result in a "best" model that has no theoretical basis.

To help decide what explanatory variables should be included in a multiple regression model **directed acyclic graphs** (DAGs) are recommended. These graphically describes the time sequence and hypothesized associations between the explanatory variables and the dependent variable. For a detailed description of DAGs, see Glymour and Greenland (2008).

A simplified example DAG is shown in Figure 6.4, which concerns the effects of high temperatures on deaths. Typical data for this research question are a sample of daily temperatures and deaths from all causes in a city. There is an arrow from "high temperatures" directly to "deaths" because of the potential stress placed on the cardiovascular system by heat. But high temperatures may cause other problems including electricity blackouts and forest fires. Fires can cause deaths directly via burning and indirectly via increased air pollution. If data were available on the days when there were blackouts and fires, and the daily levels of air pollution, these explanatory variables might be included in a multiple regression model and a variable selection technique could be used to choose the key explanatory variables. However, if interest lies in the effect of high temperatures, then including blackouts, forest fires and air pollution as explanatory variables would dilute the effect of temperature because these other variables are on the **causal pathway** between temperature and death.

If the main interest is in the effects of blackouts on deaths, then the diagram identifies high temperatures as a potential **confounder**, because high temperatures cause both deaths and blackouts. However, the reverse is not true, as blackouts are not a potential confounder of the impact that high temperatures may have on deaths, because blackouts do not cause high temperatures.

Another DAG is shown in Figure 6.5 where the dependent variable is food poisoning due to Salmonella bacteria (salmonellosis) (Stephen and Barnett

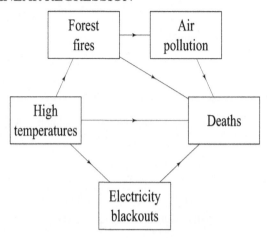

Figure 6.4 *Example DAG showing a hypothesized association between high temperatures, fires, pollution, blackouts and deaths.*

2016). The data here could be the daily number of notified cases of salmonellosis for a city, together with daily weather data. High temperatures can increase the risk of food poisoning because they promote the growth of the bacteria, and heavy rainfall can increase risk when water sources used for irrigation become contaminated. High temperatures and rainfall combine to increase humidity, but there is no reason why humidity would itself cause food poisoning. In this case it would be a mistake to add humidity to a multiple regression model because it could easily show a statistical association via its correlation with the two real risk factors (high temperatures and rainfall), but any observed association would be due to confounding and the estimates of the risks of rainfall and high temperatures would likely be biased.

The example DAG in Figure 6.5 includes the binary explanatory variable "change in test" as there was a change during the period of data collection to a more sensitive test of salmonellosis (Stephen and Barnett 2016). This explanatory variable is an important predictor of salmonellosis but is separate from the inter-related weather variables. Therefore "change in test" could be included in a multiple regression model to reduce the overall prediction error and increase the model's face validity, but not including it does not invalidate any inferences about temperature or rainfall.

Temperature and rainfall are both seasonal and there may be a thought of including season in a multiple regression model, e.g., using a categorical variable of month. However, if the interest is in the effects of the weather, then the focus should be on the more proximal explanatory variables and season

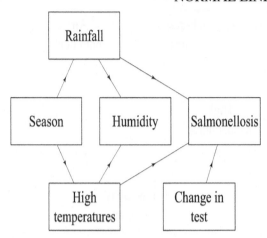

Figure 6.5 *Example DAG showing a hypothesized association between high temperatures, rainfall, humidity, and salmonellosis.*

should not be included, especially if there is no direct association between season and salmonellosis.

Sometimes a variable is not as described, but this might not become clear until a model is run. An example we encountered was for a study investigating predictors of mortality in an intensive care unit. Nutrition score was an incredibly strong predictor of mortality, but after discussions with the investigators it became clear that if a patient was thought close to death no nutrition assessment was made and the score was recorded as zero. Hence the variable is not just the nutrition score, but also includes prognostic information that is outside the scope of a typical data collection.

6.3.4 Collinearity

If some of the explanatory variables are highly correlated with one another, this is called **collinearity** or **multicollinearity**. This condition has several undesirable consequences. Firstly, the columns of the design matrix $\mathbf{X}$ may be nearly linearly dependent so that $\mathbf{X}^T\mathbf{X}$ is nearly singular and the estimating equation $\left(\mathbf{X}^T\mathbf{X}\right)\mathbf{b} = \mathbf{X}^T\mathbf{y}$ is ill-conditioned. This means that the solution $\mathbf{b}$ will be unstable in the sense that small changes in the data may cause large charges in $\mathbf{b}$ (see Section 6.2.7). Also at least some of the elements of $\sigma^2(\mathbf{X}^T\mathbf{X})^{-1}$ will be large giving large variances or covariances for elements of $\mathbf{b}$. Secondly, collinearity means that choosing the best subset of explanatory variables may be difficult.

Collinearity can be detected by calculating the **variance inflation factor** for each explanatory variable

$$\mathrm{VIF}_j = \frac{1}{1 - R^2_{(j)}}$$

where $R^2_{(j)}$ is the coefficient of determination obtained from regressing the jth explanatory variable against all the other explanatory variables. If it is uncorrelated with all the others then $\mathrm{VIF} = 1$. VIF increases as the correlation increases. It is suggested, by Montgomery et al. (2006) for example, that one should be concerned if $\mathrm{VIF} > 5$.

If several explanatory variables are highly correlated it may be impossible, on statistical grounds alone, to determine which one should be included in the model. In this case extra information from the substantive area from which the data came, an alternative specification of the model or some other non-computational approach may be needed.

6.4 Analysis of variance

Analysis of variance is the term used for statistical methods for comparing means of groups of continuous observations where the groups are defined by the levels of factors. In this case all the explanatory variables are categorical and all the elements of the design matrix $\mathbf{X}$ are dummy variables. As illustrated in Example 2.4.3, the choice of dummy variables is, to some extent, arbitrary. An important consideration is the optimal choice of specification of $\mathbf{X}$. The major issues are illustrated by two numerical examples with data from two (fictitious) designed experiments.

6.4.1 One-factor analysis of variance

The data in Table 6.9 are similar to the plant weight data in Exercise 2.1. An experiment was conducted to compare yields Y_i (measured by dried weight of plants) under a control condition and two different treatment conditions. Thus the response, dried weight, depends on one factor, growing condition, with three levels. We are interested in whether the response means differ among the groups.

More generally, if experimental units are randomly allocated to groups corresponding to J levels of a factor, this is called a **completely randomized experiment**. The data can be set out as shown in Table 6.10.

The responses at level j, $Y_{j1}, \ldots, Y_{jn_j}$, all have the same expected value and so they are called **replicates**. In general there may be different numbers of observations n_j at each level.

To simplify the discussion, suppose all the groups have the same sample size, so $n_j = K$ for $j = 1, \ldots, J$. The response $\mathbf{y}$ is the column vector of all $N = JK$ measurements

$$\mathbf{y} = [Y_{11}, Y_{12}, \ldots, Y_{1K}, Y_{21}, \ldots, Y_{2K}, \ldots, Y_{J1}, \ldots, Y_{JK}]^T.$$

We consider three different specifications of a model to test the hypothesis that the response means differ among the factor levels.

a. The simplest specification is

$$E(Y_{jk}) = \mu_j \quad \text{for } j = 1, \ldots, K. \tag{6.8}$$

This can be written as

$$E(Y_i) = \sum_{j=1}^{J} x_{ij} \mu_j, \quad i = 1, \ldots, N,$$

where $x_{ij} = 1$ if response Y_i corresponds to level A_j and $x_{ij} = 0$ otherwise. Thus, $E(\mathbf{y}) = \mathbf{X}\boldsymbol{\beta}$ with

$$\boldsymbol{\beta} = \begin{bmatrix} \mu_1 \\ \mu_2 \\ \vdots \\ \mu_J \end{bmatrix} \quad \text{and} \quad \mathbf{X} = \begin{bmatrix} 1 & 0 & \cdots & 0 \\ 0 & 1 & & \vdots \\ \vdots & & \ddots & \mathbf{O} \\ & \mathbf{O} & & \ddots & 0 \\ 0 & & & & 1 \end{bmatrix},$$

Table 6.9 *Dried weights y_i of plants from three different growing conditions.*

	Control	Treatment A	Treatment B
	4.17	4.81	6.31
	5.58	4.17	5.12
	5.18	4.41	5.54
	6.11	3.59	5.50
	4.50	5.87	5.37
	4.61	3.83	5.29
	5.17	6.03	4.92
	4.53	4.89	6.15
	5.33	4.32	5.80
	5.14	4.69	5.26
$\sum y_i$	50.32	46.61	55.26
$\sum y_i^2$	256.27	222.92	307.13

Table 6.10 *Data from a completely randomized experiment with J levels of a factor A.*

	Factor level			
	A_1	A_2	$\cdots$	A_J
	Y_{11}	Y_{21}		Y_{J1}
	Y_{12}	Y_{22}		Y_{J2}
	$\vdots$			$\vdots$
	Y_{1n_1}	Y_{2n_2}		Y_{Jn_J}
Total	$Y_{1\cdot}$	$Y_{2\cdot}$	$\cdots$	$Y_{J\cdot}$

where **0** and **1** are vectors of length K of zeros and ones, respectively, and **O** indicates that the remaining terms of the matrix are all zeros. Then $\mathbf{X}^T\mathbf{X}$ is the $J \times J$ diagonal matrix

$$\mathbf{X}^T\mathbf{X} = \begin{bmatrix} K & & & & \\ & \ddots & & \mathbf{O} & \\ & & K & & \\ & & & \ddots & \\ & \mathbf{O} & & & K \end{bmatrix} \quad \text{and} \quad \mathbf{X}^T\mathbf{y} = \begin{bmatrix} Y_{1\cdot} \\ Y_{2\cdot} \\ \vdots \\ Y_{J\bullet} \end{bmatrix}.$$

So from Equation (6.3)

$$\mathbf{b} = \frac{1}{K} \begin{bmatrix} Y_{1\cdot} \\ Y_{2\cdot} \\ \vdots \\ Y_{J\cdot} \end{bmatrix} = \begin{bmatrix} \overline{Y}_1 \\ \overline{Y}_2 \\ \vdots \\ \overline{Y}_J \end{bmatrix}$$

and

$$\mathbf{b}^T\mathbf{X}^T\mathbf{y} = \frac{1}{K}\sum_{j=1}^{J} Y_{j\cdot}^2 .$$

The fitted values are $\widehat{\mathbf{y}} = [\bar{y}_1, \bar{y}_1, \ldots, \bar{y}_1, \bar{y}_2, \ldots, \bar{y}_J]^T$. The disadvantage of this simple formulation of the model is that it cannot be extended to more than one factor. To generalize further, the model needs to be specified so that parameters for levels and combinations of levels of factors reflect differential effects beyond some average or specified response.

b. The second specification is one such formulation:

$$E(Y_{jk}) = \mu + \alpha_j, \ j = 1, \ldots, J,$$

where μ is the average effect for all levels and α_j is an additional effect due to level A_j. For this parameterization there are $J+1$ parameters.

$$
\beta = \begin{bmatrix} \mu \\ \alpha_1 \\ \vdots \\ \alpha_J \end{bmatrix}, \quad
X = \begin{bmatrix} 1 & 1 & 0 & \cdots & 0 \\ 1 & 0 & 1 & & \\ \vdots & & & O & \\ \vdots & O & & & \\ 1 & & & & 1 \end{bmatrix},
$$

where 0 and 1 are vectors of length K and O denotes a matrix of zeros. Thus,

$$
X^T y = \begin{bmatrix} Y_{..} \\ Y_{1.} \\ \vdots \\ Y_{J.} \end{bmatrix} \quad \text{and} \quad
X^T X = \begin{bmatrix} N & K & \cdots & K \\ K & K & & \\ \vdots & & O & \\ \vdots & O & & \\ K & & & K \end{bmatrix}.
$$

The first row (or column) of the $(J+1) \times (J+1)$ matrix $X^T X$ is the sum of the remaining rows (or columns), so $X^T X$ is singular and there is no unique solution of the normal equations $X^T X b = X^T y$. The general solution can be written as

$$
b = \begin{bmatrix} \widehat{\mu} \\ \widehat{\alpha}_1 \\ \vdots \\ \widehat{\alpha}_J \end{bmatrix} = \frac{1}{K} \begin{bmatrix} 0 \\ Y_{1.} \\ \vdots \\ Y_{J.} \end{bmatrix} - \lambda \begin{bmatrix} -1 \\ 1 \\ \vdots \\ 1 \end{bmatrix},
$$

where λ is an arbitrary constant. It used to be conventional to impose the additional **sum-to-zero constraint**

$$
\sum_{j=1}^{J} \alpha_j = 0
$$

so that

$$
\frac{1}{K} \sum_{j=1}^{J} Y_{j.} - J\lambda = 0,
$$

and hence,

$$
\lambda = \frac{1}{JK} \sum_{j=1}^{J} Y_{j.} = \frac{Y_{..}}{N}.
$$

This gives the solution

$$\widehat{\mu} = \frac{Y_{..}}{N} \quad \text{and} \quad \widehat{\alpha}_j = \frac{Y_{j.}}{K} - \frac{Y_{..}}{N} \quad \text{for } j = 1,\ldots,J.$$

Hence,

$$\mathbf{b}^T \mathbf{X}^T \mathbf{y} = \frac{Y_{..}^2}{N} + \sum_{j=1}^{J} Y_{j.} \left(\frac{Y_{j.}}{K} - \frac{Y_{..}}{N} \right) = \frac{1}{K} \sum_{j=1}^{J} Y_{j.}^2,$$

which is the same as for the first version of the model, and the fitted values $\widehat{\mathbf{y}} = [\bar{y}_1, \bar{y}_1, \ldots, \bar{y}_J]^T$ are also the same. Sum-to-zero constraints used to be commonly used but nowadays the version described below is the default option in most standard statistical software.

c. A third version of the model is $E(Y_{jk}) = \mu + \alpha_j$ with the constraint that $\alpha_1 = 0$. Thus μ represents the effect of the first level, and α_j measures the difference between the first level and jth level of the factor. This is called a **corner point parameterization**. For this version there are J parameters

$$\boldsymbol{\beta} = \begin{bmatrix} \mu \\ \alpha_2 \\ \vdots \\ \alpha_J \end{bmatrix}. \quad \text{Also} \quad \mathbf{X} = \begin{bmatrix} 1 & 0 & \cdots & & 0 \\ 1 & 1 & & & \\ \vdots & & \ddots & & \mathbf{O} \\ \vdots & \mathbf{O} & & & \\ 1 & & & & 1 \end{bmatrix},$$

$$\text{so } \mathbf{X}^T \mathbf{y} = \begin{bmatrix} Y_{..} \\ Y_{2.} \\ \vdots \\ Y_{J.} \end{bmatrix} \quad \text{and} \quad \mathbf{X}^T \mathbf{X} = \begin{bmatrix} N & K & \cdots & & K \\ K & K & & & \\ \vdots & & \ddots & & \mathbf{O} \\ \vdots & & & \mathbf{O} & \\ K & & & & K \end{bmatrix}.$$

The $J \times J$ matrix $\mathbf{X}^T \mathbf{X}$ is non-singular so there is a unique solution

$$\mathbf{b} = \frac{1}{K} \begin{bmatrix} Y_{1.} \\ Y_{2.} - Y_{1.} \\ \vdots \\ Y_{J.} - Y_{1.} \end{bmatrix}.$$

Also, $\mathbf{b}^T \mathbf{X}^T \mathbf{y} = \frac{1}{K} \left[Y_{..} Y_{1.} + \sum_{j=2}^{J} Y_{j.} (Y_{j.} - Y_{1.}) \right] = \frac{1}{K} \sum_{j=1}^{J} Y_{j.}^2$, and the fitted values $\widehat{\mathbf{y}} = [\bar{y}_1, \bar{y}_1, \ldots, \bar{y}_J]^T$ are the same as before.

Thus, although the three specifications of the model differ, the value of

$\mathbf{b}^T \mathbf{X}^T \mathbf{y}$ and hence

$$D_1 = \frac{1}{\sigma^2} \left(\mathbf{y}^T \mathbf{y} - \mathbf{b}^T \mathbf{X}^T \mathbf{y} \right) = \frac{1}{\sigma^2} \left[\sum_{j=1}^{J} \sum_{k=1}^{K} Y_{jk}^2 - \frac{1}{K} \sum_{j=1}^{J} Y_{j.}^2 \right]$$

is the same in each case.

These three versions of the model all correspond to the hypothesis H_1 that the response means for each level may differ. To compare this with the null hypothesis H_0 that the means are all equal, we consider the model $\mathrm{E}(Y_{jk}) = \mu$ so that $\boldsymbol{\beta} = [\mu]$ and $\mathbf{X}$ is a vector of N ones. Then $\mathbf{X}^T \mathbf{X} = N, \mathbf{X}^T \mathbf{y} = Y_{..}$, and hence, $\mathbf{b} = \widehat{\mu} = Y_{..}/N$ so that $\mathbf{b}^T \mathbf{X}^T \mathbf{y} = Y_{..}^2/N$ and

$$D_0 = \frac{1}{\sigma^2} \left[\sum_{j=1}^{J} \sum_{k=1}^{K} Y_{jk}^2 - \frac{Y_{..}^2}{N} \right].$$

To test H_0 against H_1 we assume that H_1 is correct so that $D_1 \sim \chi^2(N-J)$. If, in addition, H_0 is correct, then $D_0 \sim \chi^2(N-1)$; otherwise, D_0 has a non-central chi-squared distribution. Thus if H_0 is correct,

$$D_0 - D_1 = \frac{1}{\sigma^2} \left[\frac{1}{K} \sum_{j=1}^{J} Y_{j.}^2 - \frac{1}{N} Y_{..}^2 \right] \sim \chi^2(J-1),$$

and so

$$F = \frac{D_0 - D_1}{J-1} \Big/ \frac{D_1}{N-J} \sim F(J-1, N-J).$$

If H_0 is not correct, then F is likely to be larger than predicted from the distribution $F(J-1, N-J)$. Conventionally this hypothesis test is set out in an ANOVA table.

For the plant weight data

$$\frac{Y_{..}^2}{N} = 772.0599, \qquad \frac{1}{K} \sum_{j=1}^{J} Y_{j.}^2 = 775.8262$$

so

$$D_0 - D_1 = (775.8262 - 772.0599)/\sigma^2 = 3.7663/\sigma^2,$$

and

$$\sum_{j=1}^{J} \sum_{k=1}^{K} Y_{jk}^2 = 786.3183$$

so $D_1 = (786.3183 - 775.8262)/\sigma^2 = 10.4921/\sigma^2$. The hypothesis test is

Table 6.11 *ANOVA table for plant weight data in Table 6.9.*

Source of variation	Degrees of freedom	Sum of squares	Mean square	F
Mean	1	772.0599		
Between treatment	2	3.7663	1.883	4.85
Residual	27	10.4921	0.389	
Total	30	786.3183		

summarized in Table 6.11. Since $F = 4.85$ is significant at the 5% level when compared with the $F(2,27)$ distribution, we conclude that the group means differ.

To investigate this result further, it is convenient to use the first version of the Model (6.8), $E(Y_{jk}) = \mu_j$. The estimated means are

$$\mathbf{b} = \begin{bmatrix} \widehat{\mu}_1 \\ \widehat{\mu}_2 \\ \widehat{\mu}_3 \end{bmatrix} = \begin{bmatrix} 5.032 \\ 4.661 \\ 5.526 \end{bmatrix} .$$

If the following estimator is used

$$\widehat{\sigma}^2 = \frac{1}{N-J} (\mathbf{y} - \mathbf{Xb})^T (\mathbf{y} - \mathbf{Xb}) = \frac{1}{N-J} (\mathbf{y}^T\mathbf{y} - \mathbf{b}^T\mathbf{X}^T\mathbf{y})$$

(Equation (6.4)), we obtain $\widehat{\sigma}^2 = 10.4921/27 = 0.389$ (i.e., the residual mean square in Table 6.11). The variance–covariance matrix of $\mathbf{b}$ is $\widehat{\sigma}^2 (\mathbf{X}^T\mathbf{X})^{-1}$, where

$$\mathbf{X}^T\mathbf{X} = \begin{bmatrix} 10 & 0 & 0 \\ 0 & 10 & 0 \\ 0 & 0 & 10 \end{bmatrix} ,$$

so the standard error of each element of $\mathbf{b}$ is $\sqrt{0.389/10} = 0.197$. Now it can be seen that the significant effect is due to the mean for treatment B, $\widehat{\mu}_3 = 5.526$, being significantly larger (more than two standard deviations) than the other two means. Note that if several pairwise comparisons are made among elements of $\mathbf{b}$, the standard errors should be adjusted to take account of multiple comparisons (see, for example, Kutner et al. 2005).

If the glm (or lm) function in R is used to fit the model the command

─────────────────── R code (ANOVA) ───────────────────
```
>res.glm=glm(weight~group, family=gaussian, data=plantwt)
```

gives the corner point estimates in which treatments A and B are compared with the control condition, which is used as the reference level. These estimates (and their standard errors in brackets) are as follows: $\widehat{\mu}_1(s.e.) = 5.032(0.1971)$, $\widehat{\mu}_2 - \widehat{\mu}_1(s.e.) = -0.371(0.2788)$ and $\widehat{\mu}_3 - \widehat{\mu}_1(s.e.) = 0.494(0.2788)$. To fit the first version of the model, -1 is used to suppress the intercept to give

```
──────────────── R code (ANOVA, no intercept) ────────────────
>res.glm1=glm(weight~-1 + group, family=gaussian, data=plantwt)
```

The model corresponding to the null hypothesis of no treatment effect is

```
──────────────── R code (ANOVA, null model) ────────────────
>res.glm0=glm(weight~1, family=gaussian, data=plantwt)
```

For Stata, the command xi: needs to be used to produce the design matrix **X**. The default is the corner point parameterization so that the second and third columns of **X** have indicator variables, denoted by _Igroup_2 and _Igroup_3, for the differences of treatment conditions from the control condition. The Stata glm commands

```
──────────────── Stata code (linear model) ────────────────
.xi i.group
.glm weight _Igroup_2 _Igroup_3, family(gaussian) link(identity)
```

produces the same results as shown above.

6.4.2 Two-factor analysis of variance

Consider the fictitious data in Table 6.12 in which factor A (with $J = 3$ levels) and factor B (with $K = 2$ levels) are **crossed** so that there are JK subgroups formed by all combinations of A and B levels. In each subgroup there are $L = 2$ observations or **replicates**.

The main hypotheses are the following:

H_I: there are no interaction effects, that is, the effects of A and B are additive;

H_A: there are no differences in response associated with different levels of factor A;

H_B: there are no differences in response associated with different levels of factor B.

Thus we need to consider a **saturated model** and three **reduced models** formed by omitting various terms from the saturated model.

Table 6.12 *Fictitious data for two-factor ANOVA with equal numbers of observations in each subgroup.*

Levels of factor A	Levels of factor B		
	B_1	B_2	Total
A_1	6.8, 6.6	5.3, 6.1	24.8
A_2	7.5, 7.4	7.2, 6.5	28.6
A_3	7.8, 9.1	8.8, 9.1	34.8
Total	45.2	43.0	88.2

1. The saturated model is

$$E(Y_{jkl}) = \mu + \alpha_j + \beta_k + (\alpha\beta)_{jk}, \tag{6.9}$$

where the terms $(\alpha\beta)_{jk}$ correspond to **interaction effects** and α_j and β_k to **main effects** of the factors.

2. The **additive model** is

$$E(Y_{jkl}) = \mu + \alpha_j + \beta_k. \tag{6.10}$$

This is compared with the saturated model to test hypothesis H_I.

3. The model formed by omitting effects due to B is

$$E(Y_{jkl}) = \mu + \alpha_j. \tag{6.11}$$

This is compared with the additive model to test hypothesis H_B.

4. The model formed by omitting effects due to A is

$$E(Y_{jkl}) = \mu + \beta_k. \tag{6.12}$$

This is compared with the additive model to test hypothesis H_A.

The Models (6.9) to (6.12) have too many parameters because replicates in the same subgroup have the same expected value so there can be at most JK independent expected values, but the saturated model has $1 + J + K + JK = (J+1)(K+1)$ parameters. To overcome this difficulty (which leads to the singularity of $\mathbf{X}^T\mathbf{X}$), the extra constraints can be imposed

$$\alpha_1 + \alpha_2 + \alpha_3 = 0, \quad \beta_1 + \beta_2 = 0,$$

$$(\alpha\beta)_{11} + (\alpha\beta)_{12} = 0, \quad (\alpha\beta)_{21} + (\alpha\beta)_{22} = 0, \quad (\alpha\beta)_{31} + (\alpha\beta)_{32} = 0,$$

$$(\alpha\beta)_{11} + (\alpha\beta)_{21} + (\alpha\beta)_{31} = 0$$

(the remaining condition $(\alpha\beta)_{12} + (\alpha\beta)_{22} + (\alpha\beta)_{32} = 0$ follows from the last four equations). These are the conventional sum-to-zero constraint equations for ANOVA. An alternative is to use

$$\alpha_1 = \beta_1 = (\alpha\beta)_{11} = (\alpha\beta)_{12} = (\alpha\beta)_{21} = (\alpha\beta)_{31} = 0$$

as the corner point constraints. In either case the numbers of (linearly) independent parameters are: 1 for μ, $J-1$ for the α_j's, $K-1$ for the β_k's, and $(J-1)(K-1)$ for the $(\alpha\beta)_{jk}$'s, giving a total of JK parameters.

For simplicity all four models will be fitted using the corner point constraints. The response vector is

$$\mathbf{y} = [6.8, 6.6, 5.3, 6.1, 7.5, 7.4, 7.2, 6.5, 7.8, 9.1, 8.8, 9.1]^T,$$

and $\mathbf{y}^T\mathbf{y} = 664.10$.

For the saturated Model (6.9) with constraints

$$\alpha_1 = \beta_1 = (\alpha\beta)_{11} = (\alpha\beta)_{12} = (\alpha\beta)_{21} = (\alpha\beta)_{31} = 0,$$

$$\boldsymbol{\beta} = \begin{bmatrix} \mu \\ \alpha_2 \\ \alpha_3 \\ \beta_2 \\ (\alpha\beta)_{22} \\ (\alpha\beta)_{32} \end{bmatrix}, \quad \mathbf{X} = \begin{bmatrix} 100000 \\ 100000 \\ 100100 \\ 100100 \\ 110000 \\ 110000 \\ 110110 \\ 110110 \\ 101000 \\ 101000 \\ 101101 \\ 101101 \end{bmatrix}, \quad \mathbf{X}^T\mathbf{y} = \begin{bmatrix} Y_{...} \\ Y_{2..} \\ Y_{3..} \\ Y_{12.} \\ Y_{22.} \\ Y_{32.} \end{bmatrix} = \begin{bmatrix} 88.2 \\ 28.6 \\ 34.8 \\ 43.0 \\ 13.7 \\ 17.9 \end{bmatrix},$$

$$\mathbf{X}^T\mathbf{X} = \begin{bmatrix} 12 & 4 & 4 & 6 & 2 & 2 \\ 4 & 4 & 0 & 2 & 2 & 0 \\ 4 & 0 & 4 & 2 & 0 & 2 \\ 6 & 2 & 2 & 6 & 2 & 2 \\ 2 & 2 & 0 & 2 & 2 & 0 \\ 2 & 0 & 2 & 2 & 0 & 2 \end{bmatrix}, \quad \mathbf{b} = \begin{bmatrix} 6.7 \\ 0.75 \\ 1.75 \\ -1.0 \\ 0.4 \\ 1.5 \end{bmatrix}$$

and $\mathbf{b}^T\mathbf{X}^T\mathbf{y} = 662.62$.

For the additive Model (6.10) with the constraints $\alpha_1 = \beta_1 = 0$, the design matrix is obtained by omitting the last two columns of the design matrix for the saturated model. Thus,

$$\boldsymbol{\beta} = \begin{bmatrix} \mu \\ \alpha_2 \\ \alpha_3 \\ \beta_2 \end{bmatrix}, \quad \mathbf{X}^T\mathbf{X} = \begin{bmatrix} 12 & 4 & 4 & 6 \\ 4 & 4 & 0 & 2 \\ 4 & 0 & 4 & 2 \\ 6 & 2 & 2 & 6 \end{bmatrix}, \quad \mathbf{X}^T\mathbf{y} = \begin{bmatrix} 88.2 \\ 28.6 \\ 34.8 \\ 43.0 \end{bmatrix},$$

and hence,

$$\mathbf{b} = \begin{bmatrix} 6.383 \\ 0.950 \\ 2.500 \\ -0.367 \end{bmatrix}$$

so that $\mathbf{b}^T\mathbf{X}^T\mathbf{y} = 661.4133$.

For Model (6.11) omitting the effects of levels of factor B and using the constraint $\alpha_1 = 0$, the design matrix is obtained by omitting the last three columns of the design matrix for the saturated model. Therefore,

$$\boldsymbol{\beta} = \begin{bmatrix} \mu \\ \alpha_2 \\ \alpha_3 \end{bmatrix}, \quad \mathbf{X}^T\mathbf{X} = \begin{bmatrix} 12 & 4 & 4 \\ 4 & 4 & 0 \\ 4 & 0 & 4 \end{bmatrix}, \quad \mathbf{X}^T\mathbf{y} = \begin{bmatrix} 88.2 \\ 28.6 \\ 34.8 \end{bmatrix},$$

and hence,

$$\mathbf{b} = \begin{bmatrix} 6.20 \\ 0.95 \\ 2.50 \end{bmatrix}$$

so that $\mathbf{b}^T\mathbf{X}^T\mathbf{y} = 661.01$.

The design matrix for Model (6.12) with constraint $\beta_1 = 0$ comprises the first and fourth columns of the design matrix for the saturated model. Therefore,

$$\boldsymbol{\beta} = \begin{bmatrix} \mu \\ \beta_2 \end{bmatrix}, \quad \mathbf{X}^T\mathbf{X} = \begin{bmatrix} 12 & 6 \\ 6 & 6 \end{bmatrix}, \quad \mathbf{X}^T\mathbf{y} = \begin{bmatrix} 88.2 \\ 43.0 \end{bmatrix},$$

and hence,

$$\mathbf{b} = \begin{bmatrix} 7.533 \\ -0.367 \end{bmatrix}$$

so that $\mathbf{b}^T\mathbf{X}^T\mathbf{y} = 648.6733$.

Finally for the model with only a mean effect $E(Y_{jkl}) = \mu$, the estimate is $\mathbf{b} = [\hat{\mu}] = 7.35$ and so $\mathbf{b}^T\mathbf{X}^T\mathbf{y} = 648.27$.

The results of these calculations are summarized in Table 6.13. The subscripts S, I, B, A and M refer to the saturated model, models corresponding to H_I, H_B and H_A and the model with only the overall mean, respectively. The scaled deviances are the terms $\sigma^2 D = \mathbf{y}^T \mathbf{y} - \mathbf{b}^T \mathbf{X}^T \mathbf{y}$. The degrees of freedom (d.f.) are given by N minus the number of parameters in the model.

Table 6.13 *Summary of calculations for data in Table 6.12.*

Model	d.f.	$\mathbf{b}^T \mathbf{X}^T \mathbf{y}$	Scaled Deviance
$\mu + \alpha_j + \beta_k + (\alpha\beta)_{jk}$	6	662.6200	$\sigma^2 D_S = 1.4800$
$\mu + \alpha_j + \beta_k$	8	661.4133	$\sigma^2 D_I = 2.6867$
$\mu + \alpha_j$	9	661.0100	$\sigma^2 D_B = 3.0900$
$\mu + \beta_k$	10	648.6733	$\sigma^2 D_A = 15.4267$
μ	11	648.2700	$\sigma^2 D_M = 15.8300$

To test H_I we assume that the saturated model is correct so that $D_S \sim \chi^2(6)$. If H_I is also correct, then $D_I \sim \chi^2(8)$ so that $D_I - D_S \sim \chi^2(2)$ and

$$F = \frac{D_I - D_S}{2} \Bigg/ \frac{D_S}{6} \sim F(2, 6).$$

The value of

$$F = \frac{2.6867 - 1.48}{2\sigma^2} \Bigg/ \frac{1.48}{6\sigma^2} = 2.45$$

is not statistically significant so the data do not provide evidence against H_I. Since H_I is not rejected we proceed to test H_A and H_B. For H_B the difference in fit between the Models (6.11) and (6.10) is considered, that is, $D_B - D_I$, and compared with D_S using

$$F = \frac{D_B - D_I}{1} \Bigg/ \frac{D_S}{6} = \frac{3.09 - 2.6867}{\sigma^2} \Bigg/ \frac{1.48}{6\sigma^2} = 1.63,$$

which is not significant compared to the $F(1, 6)$ distribution, suggesting that there are no differences due to levels of factor B. The corresponding test for H_A gives $F = 25.82$, which is significant compared with $F(2, 6)$ distribution. Thus we conclude that the response means are affected only by differences in the levels of factor A. The most appropriate choice for the denominator for the F ratio, D_S or D_I, is debatable. D_S comes from a more complex model and is more likely to correspond to a central chi-squared distribution, but it has fewer degrees of freedom.

The ANOVA table for these data is shown in Table 6.14. The first number in the sum of squares column is the value of $\mathbf{b}^T \mathbf{X}^T \mathbf{y}$ corresponding to the simplest model $E(Y_{jkl}) = \mu$.

Table 6.14 *ANOVA table for data in Table 6.11.*

Source of variation	Degrees of freedom	Sum of squares	Mean square	F
Mean	1	648.2700		
Levels of A	2	12.7400	6.3700	25.82
Levels of B	1	0.4033	0.4033	1.63
Interactions	2	1.2067	0.6033	2.45
Residual	6	1.4800	0.2467	
Total	12	664.1000		

A feature of these data is that the hypothesis tests are independent in the sense that the results are not affected by which terms—other than those relating to the hypothesis in question—are also in the model. For example, the hypothesis of no differences due to factor B, $H_B : \beta_k = 0$ for all k, could equally well be tested using either models $E(Y_{jkl}) = \mu + \alpha_j + \beta_k$ and $E(Y_{jkl}) = \mu + \alpha_j$ and hence

$$\sigma^2 D_B - \sigma^2 D_I = 3.0900 - 2.6867 = 0.4033,$$

or models

$$E(Y_{jkl}) = \mu + \beta_k \quad \text{and} \quad E(Y_{jkl}) = \mu$$

and hence

$$\sigma^2 D_M - \sigma^2 D_A = 15.8300 - 15.4267 = 0.4033.$$

The reason is that the data are **balanced**, that is, there are equal numbers of observations in each subgroup. For balanced data it is possible to specify the design matrix in such a way that it is orthogonal (see Section 6.2.5 and Exercise 6.7). An example in which the hypothesis tests are not independent is given in Exercise 6.8.

The estimated sample means for each subgroup can be calculated from the values of **b**. For example, for the saturated Model (6.9) the estimated mean of the subgroup with the treatment combination A$_3$ and B$_2$ is $\widehat{\mu} + \widehat{\alpha}_3 + \widehat{\beta}_2 + (\widehat{\alpha\beta})_{32} = 6.7 + 1.75 - 1.0 + 1.5 = 8.95$.

The estimate for the same mean from the additive Model (6.10) is

$$\widehat{\mu} + \widehat{\alpha}_3 + \widehat{\beta}_2 = 6.383 + 2.5 - 0.367 = 8.516.$$

This shows the importance of deciding which model to use to summarize the data.

To assess the adequacy of an ANOVA model, residuals should be calculated and examined for unusual patterns, Normality, independence and so on, as described in Section 6.2.6. For this small data set there is no evidence of any departures from the model assumptions.

The R commands to fit the five models (Models (6.9)–(6.12) and the model with only the overall mean) are as follows:

```
———————————————————— R code (ANOVA) ————————————————
>res.glmint=glm(data~A*B, family=gaussian, data=balanced)
>res.glmadd=glm(data~A+B, family=gaussian, data=balanced)
>res.glmA=glm(data~A, family=gaussian, data=balanced)
>res.glmB=glm(data~B, family=gaussian, data=balanced)
>res.glmmean=glm(data~1, family=gaussian, data=balanced)
```

The corresponding Stata commands are

```
———————————————————— Stata code (ANOVA) ————————————————
.glm data _IfacXfac_2_2 _IfacXfac_3_2 _Ifactora_2 _Ifactora_3
 _Ifactorb_2, family(gaussian) link(identity)
.glm data _Ifactora_2 _Ifactora_3 _Ifactorb_2, family(gaussian)
 link(identity)
.glm data _Ifactora_2 _Ifactora_3, family(gaussian)
 link(identity)
.glm data _Ifactorb_2, family(gaussian) link(identity)
.glm data, family(gaussian) link(identity)
```

6.5   Analysis of covariance

Analysis of covariance is the term used for models in which some of the explanatory variables are dummy variables representing factor levels and others are continuous measurements called **covariates**. As with ANOVA, we are interested in comparing means of subgroups defined by factor levels, but recognizing that the covariates may also affect the responses, we compare the means after "adjustment" for covariate effects.

A typical example is provided by the data in Table 6.15 (Winer 1971). The responses Y_{jk} are achievement scores measured at three levels of a factor representing three different training methods, and the covariates x_{jk} are aptitude scores measured before training commenced. We want to compare the training methods, taking into account differences in initial aptitude between the three groups of subjects.

The data are plotted in Figure 6.6. There is evidence that the achievement

Table 6.15 *Achievement scores (data from Winer 1971, page 776).*

	Training method					
	A		B		C	
	y	x	y	x	y	x
	6	3	8	4	6	3
	4	1	9	5	7	2
	5	3	7	5	7	2
	3	1	9	4	7	3
	4	2	8	3	8	4
	3	1	5	1	5	1
	6	4	7	2	7	4
Total	31	15	53	24	47	19
Sum of squares	147	41	413	96	321	59
$\sum xy$	75		191		132	

scores y increase linearly with aptitude x and that the y values are generally higher for training groups B and C than for A.

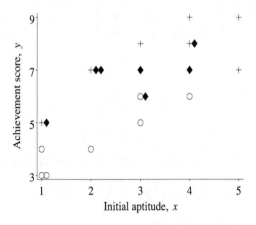

Figure 6.6 *Achievement and initial aptitude scores. Circles denote training method A, crosses denote method B and diamonds denote method C.*

To test the hypothesis that there are no differences in mean achievement scores among the three training methods, after adjustment for initial aptitude, we compare the saturated model

$$E(Y_{jk}) = \mu_j + \gamma x_{jk} \tag{6.13}$$

with the reduced model

$$E(Y_{jk}) = \mu + \gamma x_{jk}, \tag{6.14}$$

where $j = 1$ for method A, $j = 2$ for method B, $j = 3$ for method C, and $k = 1, \ldots, 7$. Let

$$\mathbf{y}_j = \begin{bmatrix} Y_{j1} \\ \vdots \\ Y_{j7} \end{bmatrix} \quad \text{and} \quad \mathbf{x}_j = \begin{bmatrix} x_{j1} \\ \vdots \\ x_{j7} \end{bmatrix}$$

so that, in matrix notation, the saturated Model (6.13) is $E(\mathbf{y}) = \mathbf{X}\boldsymbol{\beta}$ with

$$\mathbf{y} = \begin{bmatrix} \mathbf{y}_1 \\ \mathbf{y}_2 \\ \mathbf{y}_3 \end{bmatrix}, \quad \boldsymbol{\beta} = \begin{bmatrix} \mu_1 \\ \mu_2 \\ \mu_3 \\ \gamma \end{bmatrix} \quad \text{and} \quad \mathbf{X} = \begin{bmatrix} \mathbf{1} & \mathbf{0} & \mathbf{0} & \mathbf{x}_1 \\ \mathbf{0} & \mathbf{1} & \mathbf{0} & \mathbf{x}_2 \\ \mathbf{0} & \mathbf{0} & \mathbf{1} & \mathbf{x}_3 \end{bmatrix},$$

where $\mathbf{0}$ and $\mathbf{1}$ are vectors of length 7. Then

$$\mathbf{X}^T\mathbf{X} = \begin{bmatrix} 7 & 0 & 0 & 15 \\ 0 & 7 & 0 & 24 \\ 0 & 0 & 7 & 19 \\ 15 & 24 & 19 & 196 \end{bmatrix}, \quad \mathbf{X}^T\mathbf{y} = \begin{bmatrix} 31 \\ 53 \\ 47 \\ 398 \end{bmatrix},$$

and so

$$\mathbf{b} = \begin{bmatrix} 2.837 \\ 5.024 \\ 4.698 \\ 0.743 \end{bmatrix}.$$

Also, $\mathbf{y}^T\mathbf{y} = 881$ and $\mathbf{b}^T\mathbf{X}^T\mathbf{y} = 870.698$, so for the saturated Model (6.13)

$$\sigma^2 D_1 = \mathbf{y}^T\mathbf{y} - \mathbf{b}^T\mathbf{X}^T\mathbf{y} = 10.302.$$

For the reduced Model (6.14)

$$\boldsymbol{\beta} = \begin{bmatrix} \mu \\ \gamma \end{bmatrix}, \quad \mathbf{X} = \begin{bmatrix} \mathbf{1} & \mathbf{x}_1 \\ \mathbf{1} & \mathbf{x}_2 \\ \mathbf{1} & \mathbf{x}_3 \end{bmatrix}, \quad \text{so} \quad \mathbf{X}^T\mathbf{X} = \begin{bmatrix} 21 & 58 \\ 58 & 196 \end{bmatrix}$$

and

$$\mathbf{X}^T\mathbf{y} = \begin{bmatrix} 131 \\ 398 \end{bmatrix}.$$

Table 6.16 *ANCOVA table for data in Table 6.15.*

Source of variation	Degrees of freedom	Sum of squares	Mean square	F
Mean and covariate	2	853.766		
Factor levels	2	16.932	8.466	13.97
Residuals	17	10.302	0.606	
Total	21	881.000		

Hence,

$$b = \begin{bmatrix} 3.447 \\ 1.011 \end{bmatrix}, \quad b^T X^T y = 853.766, \quad \text{and so} \quad \sigma^2 D_0 = 27.234.$$

If we assume that the saturated Model (6.13) is correct, then $D_1 \sim \chi^2(17)$. If the null hypothesis corresponding to Model (6.14) is true, then $D_0 \sim \chi^2(19)$ so

$$F = \frac{D_0 - D_1}{2\sigma^2} \bigg/ \frac{D_1}{17\sigma^2} \sim F(2, 17).$$

For these data

$$F = \frac{16.932}{2} \bigg/ \frac{10.302}{17} = 13.97,$$

indicating a significant difference in achievement scores for the training methods, after adjustment for initial differences in aptitude. The usual presentation of this analysis is given in Table 6.16.

ANCOVA models are easily fitted using glm functions. For example, the saturated model with the corner point parameterization can be fitted in R using

────────────────── R code (ANCOVA) ──────────────────
```
>res.glm=glm(y~x+method, family=gaussian, data=achieve)
```

and in Stata using

────────────────── Stata code (ANCOVA) ──────────────────
```
.glm y x _Imethod_2 _Imethod_3, family(gaussian) link(identity)
```

6.6   General linear models

The term **general linear model** is used for Normal linear models with any combination of categorical and continuous explanatory variables. The factors may be **crossed**, as in Section 6.4.2, so that there are observations for each

Table 6.17 *Nested two-factor experiment.*

Hospitals	Drug A_1			Drug A_2	
	B_1	B_2	B_3	B_4	B_5
Responses	Y_{111}	Y_{121}	Y_{131}	Y_{241}	Y_{251}
	$\vdots$	$\vdots$	$\vdots$	$\vdots$	$\vdots$
	Y_{11n_1}	Y_{12n_2}	Y_{13n_3}	Y_{24n_4}	Y_{25n_5}

combination of levels of the factors. Alternatively, they may be **nested** as illustrated in the following example.

Table 6.17 shows a two-factor nested design which represents an experiment to compare two drugs (A_1 and A_2), one of which is tested in three hospitals (B_1, B_2 and B_3) and the other in two different hospitals (B_4 and B_5). We want to compare the effects of the two drugs and possible differences among hospitals using the same drug. In this case, the saturated model would be

$$E(Y_{jkl}) = \mu + \alpha_1 + \alpha_2 + (\alpha\beta)_{11} + (\alpha\beta)_{12} + (\alpha\beta)_{13} + (\alpha\beta)_{24} + (\alpha\beta)_{25},$$

subject to some constraints (the corner point constraints are $\alpha_1 = 0, (\alpha\beta)_{11} = 0$ and $(\alpha\beta)_{24} = 0$). Hospitals B_1, B_2 and B_3 can only be compared *within* drug A_1 and hospitals B_4 and B_5 *within* A_2.

Analysis for nested designs is not, in principle, different from analysis for studies with crossed factors. Key assumptions for general linear models are that the response variable has the Normal distribution, the response and explanatory variables are *linearly* related, and the variance σ^2 is the same for all responses. For the models considered in this chapter, the responses are also assumed to be independent (though this assumption is dropped in Chapter 11). All these assumptions can be examined through the use of residuals (Section 6.2.6). If they are not justified, for example, because the residuals have a skewed distribution, then it is usually worthwhile to consider transforming the response variable so that the assumption of Normality is more plausible. A useful tool, now available in many statistical programs, is the **Box–Cox transformation** (Box and Cox 1964). Let y be the original variable and y^* the transformed one, then the function

$$y^* = \begin{cases} \dfrac{y^\lambda - 1}{\lambda} & , \quad \lambda \neq 0 \\ \log y & , \quad \lambda = 0 \end{cases}$$

provides a family of transformations. For example, except for a location shift,

$\lambda = 1$ leaves y unchanged; $\lambda = \frac{1}{2}$ corresponds to taking the square root; $\lambda = -1$ corresponds to the reciprocal; and $\lambda = 0$ corresponds to the logarithmic transformation. The value of λ which produces the "most Normal" distribution can be estimated by the method of maximum likelihood.

Similarly, transformation of continuous explanatory variables may improve the linearity of associations with the response. Alternatively we can model a non-linear association as outlined in the next section.

6.7 Non-linear associations

So far we have only considered linear associations between $\mathbf{X}$ and $\mathbf{y}$, where an increase of Δ in a continuous explanatory variable x_i produces the same change β_i in $\mathbf{y}$ for all values of x_i. β_i is sometimes called a "slope" because it is a linear gradient. A **simple linear regression** equation with a single linear slope is

$$E(Y_i) = \beta_0 + \beta_1 x_i, \qquad i = 1, \dots, N. \tag{6.15}$$

There may be strong reasons to prefer a **non-linear** association because of *a priori* knowledge about the effect of x_i. For example, there is a well-known U-shaped association between daily temperature and daily mortality rates because both low and high temperatures cause physiological changes that increase the risk of death, whilst days of mild temperatures usually have lower risks (Gasparrini et al. 2015). A non-linear association may also be suggested by a plot of the residuals $\widehat{e}_i$ against x_i. For the temperature example, a plot with temperature on the x-axis and the residuals using a linear association for temperature on the y-axis would likely show a U-shaped pattern of positive residuals at low and high temperatures due to the inability of a linear slope to model the U-shaped risk.

A U-shaped association can be modelled by adding a quadratic version of the variable and an additional β parameter

$$E(Y_i) = \beta_0 + \beta_1 x_i + \beta_2 x_i^2, \qquad i = 1, \dots, N. \tag{6.16}$$

This is still **linear regression** because although the regression equation is non-linear in the $\mathbf{x}$'s it is still a linear combination of the two explanatory variables x_i and x_i^2.

In practice when using transformations such as the quadratic which can create large values of x_i's, it can be useful to **center** explanatory variables using their mean ($\bar{x}$) and **scale** them using their standard deviation (sd). For notational convenience, we first create a centred and scaled version of x_i:

$$\tilde{x}_i = (x_i - \bar{x}_i)/\text{sd}$$

Table 6.18 *Estimates for Model (6.6) using centred and scaled explanatory variables.*

Term	Estimate b_j	Standard error
Constant	37.600	1.332
Coefficient for age	−1.452	1.397
Coefficient for weight	−3.793	1.385
Coefficient for protein	4.350	1.411

and fit the model

$$E(Y_i) = \beta_0 + \beta_1 \tilde{x}_i + \beta_2 \tilde{x}_i^2.$$

This transformation improves the numerical accuracy of the matrix multiplication and inversion $(\mathbf{X}^T\mathbf{X})^{-1}$ which is less affected by the very small numbers which can occur if any explanatory variables are large (though modern computers can generally store numbers with many digits). An added advantage of centering is that the estimate for the intercept β_0 now relates the average y value to the average x value instead of the average y value when x is zero which may not be meaningful if x cannot be zero (e.g., a person's weight). Also the "slope" parameters now represent a one standard deviation change which is potentially more meaningful than a single unit change which may be very small or large. Lastly, scaling by the standard deviation makes it easier to compare the importance of variables.

The results of the previous multiple regression model on the carbohydrate diet data shown in Table 6.4 are shown using centred and scaled parameters in Table 6.18. Notice how the standard errors for all four β's are now closer to one. Also notice how the estimates for age and weight look larger than before, whereas previously it looked as if protein was the strongest explanatory variable.

Centering and scaling are mainly used to ease interpretation. They have no impact on sums of squares or hypothesis tests such as comparing Model (6.9) and Model (6.10), the exception being if there were computational problems using the original data. The scaling does not need to use the standard deviation and can be done using a unit change that may be easier to interpret, for example, 10 years for age.

6.7.1 *Example:* PLOS Medicine *journal data*

The data plotted in Figure 6.7 are from 878 journal articles published in the journal *PLOS Medicine* between 2011 and 2015. The plot shows the number of authors on the x-axis and the length of the paper's title (including spaces) on the y-axis. There were 15 papers with more than 30 authors which were truncated to 30. As the number of authors is discrete, a standard scatter plot would likely misrepresent the data as points would overlap and hence be hidden. To avoid this, the points were **jittered**, meaning a small random value was added to every point to avoid overlap.

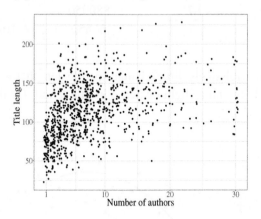

Figure 6.7 *Title length plotted against the number of authors for* PLOS *Medicine journal articles. The values have been jittered to avoid overlap.*

Table 6.19 *Parameter estimates for a linear and quadratic model fitted to the* PLOS Medicine *data.*

Parameter	Model (6.15)		Model (6.16)	
	Estimate	Standard error	Estimate	Standard error
β_0	96.11	1.76	81.40	2.48
β_1	2.18	0.17	6.07	0.51
β_2			−0.15	0.019

Table 6.19 shows the results using a linear association between author numbers and title length. We did not centre and scale the number of authors because an increase of one author is a reasonably sized difference that is easily understood. The slope for the linear model tells us that each additional author adds 2.18 characters on average to the paper's title. The slope for the quadratic model appears much steeper as it suggests an additional 6.07 characters per author, but it is hard to interpret this slope in isolation as we need to also consider the quadratic. A useful way to interpret a non-linear association is a plot as in Figure 6.8 which shows the fitted curve $\hat{\beta}_0 + \hat{\beta}_1 x + \hat{\beta}_2 x^2$ together with the mean title length for each author number from $x = 1$ to $x = 30$. The quadratic model shows a steep increase in title length for fewer than

Table 6.20 *ANOVA table for a linear and quadratic model fitted to the* PLOS Medicine *data.*

Model	Residual degrees of freedom	Residual sum of squares	F	p-value
Linear	876	947502		
Quadratic	875	880950		
Difference	1	66552	66.102	<0.001

ten authors, then relatively little change in title length for author numbers between 15 and 25, and a slight decrease in title length for the largest author numbers.

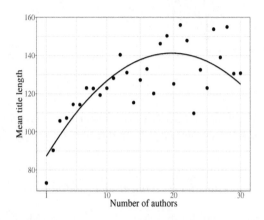

Figure 6.8 *Scatter plot of mean title length plotted against the number of authors for* PLOS Medicine *journal articles together with the fitted curve from the quadratic model.*

The ANOVA Table 6.20 shows that the quadratic model is a significant improvement in model fit compared with the linear model.

The R commands to fit the linear and quadratic models are:

```
──────────────── R code (linear vs quadratic) ────────────────
>lmodel = lm(nchar ~ authors, data=PLOS)
>qmodel = lm(nchar ~ authors + I(authors^2), data=PLOS)
>anova(lmodel, qmodel)
```

The corresponding Stata commands are

```
──────────────── Stata code (linear vs quadratic) ────────────────
.use PLOS
.generate authors2 = authors*authors
```

```
.glm nchar authors
.glm nchar authors authors2
```

6.8 Fractional polynomials

In the previous section we considered a quadratic association, but this only captures a particular non-linear shape. A quadratic function is symmetric around the inflection point, so for a U-shaped quadratic the rate of decrease as x approaches the minimum mirrors the rate of increase as x moves away from the minimum. However, there may well be occasions where the rate of increase is faster than the rate of decrease (or vice versa).

We can examine a range of non-linear associations using **fractional polynomials** (Royston et al. 1999). This approach involves power transformations of x using

$$E(Y_i) = \beta_0 + \beta_1 x_i^p, \qquad i = 1, \ldots, N. \tag{6.17}$$

where the eight candidates for p are $-2, -1, -0.5, 0, 0.5, 1, 2$ and 3 with x^0 corresponding to $\log_e(x)$. As each model has the same number of degrees of freedom, we can choose the best value of p based on the residual sum of squares.

The candidates include associations that are linear $p = 1$, quadratic $p = 2$ and reciprocal quadratic $p = -2$. The approach is similar to the Box–Cox transformation (Section 6.6) of trying multiple power transformations and selecting the best fit. In practice it can be useful to scale $\mathbf{x}$ before comparing the transformations.

Some non-linear curves using fractional polynomials are illustrated in Figure 6.9, where the rate of change in $\mathbf{y}$ can depend on $\mathbf{x}$ in different ways. The curves are also modified by the β parameter, so this approach encompasses a large number of potential non-linear associations.

Table 6.21 *Residual sum of squares for the fractional polynomial models applied to the number of authors for the* PLOS *Medicine data.*

p	Residual sum of squares	p	Residual sum of squares
-2	924,091	0.5	890,891
-1	868,424	1	943,484
-0.5	846,761	2	1,024,999
0	852,395	3	1,065,298

Table 6.21 shows the residual sum of squares for the fractional polyno-

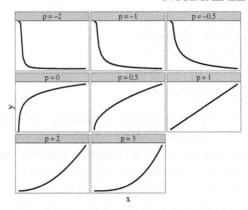

Figure 6.9 *Illustration of some of the potential non-linear associations between x and y using fractional polynomials.*

mial transformations applied to the *PLOS Medicine* data. Before fitting the models, we scaled the number of authors to $x/30$. The best transformation is the reciprocal square-root $p = -0.5$ and the best fitting curve is shown in Figure 6.10. The interpretation of this curve is quite different to the previous quadratic curve (Figure 6.8) as there is a much steeper increase in title length for author numbers from 1 to 5 and a relatively small increase in title length from 10 authors onwards.

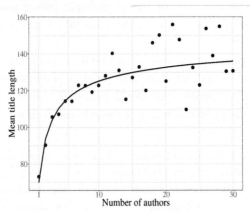

Figure 6.10 *Scatter plot of mean title length plotted against the number of authors for* PLOS Medicine *journal articles together with the best fitting fractional polynomial curve of* $p = -0.5$.

6.9   Exercises

6.1 Table 6.22 shows the average apparent per capita consumption of sugar (in kg per year) in Australia, as refined sugar and in manufactured foods (from Australian Bureau of Statistics, 1998).

Table 6.22 *Australian sugar consumption.*

Period	Refined sugar	Sugar in manufactured food
1936–39	32.0	16.3
1946–49	31.2	23.1
1956–59	27.0	23.6
1966–69	21.0	27.7
1976–79	14.9	34.6
1986–89	8.8	33.9

a. Plot sugar consumption against time separately for refined sugar and sugar in manufactured foods. Fit simple linear regression models to summarize the pattern of consumption of each form of sugar. Calculate 95% confidence intervals for the average annual change in consumption for each form.

b. Calculate the total average sugar consumption for each period and plot these data against time. Using suitable models, test the hypothesis that total sugar consumption did not change over time.

6.2 Table 6.23 shows response of a grass and legume pasture system to various quantities of phosphorus fertilizer (data from D. F. Sinclair; the results were reported in Sinclair and Probert, 1986). The total yield, of grass and legume together, and amount of phosphorus (K) are both given in kilograms per hectare. Find a suitable model for describing the association between yield and quantity of fertilizer.

a. Plot yield against phosphorus to obtain an approximately linear association (you may need to try several transformations of either or both variables in order to achieve approximate linearity).

b. Use the results of (a) to specify a possible model. Fit the model.

c. Calculate the standardized residuals for the model and use appropriate plots to check for any systematic effects that might suggest alternative models and to investigate the validity of any assumptions made.

Table 6.23 *Yield of grass and legume pasture and phosphorus levels (K).*

K	Yield	K	Yield	K	Yield
0	1753.9	15	3107.7	10	2400.0
40	4923.1	30	4415.4	5	2861.6
50	5246.2	50	4938.4	40	3723.0
5	3184.6	5	3046.2	30	4892.3
10	3538.5	0	2553.8	40	4784.6
30	4000.0	10	3323.1	20	3184.6
15	4184.6	40	4461.5	0	2723.1
40	4692.3	20	4215.4	50	4784.6
20	3600.0	40	4153.9	15	3169.3

6.3 Analyze the carbohydrate data in Table 6.3 using appropriate software (or, preferably, repeat the analyses using several different regression programs and compare the results).

   a. Plot the responses y against each of the explanatory variables x_1, x_2 and x_3 to see if y appears to be linearly related to them.

   b. Fit the Model (6.6) and examine the residuals to assess the adequacy of the model and the assumptions.

   c. Fit the models

$$E(Y_i) = \beta_0 + \beta_1 x_{i1} + \beta_3 x_{i3}$$

   and

$$E(Y_i) = \beta_0 + \beta_3 x_{i3}$$

   (note the variable x_2, relative weight, is omitted from both models), and use these to test the hypothesis: $\beta_1 = 0$. Compare your results with Table 6.5.

6.4 It is well known that the concentration of cholesterol in blood serum increases with age, but it is less clear whether cholesterol level is also associated with body weight. Table 6.24 shows for thirty women serum cholesterol (millimoles per liter), age (years) and body mass index (weight divided by height squared, where weight was measured in kilograms and height in meters). Use multiple regression to test whether serum cholesterol is associated with body mass index when age is already included in the model.

6.5 Table 6.25 shows plasma inorganic phosphate levels (mg/dl) one hour after a standard glucose tolerance test for obese subjects, with or without hyperinsulinemia, and controls (data from Jones, 1987).

Table 6.24 *Cholesterol (CHOL), age and body mass index (BMI) for thirty women.*

CHOL	Age	BMI	CHOL	Age	BMI
5.94	52	20.7	6.48	65	26.3
4.71	46	21.3	8.83	76	22.7
5.86	51	25.4	5.10	47	21.5
6.52	44	22.7	5.81	43	20.7
6.80	70	23.9	4.65	30	18.9
5.23	33	24.3	6.82	58	23.9
4.97	21	22.2	6.28	78	24.3
8.78	63	26.2	5.15	49	23.8
5.13	56	23.3	2.92	36	19.6
6.74	54	29.2	9.27	67	24.3
5.95	44	22.7	5.57	42	22.0
5.83	71	21.9	4.92	29	22.5
5.74	39	22.4	6.72	33	24.1
4.92	58	20.2	5.57	42	22.7
6.69	58	24.4	6.25	66	27.3

Table 6.25 *Plasma phosphate levels in obese and control subjects.*

Hyperinsulinemic obese	Non-hyperinsulinemic obese	Controls
2.3	3.0	3.0
4.1	4.1	2.6
4.2	3.9	3.1
4.0	3.1	2.2
4.6	3.3	2.1
4.6	2.9	2.4
3.8	3.3	2.8
5.2	3.9	3.4
3.1		2.9
3.7		2.6
3.8		3.1
		3.2

Table 6.26 *Weights of machine components made by workers on different days.*

| | \multicolumn{4}{c}{Workers} | | | |
	1	2	3	4
Day 1	35.7	38.4	34.9	37.1
	37.1	37.2	34.3	35.5
	36.7	38.1	34.5	36.5
	37.7	36.9	33.7	36.0
	35.3	37.2	36.2	33.8
Day 2	34.7	36.9	32.0	35.8
	35.2	38.5	35.2	32.9
	34.6	36.4	33.5	35.7
	36.4	37.8	32.9	38.0
	35.2	36.1	33.3	36.1

a. Perform a one-factor analysis of variance to test the hypothesis that there are no mean differences among the three groups. What conclusions can you draw?

b. Obtain a 95% confidence interval for the difference in means between the two obese groups.

c. Using an appropriate model, examine the standardized residuals for all the observations to look for any systematic effects and to check the Normality assumption.

6.6 The weights (in grams) of machine components of a standard size made by four different workers on two different days are shown in Table 6.26; five components were chosen randomly from the output of each worker on each day. Perform an analysis of variance to test for differences among workers, among days, and possible interaction effects. What are your conclusions?

6.7 For the balanced data in Table 6.12, the analyses in Section 6.4.2 showed that the hypothesis tests were independent. An alternative specification of the design matrix for the saturated Model (6.9) with the corner point constraints $\alpha_1 = \beta_1 = (\alpha\beta)_{11} = (\alpha\beta)_{12} = (\alpha\beta)_{21} = (\alpha\beta)_{31} = 0$ so that

$$
\beta = \begin{bmatrix} \mu \\ \alpha_2 \\ \alpha_3 \\ \beta_2 \\ (\alpha\beta)_{22} \\ (\alpha\beta)_{32} \end{bmatrix} \quad \text{is} \quad X = \begin{bmatrix}
1 & -1 & -1 & -1 & 1 & 1 \\
1 & -1 & -1 & -1 & 1 & 1 \\
1 & -1 & -1 & 1 & -1 & -1 \\
1 & -1 & -1 & 1 & -1 & -1 \\
1 & 1 & 0 & -1 & -1 & 0 \\
1 & 1 & 0 & -1 & -1 & 0 \\
1 & 1 & 0 & 1 & 1 & 0 \\
1 & 1 & 0 & 1 & 1 & 0 \\
1 & 0 & 1 & -1 & 0 & -1 \\
1 & 0 & 1 & -1 & 0 & -1 \\
1 & 0 & 1 & 1 & 0 & 1 \\
1 & 0 & 1 & 1 & 0 & 1
\end{bmatrix},
$$

where the columns of X corresponding to the terms $(\alpha\beta)_{jk}$ are the products of columns corresponding to terms α_j and β_k.

a. Show that $X^T X$ has the block diagonal form described in Section 6.2.5. Fit the Model (6.9) and also Models (6.10) to (6.12) and verify that the results in Table 6.11 are the same for this specification of X.

b. Show that the estimates for the mean of the subgroup with treatments A_3 and B_2 for two different models are the same as the values given at the end of Section 6.4.2.

6.8 Table 6.27 shows the data from a fictitious two-factor experiment.

a. Test the hypothesis that there are no interaction effects.

b. Test the hypothesis that there is no effect due to Factor A

(i) by comparing the models

$$E(Y_{jkl}) = \mu + \alpha_j + \beta_k \quad \text{and} \quad E(Y_{jkl}) = \mu + \beta_k;$$

(ii) by comparing the models

$$E(Y_{jkl}) = \mu + \alpha_j \quad \text{and} \quad E(Y_{jkl}) = \mu.$$

Explain the results.

6.9 Examine if there is a non-linear association between age and cholesterol using the fractional polynomial approach for the data in Table 6.24.

Table 6.27 *Two-factor experiment with unbalanced data.*

Factor A	Factor B	
	B_1	B_2
A_1	5	3, 4
A_2	6, 4	4, 3
A_3	7	6, 8

## Chapter 7

Binary Variables and Logistic Regression

7.1 Probability distributions

In this chapter we consider generalized linear models in which the outcome variables are measured on a binary scale. For example, the responses may be alive or dead, or present or absent. *Success* and *failure* are used as generic terms of the two categories.

First, we define the **binary random variable**

$$Z = \begin{cases} 1 \text{ if the outcome is a success} \\ 0 \text{ if the outcome is a failure} \end{cases}$$

with probabilities $\Pr(Z = 1) = \pi$ and $\Pr(Z = 0) = 1 - \pi$, which is the Bernoulli distribution $B(\pi)$. If there are n such random variables $Z_1, \ldots, Z_n$, which are independent with $\Pr(Z_j = 1) = \pi_j$, then their joint probability is

$$\prod_{j=1}^{n} \pi_j^{z_j}(1 - \pi_j)^{1-z_j} = \exp\left[\sum_{j=1}^{n} z_j \log\left(\frac{\pi_j}{1 - \pi_j}\right) + \sum_{j=1}^{n} \log(1 - \pi_j)\right], \quad (7.1)$$

which is a member of the exponential family (see Equation (3.3)).

Next, for the case where the π_j's are all equal, we can define

$$Y = \sum_{j=1}^{n} Z_j$$

so that Y is the number of successes in n "trials." The random variable Y has the distribution $\mathrm{Bin}(n, \pi)$:

$$\Pr(Y = y) = \binom{n}{y} \pi^y (1 - \pi)^{n-y}, \quad y = 0, 1, \ldots, n. \quad (7.2)$$

Finally, we consider the general case of N independent random variables

$Y_1, Y_2, \ldots, Y_N$ corresponding to the numbers of successes in N different subgroups or strata (Table 7.1). If $Y_i \sim \text{Bin}(n_i, \pi_i)$, the log-likelihood function is

$$l(\pi_1, \ldots, \pi_N; y_1, \ldots, y_N)$$
$$= \sum_{i=1}^{N} \left[y_i \log \left(\frac{\pi_i}{1 - \pi_i} \right) + n_i \log(1 - \pi_i) + \log \binom{n_i}{y_i} \right]. \qquad (7.3)$$

Table 7.1 *Frequencies for N Binomial distributions.*

	Subgroups			
	1	2	...	N
Successes	Y_1	Y_2	...	Y_N
Failures	$n_1 - Y_1$	$n_2 - Y_2$	...	$n_N - Y_N$
Totals	n_1	n_2	...	n_N

7.2 Generalized linear models

We want to describe the proportion of successes, $P_i = Y_i/n_i$, in each subgroup in terms of factor levels and other explanatory variables which characterize the subgroup. As $E(Y_i) = n_i \pi_i$ and so $E(P_i) = \pi_i$, we model the probabilities π_i as

$$g(\pi_i) = \mathbf{x}_i^T \boldsymbol{\beta},$$

where $\mathbf{x}_i$ is a vector of explanatory variables (dummy variables for factor levels and measured values for covariates), $\boldsymbol{\beta}$ is a vector of parameters and g is a link function.

The simplest case is the **linear model**

$$\pi = \mathbf{x}^T \boldsymbol{\beta}.$$

This is used in some practical applications, but it has the disadvantage that although π is a probability, the fitted values $\mathbf{x}^T \mathbf{b}$ may be less than zero or greater than one.

To ensure that π is restricted to the interval $[0,1]$ it is often modelled using a cumulative probability distribution

$$\pi = \int_{-\infty}^{t} f(s) ds,$$

where $f(s) \geqslant 0$ and $\int_{-\infty}^{\infty} f(s) ds = 1$. The probability density function $f(s)$ is called the **tolerance distribution**. Some commonly used examples are considered in Section 7.3.

7.3   Dose response models

Historically, one of the first uses of regression-like models for Binomial data was for bioassay results (Finney 1973). Responses were the proportions or percentages of "successes"; for example, the proportion of experimental animals killed by various dose levels of a toxic substance. Such data are sometimes called **quantal responses**. The aim is to describe the probability of "success", π, as a function of the dose, x; for example, $g(\pi) = \beta_1 + \beta_2 x$.

If the tolerance distribution $f(s)$ is the Uniform distribution on the interval $[c_1, c_2]$

$$f(s) = \begin{cases} \dfrac{1}{c_2 - c_1} & \text{if } c_1 \leqslant s \leqslant c_2 \\ 0 & \text{otherwise} \end{cases},$$

then π is cumulative

$$\pi = \int_{c_1}^x f(s)ds = \frac{x - c_1}{c_2 - c_1} \qquad \text{for } c_1 \leqslant x \leqslant c_2$$

(see Figure 7.1). This equation has the form $\pi = \beta_1 + \beta_2 x$, where

$$\beta_1 = \frac{-c_1}{c_2 - c_1} \text{ and } \beta_2 = \frac{1}{c_2 - c_1}.$$

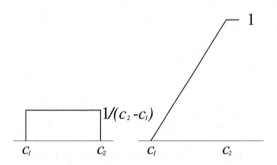

Figure 7.1 *Uniform distribution:* $f(s)$ *and* π.

This **linear model** is equivalent to using the identity function as the link function g and imposing conditions on x, β_1 and β_2 corresponding to $c_1 \leq x \leq c_2$. These extra conditions mean that the standard methods for estimating β_1 and β_2 for generalized linear models cannot be directly applied. In practice, this model is not widely used.

One of the original models used for bioassay data is called the **probit model**. The Normal distribution is used as the tolerance distribution (see Figure 7.2).

$$\pi \; = \; \frac{1}{\sigma\sqrt{2\pi}} \int_{-\infty}^{x} \exp\left[-\frac{1}{2}\left(\frac{s-\mu}{\sigma} \right)^2 \right] ds$$

$$= \; \Phi\left(\frac{x-\mu}{\sigma} \right),$$

where Φ denotes the cumulative probability function for the standard Normal distribution $N(0,1)$. Thus,

$$\Phi^{-1}(\pi) = \beta_1 + \beta_2 x$$

where $\beta_1 = -\mu/\sigma$ and $\beta_2 = 1/\sigma$ and the link function g is the inverse cumulative Normal probability function Φ^{-1}. Probit models are used in several areas of biological and social sciences in which there are natural interpretations of the model; for example, $x = \mu$ is called the **median lethal dose** LD(50) because it corresponds to the dose that can be expected to kill half of the animals.

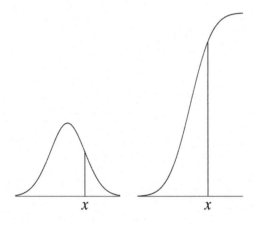

Figure 7.2 *Normal distribution:* $f(s)$ *and* π.

Another model that gives numerical results very much like those from the probit model, but which computationally is somewhat easier, is the **logistic** or **logit model**. The tolerance distribution is

$$f(s) = \frac{\beta_2 \exp(\beta_1 + \beta_2 s)}{[1 + \exp(\beta_1 + \beta_2 s)]^2},$$

so

$$\pi = \int_{-\infty}^{x} f(s)ds = \frac{\exp(\beta_1 + \beta_2 x)}{1 + \exp(\beta_1 + \beta_2 x)}.$$

This gives the link function

$$\log\left(\frac{\pi}{1-\pi}\right) = \beta_1 + \beta_2 x.$$

The term $\log[\pi/(1-\pi)]$ is sometimes called the **logit function** and it has a natural interpretation as the logarithm of odds (see Exercise 7.2). The logistic model is widely used for Binomial data and is implemented in many statistical programs. The shapes of the functions $f(s)$ and $\pi(x)$ are similar to those for the probit model (Figure 7.2) except in the tails of the distributions (see Cox and Snell, 1989).

Several other models are also used for dose response data. For example, if the **extreme value distribution**

$$f(s) = \beta_2 \exp\left[(\beta_1 + \beta_2 s) - \exp(\beta_1 + \beta_2 s)\right]$$

is used as the tolerance distribution, then

$$\pi = 1 - \exp\left[-\exp(\beta_1 + \beta_2 x)\right],$$

and so $\log[-\log(1-\pi)] = \beta_1 + \beta_2 x$. This link, $\log[-\log(1-\pi)]$, is called the **complementary log-log function**. The model is similar to the logistic and probit models for values of π near 0.5 but differs from them for π near 0 or 1. These models are illustrated in the following example.

Table 7.2 *Beetle mortality data.*

Dose, x_i ($\log_{10} CS_2 mgl^{-1}$)	Number of beetles, n_i	Number killed, y_i
1.6907	59	6
1.7242	60	13
1.7552	62	18
1.7842	56	28
1.8113	63	52
1.8369	59	53
1.8610	62	61
1.8839	60	60

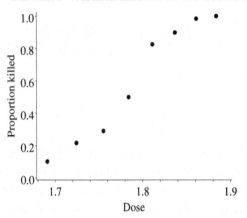

Figure 7.3 *Beetle mortality data from Table 7.2: proportion killed, $p_i = y_i/n_i$, plotted against dose, x_i ($log_{10}CS_2mgl^{-1}$).*

7.3.1   Example: Beetle mortality

Table 7.2 shows numbers of beetles dead after five hours of exposure to gaseous carbon disulphide at various concentrations (data from Bliss, 1935). Figure 7.3 shows the proportions $p_i = y_i/n_i$ plotted against dose x_i (actually x_i is the logarithm of the quantity of carbon disulphide). We begin by fitting the logistic model

$$\pi_i = \frac{\exp(\beta_1 + \beta_2 x_i)}{1 + \exp(\beta_1 + \beta_2 x_i)}$$

so

$$\log\left(\frac{\pi_i}{1 - \pi_i}\right) = \beta_1 + \beta_2 x_i$$

and

$$\log(1 - \pi_i) = -\log[1 + \exp(\beta_1 + \beta_2 x_i)].$$

Therefore from Equation (7.3) the log-likelihood function is

$$l = \sum_{i=1}^{N}\left[y_i(\beta_1 + \beta_2 x_i) - n_i \log[1 + \exp(\beta_1 + \beta_2 x_i)] + \log\binom{n_i}{y_i}\right],$$

and the scores with respect to β_1 and β_2 are

$$
\begin{aligned}
U_1 &= \frac{\partial l}{\partial \beta_1} = \sum\left\{y_i - n_i\left[\frac{\exp(\beta_1 + \beta_2 x_i)}{1 + \exp(\beta_1 + \beta_2 x_i)}\right]\right\} = \sum(y_i - n_i\pi_i) \\
U_2 &= \frac{\partial l}{\partial \beta_2} = \sum\left\{y_i x_i - n_i x_i\left[\frac{\exp(\beta_1 + \beta_2 x_i)}{1 + \exp(\beta_1 + \beta_2 x_i)}\right]\right\} \\
&= \sum x_i(y_i - n_i\pi_i).
\end{aligned}
$$

Similarly the information matrix is

$$
\mathfrak{I} = \begin{bmatrix} \sum n_i \pi_i (1 - \pi_i) & \sum n_i x_i \pi_i (1 - \pi_i) \\[2mm] \sum n_i x_i \pi_i (1 - \pi_i) & \sum n_i x_i^2 \pi_i (1 - \pi_i) \end{bmatrix}.
$$

Maximum likelihood estimates are obtained by solving the iterative equation

$$
\mathfrak{I}^{(m-1)} \mathbf{b}^m = \mathfrak{I}^{(m-1)} \mathbf{b}^{(m-1)} + \mathbf{U}^{(m-1)}
$$

(from (4.22)) where the superscript (m) indicates the mth approximation and $\mathbf{b}$ is the vector of estimates. Starting with $b_1^{(0)} = 0$ and $b_2^{(0)} = 0$, successive approximations are shown in Table 7.3. The estimates converge by the sixth iteration. The table also shows the increase in values of the log-likelihood function (7.3), omitting the constant term $\log \binom{n_i}{y_i}$. The fitted values are $\widehat{y}_i = n_i \widehat{\pi}_i$ calculated at each stage (initially $\widehat{\pi}_i = 0.5$ for all i).

For the final approximation, the estimated variance–covariance matrix for $\mathbf{b}$, $[\mathfrak{I}(\mathbf{b})^{-1}]$, is shown at the bottom of Table 7.3 together with the deviance

$$
D = 2 \sum_{i=1}^{N} \left[y_i \log \left(\frac{y_i}{\widehat{y}_i} \right) + (n_i - y_i) \log \left(\frac{n - y_i}{n - \widehat{y}_i} \right) \right]
$$

(from Section 5.6.1).

The estimates and their standard errors are

$$
b_1 = -60.72, \quad \text{standard error} = \sqrt{26.840} = 5.18
$$
$$
\text{and} \quad b_2 = 34.27, \quad \text{standard error} = \sqrt{8.481} = 2.91.
$$

If the model is a good fit of the data, the deviance should approximately have the distribution $\chi^2(6)$ because there are $N = 8$ covariate patterns (i.e., different values of x_i) and $p = 2$ parameters. But the calculated value of D is almost twice the "expected" value of 6 and is almost as large as the upper 5% point of the $\chi^2(6)$ distribution, which is 12.59. This suggests that the model does not fit particularly well.

Statistical software for fitting generalized linear models for dichotomous responses often differs between the case when the data are grouped as counts of successes y and failures $n - y$ in n trials with the same covariate pattern and the case when the data are binary (0 or 1) responses (see later example with data in Table 7.8). For Stata the logistic regression model for the grouped data for beetle mortality in Table 7.2 can be fitted using the following command

──────────── Stata code (logistic regression) ────────────
```
.glm y x, family(binomial n) link(logit)
```

Table 7.3 *Fitting a linear logistic model to the beetle mortality data.*

	Initial estimate	First	Second	Sixth
		Approximation		
β_1	0	−37.856	−53.853	−60.717
β_2	0	21.337	30.384	34.270
log-likelihood	−333.404	−200.010	−187.274	−186.235

Observations		Fitted values		
y_1	6	29.5	8.505	4.543	3.458
y_2	13	30.0	15.366	11.254	9.842
y_3	18	31.0	24.808	23.058	22.451
y_4	28	28.0	30.983	32.947	33.898
y_5	52	31.5	43.362	48.197	50.096
y_6	53	29.5	46.741	51.705	53.291
y_7	61	31.0	53.595	58.061	59.222
y_8	60	30.0	54.734	58.036	58.743

$$[\mathcal{J}(\mathbf{b})]^{-1} = \begin{bmatrix} 26.840 & -15.082 \\ -15.082 & 8.481 \end{bmatrix}, \quad D = 11.23$$

The estimated variance–covariance matrix can be obtained by selecting the option to display the negative Hessian matrix using

────────── Stata code (logistic regression) ──────────
```
.glm y x, family(binomial n) link(logit) hessian
```

Evaluated for the final estimates, this matrix is given as $\begin{bmatrix} 58.484 & 104.011 \\ 104.011 & 185.094 \end{bmatrix}$,

which can be inverted to obtain $\begin{bmatrix} 26.8315 & -15.0775 \\ -15.0775 & 8.4779 \end{bmatrix}$, which is the

same as in Table 7.3 (except for rounding effects). The value for the log-likelihood shown by Stata does not include the term $\sum_{i=1}^{N} \log \binom{n_i}{y_i}$ in (7.3). For the beetle mortality data the value of this term in -167.5203 so, compared with the value of -186.235 in Table 7.3, the log-likelihood value shown by Stata is $-186.235 - (-167.5203) = -18.715$. Fitted values can be obtained after the model is fitted using the command

────────── Stata code (fitted values) ──────────
```
.predict fit, mu
```

To use R to fit generalized linear models to grouped dichotomous data, it is necessary to construct a response matrix with two columns, y and $(n - y)$, as shown below for the beetle mortality data using the S-PLUS lists denoted by c.

```
—————————— R code (data entry and manipulation) ——————————
>y=c(6,13,18,28,52,53,61,60)
>n=c(59,60,62,56,63,59,62,60)
>x=c(1.6907,1.7242,1.7552,1.7842,1.8113,1.8369,1.8610,1.8839)
>n_y=n-y
>beetle.mat=cbind(y,n_y)
```

The logistic regression model can then be fitted using the command

```
—————————— R code (logistic regression) ——————————
>res.glm1=glm(beetle.mat~x, family=binomial(link="logit"))
```

The fitted values obtained from

```
—————————— R code (fitted values) ——————————
>fitted.values(res.glm1)
```

are estimated proportions of deaths in each group and so the fitted values for **y** need to be calculated as follows:

```
—————————— R code (fitted values) ——————————
>fit_p=c(fitted.values(res.glm1))
>fit_y=n*fit_p
```

Several alternative models can be fitted to the beetle mortality data. The results are shown in Table 7.4. Among these models the extreme value model appears to fit the data best. For Stata the relevant commands are

```
—————————— Stata code (probit model) ——————————
.glm y x, family(binomial n) link(probit)
```

and

```
—————————— Stata code (extreme value model) ——————————
.glm y x, family(binomial n) link(cloglog)
```

For R they are

```
—————————— R code (probit model) ——————————
>res.glm2=glm(beetle.mat~x, family=binomial(link="probit"))
```

and

Table 7.4 *Comparison of observed numbers killed with fitted values obtained from various dose-response models for the beetle mortality data. Deviance statistics are also given.*

Observed value of Y	Logistic model	Probit model	Extreme value model
6	3.46	3.36	5.59
13	9.84	10.72	11.28
18	22.45	23.48	20.95
28	33.90	33.82	30.37
52	50.10	49.62	47.78
53	53.29	53.32	54.14
61	59.22	59.66	61.11
60	58.74	59.23	59.95
D	11.23	10.12	3.45
$b_1(s.e.)$	$-60.72(5.18)$	$-34.94(2.64)$	$-39.57(3.23)$
$b_2(s.e.)$	$34.27(2.91)$	$19.73(1.48)$	$22.04(1.79)$

```
———————————— R code (extreme value model) ——————————
>res.glm3=glm(beetle.mat~x, family=binomial(link="cloglog"))
```

7.4   General logistic regression model

The simple linear logistic model $\log[\pi_i/(1 - \pi_i)] = \beta_1 + \beta_2 x_i$ used in Example 7.3.1 is a special case of the general logistic regression model

$$\text{logit } \pi_i = \log\left(\frac{\pi_i}{1 - \pi_i}\right) = \mathbf{x}_i^T \boldsymbol{\beta},$$

where $\mathbf{x}_i$ is a vector of continuous measurements corresponding to covariates and dummy variables corresponding to factor levels and $\boldsymbol{\beta}$ is the parameter vector. This model is very widely used for analyzing data involving binary or Binomial responses and several explanatory variables. It provides a powerful technique analogous to multiple regression and ANOVA for continuous responses.

Maximum likelihood estimates of the parameters $\boldsymbol{\beta}$, and consequently of the probabilities $\pi_i = g^{-1}(\mathbf{x}_i^T \boldsymbol{\beta})$, are obtained by maximizing the log-likelihood function

$$l(\boldsymbol{\pi}; \mathbf{y}) = \sum_{i=1}^{N} [y_i \log \pi_i + (n_i - y_i) \log(1 - \pi_i) + \log\binom{n_i}{y_i}] \qquad (7.4)$$

using the methods described in Chapter 4.

The estimation process is essentially the same whether the data are grouped as frequencies for each **covariate pattern** (i.e., observations with the same values of all the explanatory variables) or each observation is coded 0 or 1 and its covariate pattern is listed separately. If the data can be grouped, the response Y_i, the number of "successes" for covariate pattern i, may be modelled by the Binomial distribution. If each observation has a different covariate pattern, then $n_i = 1$ and the response Y_i is binary.

The deviance, derived in Section 5.6.1, is

$$D = 2\sum_{i=1}^{N}\left[y_i\log\left(\frac{y_i}{\widehat{y}_i}\right) + (n_i - y_i)\log\left(\frac{n_i - y_i}{n_i - \widehat{y}_i}\right)\right]. \qquad (7.5)$$

This has the form

$$D = 2\sum o\log\frac{o}{e}$$

where o denotes the observed "successes" y_i and "failures" $(n_i - y_i)$ from the cells of Table 7.1 and e denotes the corresponding estimated expected frequencies or fitted values $\widehat{y}_i = n_i\widehat{\pi}_i$ and $(n_i - \widehat{y}_i) = (n_i - n_i\widehat{\pi}_i)$. Summation is over all $2 \times N$ cells of the table.

Notice that D does not involve any nuisance parameters (like σ^2 for Normal response data), so goodness of fit can be assessed and hypotheses can be tested directly using the approximation

$$D \sim \chi^2(N - p),$$

where p is the number of parameters estimated and N the number of covariate patterns.

The estimation methods and sampling distributions used for inference depend on asymptotic results. For small studies or situations where there are few observations for each covariate pattern, the asymptotic results may be poor approximations. However software, such as StatXact and LogXact, has been developed using "exact" methods so that the methods described in this chapter can be used even when sample sizes are small.

7.4.1   Example: Embryogenic anthers

The data in Table 7.5, cited by Wood (1978), are taken from Sangwan-Norrell (1977). They are numbers y_{jk} of embryogenic anthers of the plant species *Datura innoxia* Mill. obtained when numbers n_{jk} of anthers were prepared under several different conditions. There is one qualitative factor with two

Table 7.5 *Embryogenic anther data.*

Storage condition		Centrifuging force (g)		
		40	150	350
Control	y_{1k}	55	52	57
	n_{1k}	102	99	108
Treatment	y_{2k}	55	50	50
	n_{2k}	76	81	90

levels, a treatment consisting of storage at 3°C for 48 hours or a control storage condition, and one continuous explanatory variable represented by three values of centrifuging force. We will compare the treatment and control effects on the proportions after adjustment (if necessary) for centrifuging force.

The proportions $p_{jk} = y_{jk}/n_{jk}$ in the control and treatment groups are plotted against x_k, the logarithm of the centrifuging force, in Figure 7.4. The response proportions appear to be higher in the treatment group than in the control group, and at least for the treated group, the response decreases with centrifuging force.

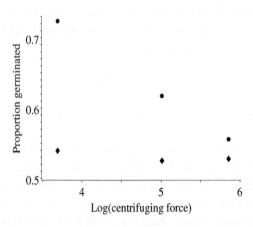

Figure 7.4 *Anther data from Table 7.5: proportion that germinated $p_{jk} = y_{jk}/n_{jk}$ plotted against $\log_e$(centrifuging force); dots represent the treatment condition and diamonds represent the control condition.*

We will compare three logistic models for π_{jk}, the probability of the anthers being embryogenic, where $j = 1$ for the control group and $j = 2$ for the treatment group and $x_1 = \log_e 40 = 3.689$, $x_2 = \log_e 150 = 5.011$, and $x_3 = \log_e 350 = 5.858$.

Model 1: logit $\pi_{jk} = \alpha_j + \beta_j x_k$ (i.e., different intercepts and slopes);

Model 2: logit $\pi_{jk} = \alpha_j + \beta x_k$ (i.e., different intercepts but the same slope);

Model 3: logit $\pi_{jk} = \alpha + \beta x_k$ (i.e., same intercept and slope).

These models were fitted by the method of maximum likelihood. The results are summarized in Table 7.6. To test the null hypothesis that the slope is the same for the treatment and control groups, we use $D_2 - D_1 = 2.591$. From the tables for the $\chi^2(1)$ distribution, the significance level is between 0.1 and 0.2, and so we could conclude that the data provide little evidence against the null hypothesis of equal slopes. On the other hand, the power of this test is very low and both Figure 7.4 and the estimates for Model 1 suggest that although the slope for the control group may be zero, the slope for the treatment group is negative. Comparison of the deviances from Models 2 and 3 gives a test for equality of the control and treatment effects after a common adjustment for centrifuging force: $D_3 - D_2 = 5.473$, which indicates that the storage effects are different. The observed proportions and the corresponding fitted values for Models 1, 2 and 3 are shown in Table 7.7. Obviously, Model 1 fits the data very well but this is hardly surprising since four parameters have been used to describe six data points—such "over-fitting" is not recommended!

Table 7.6 *Maximum likelihood estimates and deviances for logistic models for the embryogenic anther data (standard errors of estimates in brackets).*

Model 1	Model 2	Model 3
$a_1 = 0.234(0.628)$	$a_1 = 0.877(0.487)$	$a = 1.021(0.481)$
$a_2 - a_1 = 1.977(0.998)$	$a_2 - a_1 = 0.407(0.175)$	$b = -0.148(0.096)$
$b_1 = -0.023(0.127)$	$b = -0.155(0.097)$	
$b_2 - b_1 = -0.319(0.199)$		
$D_1 = 0.028$	$D_2 = 2.619$	$D_3 = 8.092$

These results can be reproduced using Stata. If the control and treatment groups are recoded to 0 and 1, respectively (in a variable called newstor $= j - 1$), and an interaction term is created by multiplying this variable and the **x** vector, then the models can be fitted using the following commands:

```
————————————— Stata code (logistic models) —————————————
.glm y newstor x interaction, family(binomial n) link(logit)
.glm y newstor x, family(binomial n) link(logit)
.glm y x, family(binomial n) link(logit)
```

For R it is necessary to use columns of **y** and $(\mathbf{n} - \mathbf{y})$ and then the commands are similar to those for Stata. For example, for Model 1 the interaction term can be explicitly used and it includes the main effects

—————————————— R code (logistic model) ——————————————

```
>data(anthers)
>res.glm3=glm(cbind(y,n-y)~newstor*x,family=binomial(link=
  "logit"),data=anthers)
```

Table 7.7 *Observed and expected frequencies for the embryogenic anther data for various models.*

Storage condition	Covariate value	Observed frequency	Expected frequencies		
			Model 1	Model 2	Model 3
Control	x_1	55	54.82	58.75	62.91
	x_2	52	52.47	52.03	56.40
	x_3	57	56.72	53.22	58.18
Treatment	x_1	55	54.83	51.01	46.88
	x_2	50	50.43	50.59	46.14
	x_3	50	49.74	53.40	48.49

7.5   Goodness of fit statistics

Instead of using maximum likelihood estimation, we could estimate the parameters by minimizing the weighted sum of squares

$$S_w = \sum_{i=1}^{N} \frac{(y_i - n_i \pi_i)^2}{n_i \pi_i (1 - \pi_i)}$$

since $E(Y_i) = n_i \pi_i$ and $\text{var}(Y_i) = n_i \pi_i (1 - \pi_i)$.

This is equivalent to minimizing the **Pearson chi-squared statistic**

$$X^2 = \sum \frac{(o - e)^2}{e},$$

where o represents the observed frequencies in Table 7.1, e represents the expected frequencies and summation is over all $2 \times N$ cells of the table. The reason is that

$$X^2 = \sum_{i=1}^{N} \frac{(y_i - n_i \pi_i)^2}{n_i \pi_i} + \sum_{i=1}^{N} \frac{[(n_i - y_i) - n_i(1 - \pi_i)]^2}{n_i(1 - \pi_i)}$$

$$= \sum_{i=1}^{N} \frac{(y_i - n_i \pi_i)^2}{n_i \pi_i (1 - \pi_i)} (1 - \pi_i + \pi_i) = S_w.$$

When X^2 is evaluated at the estimated expected frequencies, the statistic is

$$X^2 = \sum_{i=1}^{N} \frac{(y_i - n_i \widehat{\pi}_i)^2}{n_i \widehat{\pi}_i (1 - \widehat{\pi}_i)} \tag{7.6}$$

which is asymptotically equivalent to the deviances in (7.5),

$$D = 2 \sum_{i=1}^{N} \left[y_i \log \left(\frac{y_i}{n_i \widehat{\pi}_i} \right) + (n_i - y_i) \log \left(\frac{n_i - y_i}{n_i - n_i \widehat{\pi}_i} \right) \right].$$

The proof of the relationship between X^2 and D uses the Taylor series expansion of $s \log(s/t)$ about $s = t$, namely,

$$s \log \frac{s}{t} = (s-t) + \frac{1}{2} \frac{(s-t)^2}{t} + \cdots \quad .$$

Thus,

$$
\begin{aligned}
D &= 2 \sum_{i=1}^{N} \left\{ (y_i - n_i \widehat{\pi}_i) + \frac{1}{2} \frac{(y_i - n_i \widehat{\pi}_i)^2}{n_i \widehat{\pi}_i} + [(n_i - y_i) - (n_i - n_i \widehat{\pi}_i)] \right. \\
&\quad \left. + \frac{1}{2} \frac{[(n_i - y_i) - (n_i - n_i \widehat{\pi}_i)]^2}{n_i - n_i \widehat{\pi}_i} + \cdots \right\} \\
&\cong \sum_{i=1}^{N} \frac{(y_i - n_i \widehat{\pi}_i)^2}{n_i \widehat{\pi}_i (1 - \widehat{\pi}_i)} = X^2.
\end{aligned}
$$

The asymptotic distribution of D, under the hypothesis that the model is correct, is $D \sim \chi^2(N-p)$, therefore, approximately $X^2 \sim \chi^2(N-p)$. The choice between D and X^2 depends on the adequacy of the approximation to the $\chi^2(N-p)$ distribution. There is some evidence to suggest that X^2 is often better than D because D is unduly influenced by very small frequencies (Cressie and Read 1989). Both the approximations are likely to be poor, however, if the expected frequencies are too small (e.g., less than 1).

In particular, if each observation has a different covariate pattern so y_i is zero or one, then neither D nor X^2 provides a useful measure of fit. This can happen if the explanatory variables are continuous, for example. The most commonly used approach in this situation is due to Hosmer and Lemeshow (1980). Their idea was to group observations into categories on the basis of their predicted probabilities. Typically about 10 groups are used with approximately equal numbers of observations in each group. The observed numbers of successes and failures in each of the g groups are summarized as shown in Table 7.1. Then the Pearson chi-squared statistic for a $g \times 2$ contingency table is calculated and used as a measure of fit. This **Hosmer–Lemeshow statistic** is denoted by X_{HL}^2. The sampling distribution of X_{HL}^2 has been found by simulation to be approximately $\chi^2(g-2)$. The use of this statistic is illustrated in the example in Section 7.8.

Sometimes the log-likelihood function for the fitted model is compared with the log-likelihood function for a minimal model, in which the values π_i are all equal (in contrast to the saturated model which is used to define the deviance). Under the **minimal model** $\tilde{\pi} = (\Sigma y_i)/(\Sigma n_i)$. Let $\hat{\pi}_i$ denote the estimated probability for Y_i under the model of interest (so the fitted value is $\hat{y}_i = n_i \hat{\pi}_i$). The statistic is defined by

$$C = 2\left[l\left(\hat{\boldsymbol{\pi}};\mathbf{y}\right) - l\left(\tilde{\boldsymbol{\pi}};\mathbf{y}\right)\right],$$

where l is the log-likelihood function given by (7.4). Thus,

$$C = 2\sum\left[y_i\log\left(\frac{\hat{y}_i}{n_i\tilde{\pi}_i}\right) + (n_i - y_i)\log\left(\frac{n_i - \hat{y}_i}{n_i - n_i\tilde{\pi}_i}\right)\right].$$

From the results in Section 5.2, the approximate sampling distribution for C is $\chi^2(p-1)$ if all the p parameters except the intercept term β_1 are zero (see Exercise 7.4). Otherwise C will have a non-central distribution. Thus C is a test statistic for the hypothesis that none of the explanatory variables is needed for a parsimonious model. C is sometimes called the **likelihood ratio chi-squared statistic**.

By analogy with R^2 for multiple linear regression (see Section 6.3.2) another statistic sometimes used is

$$\text{pseudo } R^2 = \frac{l\left(\tilde{\boldsymbol{\pi}};\mathbf{y}\right) - l\left(\hat{\boldsymbol{\pi}};\mathbf{y}\right)}{l\left(\tilde{\boldsymbol{\pi}};\mathbf{y}\right)},$$

which represents the proportional improvement in the log-likelihood function due to the terms in the model of interest, compared with the minimal model. This statistic is produced by some statistical programs as a measure of goodness of fit. As for R^2, the sampling distribution of pseudo R^2 is not readily determined (so p-values cannot be obtained), and it increases as more parameters are added to the model. Therefore, various modifications of pseudo R^2 are used to adjust for the number of parameters (see, for example, Liao and McGee, 2003). For logistic regression R^2-type measures often appear alarmingly small even when other measures suggest that the model fits the data well. The reason is that pseudo R^2 is a measure of the predictability of individual outcomes Y_i rather than the predictability of all the event rates (Mittlbock and Heinzl 2001).

The **Akaike information criterion** *AIC* and the Schwartz or **Bayesian information criterion** *BIC* are other goodness of fit statistics based on the log-likelihood function with adjustment for the number of parameters estimated and for the amount of data. These statistics are usually defined as follows:

$$AIC = -2l\left(\widehat{\pi};\mathbf{y}\right) + 2p \tag{7.7}$$

$$BIC = -2l\left(\widehat{\pi};\mathbf{y}\right) + p \times \ln(\text{number of observations}),$$

where p is the number of parameters estimated. The statistical software R uses this definition of *AIC*, for example. However, the versions used by Stata are somewhat different:

$$AIC_{Stata} = \left(-2l\left(\widehat{\pi};\mathbf{y}\right) + 2p\right)/N,$$

where N is the number of subgroups in Table 7.1, and

$$BIC_{Stata} = D - (N - p)\ln\sum_{i=1}^{N} n_i,$$

where D is the deviance for the model, $(N - p)$ is the corresponding degrees of freedom and $\sum_{i=1}^{N} n_i$ is the total number of observations.

Note that the statistics (except for pseudo R^2) discussed in this section summarize how well a particular model fits the data. So a small value of the statistic and, hence, a large p-value, indicates that the model fits well. These statistics are not usually appropriate for testing hypotheses about the parameters of nested models, but they can be particularly useful for comparing models that are not nested.

For the logistic model for the beetle mortality example (Section 7.3.1), the log-likelihood for the model with no explanatory variable is $l\left(\widetilde{\pi};\mathbf{y}\right) = (-167.5203 - 155.2002) = -322.7205$, while $l\left(\widehat{\pi};\mathbf{y}\right) = (-167.5203 - 18.7151) = -186.2354$, so the statistic $C = 2 \times (-186.2354 - (-322.7205)) = 272.970$ with one degree of freedom, indicating that the slope parameter β_1 is definitely needed! The pseudo R^2 value is $2 \times (-322.72051 - (-186.23539))/2 \times (-322.72051) = 0.4229$ indicating reasonable but not excellent fit. The usual value for $AIC = -2 \times (-18.7151) + 2 \times 2 = 41.430$—this is the value given by R, for example. Stata gives $AIC_{Stata} = 41.430/8 = 5.179$ and $BIC_{Stata} = 11.2322 - (8 - 2) \times \ln 481 = -25.823$. All these measures show marked improvements when the extreme value model is fitted compared with the logistic model.

7.6   Residuals

For logistic regression there are two main forms of residuals corresponding to the goodness of fit measures D and X^2. If there are m different covariate patterns, then m residuals can be calculated. Let Y_k denote the number of successes, n_k the number of trials and $\widehat{\pi}_k$ the estimated probability of success for the kth covariate pattern.

The **Pearson**, or **chi-squared**, **residual** is

$$X_k = \frac{(y_k - n_k\widehat{\pi}_k)}{\sqrt{n_k\widehat{\pi}_k(1-\widehat{\pi}_k)}} \quad , k = 1, \dots, m. \tag{7.8}$$

From (7.6), $\sum_{k=1}^{m} X_k^2 = X^2$, the Pearson chi-squared goodness of fit statistic. The **standardized Pearson residuals** are

$$r_{Pk} = \frac{X_k}{\sqrt{1-h_k}},$$

where h_k is the leverage, which is obtained from the hat matrix (see Section 6.2.6).

Deviance residuals can be defined similarly,

$$d_k = \text{sign}(y_k - n_k\widehat{\pi}_k)\left\{2\left[y_k \log\left(\frac{y_k}{n_k\widehat{\pi}_k}\right) + (n_k - y_k)\log\left(\frac{n_k - y_k}{n_k - n_k\widehat{\pi}_k}\right)\right]\right\}^{1/2} \tag{7.9}$$

where the term $\text{sign}(y_k - n_k\widehat{\pi}_k)$ ensures that d_k has the same sign as X_k. From Equation (7.5), $\sum_{k=1}^{m} d_k^2 = D$, the deviance. Also standardized deviance residuals are defined by

$$r_{Dk} = \frac{d_k}{\sqrt{1-h_k}}.$$

Pearson and deviance residuals can be used for checking the adequacy of a model, as described in Section 2.3.4. For example, they should be plotted against each continuous explanatory variable in the model to check if the assumption of linearity is appropriate and against other possible explanatory variables not included in the model. They should be plotted in the order of the measurements, if applicable, to check for serial correlation. Normal probability plots can also be used because the standardized residuals should have, approximately, the standard Normal distribution $N(0,1)$, provided the numbers of observations for each covariate pattern are not too small.

If the data are binary, or if n_k is small for most covariate patterns, then there are few distinct values of the residuals and the plots may be relatively uninformative. In this case, it may be necessary to rely on the aggregated goodness of fit statistics X^2 and D and other diagnostics (see Section 7.7).

For more details about the use of residuals for Binomial and binary data, see Chapter 5 of Collett (2003), for example.

7.7 Other diagnostics

By analogy with the statistics used to detect influential observations in multiple linear regression, the statistics delta-beta, delta-chi-squared and delta-deviance are also available for logistic regression (see Section 6.2.7).

For binary or Binomial data there are additional issues to consider. The first is to check the choice of the link function. Brown (1982) developed a test for the logit link which is implemented in some software. The approach suggested by Aranda-Ordaz (1981) is to consider a more general family of link functions

$$g(\pi, \alpha) = \log \left[\frac{(1 - \pi)^{-\alpha} - 1}{\alpha} \right].$$

If $\alpha = 1$, then $g(\pi) = \log[\pi/(1 - \pi)]$, the logit link. As $\alpha \to 0$, then $g(\pi) \to \log[-\log(1 - \pi)]$, the complementary log-log link. In principle, an optimal value of α can be estimated from the data, but the process requires several steps. In the absence of suitable software to identify the best link function, it is advisable to experiment with several alternative links.

The second issue in assessing the adequacy of models for binary or Binomial data is **overdispersion**. Observations Y_i may have observed variance greater than Binomial variance $n_i \pi_i (1 - \pi_i)$, or equivalently var($\widehat{\pi}_i$) may be greater than $\pi_i (1 - \pi_i)/n_i$. There is an indicator of this problem if the deviance D is much greater than the expected value of $N - p$. This could be due to inadequate specification of the model (e.g., relevant explanatory variables have been omitted or the link function is incorrect) or to a more complex structure (see Exercise 7.5). One approach is to include an extra parameter ϕ in the model so that var(Y_i) $= n_i \pi_i (1 - \pi_i)\phi$. This is implemented in various ways in statistical software. For example, in R there is an option in glm to specify a quasibinomial distribution instead of a Binomial distribution. Another possible explanation for overdispersion is that the Y_i's are not independent. For example if the binary responses counted by Y_i are not independent, the effective number of trials n', will be less than n so that var($\widehat{\pi}_i$) $= \pi_i (1 - \pi_i)/n'_i > \pi_i (1 - \pi_i)/n_i$. Methods for modelling correlated data are outlined in Chapter 11. For a detailed discussion of overdispersion for Binomial data, see, for example, Collett (2003, Chapter 6).

7.8   Example: Senility and WAIS

Table 7.8 *Symptoms of senility (s=1 if symptoms are present and s=0 otherwise) and WAIS scores (x) for N=54 people.*

x	s	x	s	x	s	x	s	x	s
9	1	7	1	7	0	17	0	13	0
13	1	5	1	16	0	14	0	13	0
6	1	14	1	9	0	19	0	9	0
8	1	13	0	9	0	9	0	15	0
10	1	16	0	11	0	11	0	10	0
4	1	10	0	13	0	14	0	11	0
14	1	12	0	15	0	10	0	12	0
8	1	11	0	13	0	16	0	4	0
11	1	14	0	10	0	10	0	14	0
7	1	15	0	11	0	16	0	20	0
9	1	18	0	6	0	14	0		

A sample of elderly people was given a psychiatric examination to determine whether symptoms of senility were present. Other measurements taken at the same time included the score on a subset of the Wechsler Adult Intelligent Scale (WAIS). The data are shown in Table 7.8. The data in Table 7.8 are binary although some people have the same WAIS scores and so there are $m = 17$ different covariate patterns (see Table 7.9). Let Y_i denote the number of people with symptoms among n_i people with the ith covariate pattern. The logistic regression model

$$\log\left(\frac{\pi_i}{1-\pi_i}\right) = \beta_1 + \beta_2 x_i; \quad Y_i \sim \text{Bin}(n_i, \pi_i) \quad i = 1, \ldots, m,$$

was fitted with the following results:

$b_1 = 2.404$, standard error $(b_1) = 1.192$;
$b_2 = -0.3235$, standard error $(b_2) = 0.1140$;
$X^2 = \sum X_i^2 = 8.083$ and $D = \sum d_i^2 = 9.419$.

As there are $m = 17$ covariate patterns (different values of x, in this example) and $p = 2$ parameters, X^2 and D can be compared with $\chi^2(15)$ (by these criteria the model appears to fit well).

Figure 7.5 shows the observed relative frequencies y_i/n_i for each covariate pattern and the fitted probabilities $\hat{\pi}_i$ plotted against WAIS score, x (for $i = 1, \ldots, m$). The model appears to fit better for higher values of x.

Table 7.9 *Covariate patterns and responses, estimated probabilities ($\widehat{\pi}$), Pearson residuals (X) and deviance residuals (d) for senility and WAIS.*

x	y	n	$\widehat{\pi}$	X	d
4	1	2	0.752	−0.826	−0.766
5	1	1	0.687	0.675	0.866
6	1	2	0.614	−0.330	−0.326
7	2	3	0.535	0.458	0.464
8	2	2	0.454	1.551	1.777
9	2	6	0.376	−0.214	−0.216
10	1	6	0.303	−0.728	−0.771
11	1	6	0.240	−0.419	−0.436
12	0	2	0.186	−0.675	−0.906
13	1	6	0.142	0.176	0.172
14	2	7	0.107	1.535	1.306
15	0	3	0.080	−0.509	−0.705
16	0	4	0.059	−0.500	−0.696
17	0	1	0.043	−0.213	−0.297
18	0	1	0.032	−0.181	−0.254
19	0	1	0.023	−0.154	−0.216
20	0	1	0.017	−0.131	−0.184
Sum	14	54			
			Sum of squares	8.084*	9.418*

* Sums of squares differ slightly from the goodness of fit statistics X^2 and D mentioned in the text due to rounding errors.

Table 7.9 shows the covariate patterns, estimates $\widehat{\pi}_i$ and the corresponding chi-squared and deviance residuals calculated using Equations (7.8) and (7.9), respectively.

The residuals and associated residual plots (not shown) do not suggest that there are any unusual observations but the small numbers of observations for each covariate value make the residuals difficult to assess. The Hosmer–Lemeshow approach provides some simplification; Table 7.10 shows the data in categories defined by grouping values of $\widehat{\pi}_i$ so that the total numbers of observations per category are approximately equal. For this illustration, $g = 3$ categories were chosen. The expected frequencies are obtained from the values in Table 7.9; there are $\sum n_i \widehat{\pi}_i$ with symptoms and $\sum n_i (1 - \widehat{\pi}_i)$ without symptoms for each category. The Hosmer–Lemeshow statistic X^2_{HL} is obtained by calculating $X^2 = \Sigma \left[(o - e)^2/e\right]$, where the observed frequencies, o, and expected frequencies, e, are given in Table 7.10 and summation is over

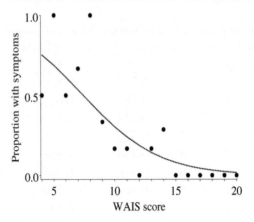

Figure 7.5 *Association between presence of symptoms and WAIS score from data in Tables 7.8 and 7.9; dots represent observed proportions and the dotted line represents estimated probabilities.*

all 6 cells of the table; $X^2_{HL} = 1.15$, which is not significant when compared with the $\chi^2(1)$ distribution.

Table 7.10 *Hosmer–Lemeshow test for data in Table 7.9: observed frequencies (o) and expected frequencies (e) for numbers of people with or without symptoms, grouped by values of $\widehat{\pi}$.*

Values of $\widehat{\pi}$		≤ 0.107	0.108–0.303	> 0.303
Corresponding values of x		14–20	10–13	4–9
Number of people	o	2	3	9
with symptoms	e	1.335	4.479	8.186
Number of people	o	16	17	7
without symptoms	e	16.665	15.521	7.814
Total number of people		18	20	16

For the minimal model, without x, the maximum value of the log-likelihood function is $l(\widetilde{\pi}, y) = -30.9032$. For the model with x, the corresponding value is $l(\widehat{\pi}, y) = -25.5087$. Therefore, from Section 7.5, $C = 10.789$, which is highly significant compared with $\chi^2(1)$, showing that the slope parameter is non-zero. Also pseudo $R^2 = 0.17$ which suggests the model does not predict the outcomes for individuals particularly well even though the residuals for all the covariate patterns are small and the other summary measures suggest the fit is good.

These data illustrate the differences between fitting the model to binary (ungrouped) and Binomial (grouped) data. The relevant Stata commands are

```
————————————— Stata code (binary model) —————————————
.glm s x, family(binomial 1) link(logit)
```

and

```
————————————— Stata code (Binomial model) —————————————
.gen n=1
.collapse (sum) n s, by(x)
.glm s x, family(binomial n) link(logit)
```

Using R the command for the ungrouped data is

```
————————————— R code (binary model) —————————————
>res.glm=glm(s~x,family=binomial(link="logit"),data=senility)
```

For grouped data a matrix of columns of "successes" and "failures" has to be constructed. The code below shows this construction using the 'doBy' library (Højsgaard and Halekoh 2016) followed by the GLM command

```
————————————— R code (Binomial model) —————————————
>library(doBy)
>waisgrp=summaryBy(s~x,data=senility,FUN=c(sum,length))
>names(waisgrp)=c('x','y','n')
>res.glm=glm(cbind(y, n-y)~x,family=binomial(link="logit")
 ,data=waisgrp)
```

For either form of the data the values of the estimates and their standard errors are the same, but the measures of goodness of fit differ as shown in Table 7.11. In this table M_0 refers to a model with only a constant (i.e., no effect of WAIS scores x), M_1 refers to the model with a constant and an effect of x, and M_S refers to the saturated model with a parameter for every observation. The statistics pseudo R^2 and AIC can be interpreted to indicate that the model M_1 is better able to predict the group outcomes (i.e., event rates) than to predict individual outcomes. However, the differences caused by the form in which the data are analyzed indicate that caution is needed when assessing adequacy of logistic regression models using these measures.

7.9 Odds ratios and prevalence ratios

Logistic regression models commonly use the logit link and this is the default link function in most statistical software packages. The parameter estimates

Table 7.11 *Measures of goodness of fit for models for data in Table 7.9 obtained using ungrouped and grouped observations.*

	Ungrouped	Grouped
Number of observations	54	17
M_0 log-likelihood	-30.90316	-17.29040
M_1 log-likelihood	-25.50869	-11.89593
M_S log-likelihood	0.0	-7.18645
M_0 deviance	61.80632	20.20791
M_1 deviance	51.01738	9.14897
$M_0 - M_1$ deviance	10.7889	10.7889
M_1 pseudo R^2	0.1746	0.3120
M_1 AIC	55.0173	27.7919
M_1 AIC_{Stata}	1.01884	1.6348

using the logit link are odds ratios (see Exercise 7.2), but these are often misinterpreted in terms of probability. For example, if the probability of disease in one group is 0.8 and in another is 0.2 then the odds ratio (OR) is:

$$OR = \frac{0.8/(1-0.8)}{0.2/(1-0.2)} = \frac{4}{0.25} = 16$$

whereas the prevalence ratio or relative risk (PR) is the much smaller

$$PR = \frac{0.8}{0.2} = 4.$$

Both statistics are correct, but odds ratios are often reported as if they were prevalence ratios with incorrect phrases such as "16 times more likely."

In principle, prevalence ratios can be estimated by using a log link in place of a logit link in a binomial regression model (Deddens and Petersen 2008). However, in practice there are two main problems. Firstly, models using the log link frequently fail to converge. This is because the transformed probabilities for a logit link can be positive or negative, whereas the log-transformed probabilities cannot be positive. Secondly, while prevalence ratios are easier to interpret correctly than odds ratios are, this is only true if the explanatory variables are all categorical. For a continuous explanatory variable the prevalence ratio is not linearly related to changes in the explanatory variable so it is necessary to state the values of the variable (such as a 'dose' level) that the prevalence ratio applies to. Both these problems are explained in more detail in Section 14.2.1.

The following example shows the relationships between odds ratios and prevalence ratios using the anther data from Table 7.5. For this example the centrifuging force is ignored as the models summarized in Table 7.6 do not provide much evidence that centrifuging force had an effect. Table 7.12 shows the resulting 2×2 table.

Table 7.12 *Summary table for the anther data ignoring centrifuging force; the y-values in Table 7.5 denote 'success.'*

Storage condition	Success	Failure	Total
Control	164	145	309
Treatment	155	92	247

From Table 7.12 the prevalence ratio of success under the treatment condition relative to the control condition is $(155/247)/(164/309) = 1.182$ and the odds ratio is $(155/92)/(164/145) = 1.490$.

Table 7.13 shows the results from logistic regression and from fitting a binomial model using the log link function. The relevant commands for Stata are shown below.

— Stata code (logistic & log-binomial regression models) —
```
. glm y i.storage, family(binomial n) link(logit) eform
. glm y i.storage, family(binomial n) link(log) eform
```

The R commands are shown below; the first line creates the summed data in Table 7.12.

—— R code (logistic & log-binomial regression models) ——
```
>anthers.sum <- aggregate(anthers[c("n","y")],
 by = anthers[c("storage")], FUN=sum)
>summary(glm(cbind(y, n-y) ~ storage, data=anthers.sum,
 family=binomial(link='logit')))
>summary(glm(cbind(y, n-y) ~ storage, data=anthers.sum,
 family=binomial(link='log')))
```

Table 7.13 *Point estimates and 95% confidence intervals (CIs) from two models fitted to the anther data in Table 7.12.*

	Logit model odds ratios	Log model prevalence ratios
Constant	1.131 (0.905, 1.414)	0.531 (0.478, 0.589)
Treatment vs. control	1.490 (1.059, 2.095)	1.182 (1.026, 1.363)

The prevalence of 'success' is about 20% higher under the treatment

storage condition than the control condition, PR = 1.182 (1.026, 1.363). However the effect size may appear larger if reported as almost 50% higher odds,
OR = 1.490 (1.059, 2.095).

If the explanatory variable centrifuge force is added to the model as a
categorical variable with three levels, then the estimated odds ratio for the
treatment condition relative to the control condition from the logistic model
is 1.502 (95% CI 1.067, 2.115) which corresponds to the estimate for the
difference $a_2 - a_1 = 0.407$ shown in Table 7.6 because $\exp(0.407) = 1.502$.

7.10 Exercises

7.1 The number of deaths from leukemia and other cancers among survivors
of the Hiroshima atom bomb are shown in Table 7.14, classified by the
radiation dose received. The data refer to deaths during the period 1950–
1959 among survivors who were aged 25 to 64 years in 1950 (from data
set 13 of Cox and Snell, 1981, attributed to Otake, 1979).

a. Obtain a suitable model to describe the dose–response association
between radiation and the proportional cancer mortality rates for
leukemia.

b. Examine how well the model describes the data.

c. Interpret the results.

Table 7.14 *Deaths from leukemia and other cancers classified by radiation dose received from the Hiroshima atomic bomb.*

	\multicolumn{6}{c}{Radiation dose (rads)}					
Deaths	0	1–9	10–49	50–99	100–199	200+
Leukemia	13	5	5	3	4	18
Other cancers	378	200	151	47	31	33
Total cancers	391	205	156	50	35	51

7.2 **Odds ratios.** Consider a 2×2 contingency table from a prospective study
in which people who were or were not exposed to some pollutant are followed up and, after several years, categorized according to the presence or
absence of a disease. Table 7.15 shows the probabilities for each cell. The
odds of disease for either exposure group is $O_i = \pi_i/(1 - \pi_i)$, for $i = 1, 2$,
and so the odds ratio

$$\phi = \frac{O_1}{O_2} = \frac{\pi_1(1 - \pi_2)}{\pi_2(1 - \pi_1)}$$

is a measure of the relative likelihood of disease for the exposed and not
exposed groups.

Table 7.15 *2×2 table for a prospective study of exposure and disease outcome.*

	Diseased	Not diseased
Exposed	π_1	$1 - \pi_1$
Not exposed	π_2	$1 - \pi_2$

a. For the simple logistic model $\pi_i = e^{\beta_i}/(1 + e^{\beta_i})$, show that if there is no difference between the exposed and not exposed groups (i.e., $\beta_1 = \beta_2$), then $\phi = 1$.

b. Consider J 2×2 tables like Table 7.15, one for each level x_j of a factor, such as age group, with $j = 1, \ldots, J$. For the logistic model

$$\pi_{ij} = \frac{\exp(\alpha_i + \beta_i x_j)}{1 + \exp(\alpha_i + \beta_i x_j)}, \qquad i = 1, 2, \quad j = 1, \ldots, J.$$

Show that $\log \phi$ is constant over all tables if $\beta_1 = \beta_2$ (McKinlay 1978).

7.3 Tables 7.16 and 7.17 show the survival 50 years after graduation of men and women who graduated each year from 1938 to 1947 from various faculties of the University of Adelaide (data compiled by J.A. Keats). The columns labelled S contain the number of graduates who survived and the columns labelled T contain the total number of graduates. There were insufficient women graduates from the faculties of Medicine and Engineering to warrant analysis.

a. Are the proportions of graduates who survived for 50 years after graduation the same all years of graduation?

b. Are the proportions of male graduates who survived for 50 years after graduation the same for all Faculties?

c. Are the proportions of female graduates who survived for 50 years after graduation the same for Arts and Science?

d. Is the difference between men and women in the proportion of graduates who survived for 50 years after graduation the same for Arts and Science?

7.4 Let $l(\mathbf{b}_{\min})$ denote the maximum value of the log-likelihood function for the minimal model with linear predictor $\mathbf{x}^T \boldsymbol{\beta} = \beta_1$, and let $l(\mathbf{b})$ be the corresponding value for a more general model $\mathbf{x}^T \boldsymbol{\beta} = \beta_1 + \beta_2 x_1 + \ldots + \beta_p x_{p-1}$.

a. Show that the likelihood ratio chi-squared statistic is

$$C = 2[l(\mathbf{b}) - l(\mathbf{b}_{\min})] = D_0 - D_1,$$

Table 7.16 *Fifty years survival for men after graduation from the University of Adelaide.*

Year of graduation	Faculty							
	Medicine		Arts		Science		Engineering	
	S	T	S	T	S	T	S	T
1938	18	22	16	30	9	14	10	16
1939	16	23	13	22	9	12	7	11
1940	7	17	11	25	12	19	12	15
1941	12	25	12	14	12	15	8	9
1942	24	50	8	12	20	28	5	7
1943	16	21	11	20	16	21	1	2
1944	22	32	4	10	25	31	16	22
1945	12	14	4	12	32	38	19	25
1946	22	34			4	5		
1947	28	37	13	23	25	31	25	35
Total	177	275	92	168	164	214	100	139

Table 7.17 *Fifty years survival for women after graduation from the University of Adelaide.*

Year of graduation	Faculty			
	Arts		Science	
	S	T	S	T
1938	14	19	1	1
1939	11	16	4	4
1940	15	18	6	7
1941	15	21	3	3
1942	8	9	4	4
1943	13	13	8	9
1944	18	22	5	5
1945	18	22	16	17
1946	1	1	1	1
1947	13	16	10	10
Total	126	157	58	61

where D_0 is the deviance for the minimal model and D_1 is the deviance for the more general model.

b. Deduce that if $\beta_2 = \ldots = \beta_p = 0$, then C has the central chi-squared distribution with $(p-1)$ degrees of freedom.

7.5 Let Y_i be the number of successes in n_i trials with

$$Y_i \sim \text{Bin}(n_i, \pi_i),$$

where the probabilities π_i have a Beta distribution

$$\pi_i \sim \text{Be}(\alpha, \beta).$$

The probability density function for the Beta distribution is $f(x; \alpha, \beta) = x^{\alpha-1}(1-x)^{(\beta-1)}/B(\alpha, \beta)$ for x in [0, 1], $\alpha > 0, \beta > 0$ and the beta function $B(\alpha, \beta)$ defining the normalizing constant required to ensure that $\int_0^1 f(x; \alpha, \beta)dx = 1$. It can be shown that $E(X) = \alpha/(\alpha + \beta)$ and $\text{var}(X) = \alpha\beta/[(\alpha+\beta)^2(\alpha+\beta+1)]$. Let $\theta = \alpha/(\alpha+\beta)$, and hence, show that

a. $E(\pi_i) = \theta$

b. $\text{var}(\pi_i) = \theta(1-\theta)/(\alpha+\beta+1) = \phi\theta(1-\theta)$

c. $E(Y_i) = n_i\theta$

d. $\text{var}(Y_i) = n_i\theta(1-\theta)[1+(n_i-1)\phi]$ so that $\text{var}(Y_i)$ is larger than the Binomial variance (unless $n_i = 1$ or $\phi = 0$).

Chapter 8

Nominal and Ordinal Logistic Regression

8.1 Introduction

If the response variable is categorical, with more than two categories, then there are two options for generalized linear models. One relies on generalizations of logistic regression from dichotomous responses, described in Chapter 7, to nominal or ordinal responses with more than two categories. This first approach is the subject of this chapter. The other option is to model the frequencies or counts for the covariate patterns as the response variables with Poisson distributions. The second approach, called **log-linear modelling**, is covered in Chapter 9.

For nominal or ordinal logistic regression, one of the measured or observed categorical variables is regarded as the response, and all other variables are explanatory variables. For log-linear models, all the variables are treated alike. The choice of which approach to use in a particular situation depends on whether one variable is clearly a "response" (for example, the outcome of a prospective study) or several variables have the same status (as may be the situation in a cross-sectional study). Additionally, the choice may depend on how the results are to be presented and interpreted. Nominal and ordinal logistic regression yield odds ratio estimates which are relatively easy to interpret if there are no interactions (or only fairly simple interactions). Log-linear models are good for testing hypotheses about complex interactions, but the parameter estimates are less easily interpreted.

This chapter begins with the Multinomial distribution which provides the basis for modelling categorical data with more than two categories. Then the various formulations for nominal and ordinal logistic regression models are discussed, including the interpretation of parameter estimates and methods for checking the adequacy of a model. A numerical example is used to illustrate the methods.

8.2   Multinomial distribution

Consider a random variable Y with J categories. Let $\pi_1, \pi_2, \ldots, \pi_J$ denote the respective probabilities, with $\pi_1 + \pi_2 + \ldots + \pi_J = 1$. If there are n independent observations of Y which result in y_1 outcomes in category 1, y_2 outcomes in category 2, and so on, then let

$$\mathbf{y} = \begin{bmatrix} y_1 \\ y_2 \\ \vdots \\ y_J \end{bmatrix}, \quad \text{with} \sum_{j=1}^{J} y_j = n.$$

The Multinomial distribution is

$$f(\mathbf{y} \,|\, n) = \frac{n!}{y_1! y_2! \ldots, y_J!} \pi_1^{y_1} \pi_2^{y_2} \ldots \pi_J^{y_J}, \tag{8.1}$$

it is denoted by $M(n, \pi_1, \ldots, \pi_j)$. If $J = 2$, then $\pi_2 = 1 - \pi_1$, $y_2 = n - y_1$ and (8.1) is the Binomial distribution $B(n, \pi)$; see (7.1). In general, (8.1) does not satisfy the requirements for being a member of the exponential family of distributions (3.3). However, the following relationship with the Poisson distribution ensures that generalized linear modelling is appropriate.

Let $Y_1, \ldots, Y_J$ denote independent random variables with distributions $Y_j \sim \text{Po}(\lambda_j)$. Their joint probability distribution is

$$f(\mathbf{y}) = \prod_{j=1}^{J} \frac{\lambda_j^{y_j} e^{-\lambda_j}}{y_j!}, \tag{8.2}$$

where

$$\mathbf{y} = \begin{bmatrix} y_1 \\ \vdots \\ y_J \end{bmatrix}.$$

Let $n = Y_1 + Y_2 + \ldots + Y_J$, then n is a random variable with the distribution $n \sim \text{Po}(\lambda_1 + \lambda_2 + \ldots + \lambda_J)$ (see, for example, Forbes, Evans, Hastings, and Peacock, 2010). Therefore, the distribution of $\mathbf{y}$ conditional on n is

$$f(\mathbf{y} \,|\, n) = \left[\prod_{j=1}^{J} \frac{\lambda_j^{y_j} e^{-\lambda_j}}{y_j!} \right] \Big/ \frac{(\lambda_1 + \ldots + \lambda_J)^n e^{-(\lambda_1 + \ldots + \lambda_J)}}{n!},$$

which can be simplified to

$$f(\mathbf{y} \,|\, n) = \left(\frac{\lambda_1}{\sum \lambda_k} \right)^{y_1} \ldots \left(\frac{\lambda_J}{\sum \lambda_k} \right)^{y_J} \frac{n!}{y_1! \ldots, y_J!}. \tag{8.3}$$

If $\pi_j = \lambda_j / \left(\sum_{k=1}^{K} \lambda_k \right)$, for $j = 1, \ldots, J$, then (8.3) is the same as (8.1) and $\sum_{j=1}^{J} \pi_j = 1$, as required. Therefore, the Multinomial distribution can be regarded as the joint distribution of Poisson random variables, conditional upon their sum n. This result provides a justification for the use of generalized linear modelling.

For the Multinomial distribution (8.1) it can be shown that $E(Y_j) = n\pi_j$, $\text{var}(Y_j) = n\pi_j(1 - \pi_j)$ and $\text{cov}(Y_j, Y_k) = -n\pi_j\pi_k$ (see, for example, Forbes, Evans, Hastings, and Peacock, 2010).

In this chapter models based on the Binomial distribution are considered, because pairs of response categories are compared, rather than all J categories simultaneously.

8.3   Nominal logistic regression

Nominal logistic regression models are used when there is no natural order among the response categories. One category is arbitrarily chosen as the **reference category**. Suppose this is the first category. Then the logits for the other categories are defined by

$$\text{logit}(\pi_j) = \log\left(\frac{\pi_j}{\pi_1}\right) = \mathbf{x}_j^T \boldsymbol{\beta}_j, \quad \text{for } j = 2, \ldots, J. \tag{8.4}$$

The $(J-1)$ logit equations are used simultaneously to estimate the parameters $\boldsymbol{\beta}_j$. Once the parameter estimates $\mathbf{b}_j$ have been obtained, the linear predictors $\mathbf{x}_j^T \mathbf{b}_j$ can be calculated. From (8.4)

$$\widehat{\pi}_j = \widehat{\pi}_1 \exp\left(\mathbf{x}_j^T \mathbf{b}_j\right) \quad \text{for } j = 2, \ldots, J.$$

But $\widehat{\pi}_1 + \widehat{\pi}_2 + \ldots + \widehat{\pi}_J = 1$, so

$$\widehat{\pi}_1 = \frac{1}{1 + \sum_{j=2}^{J} \exp\left(\mathbf{x}_j^T \mathbf{b}_j\right)}$$

and

$$\widehat{\pi}_j = \frac{\exp\left(\mathbf{x}_j^T \mathbf{b}_j\right)}{1 + \sum_{j=2}^{J} \exp\left(\mathbf{x}_j^T \mathbf{b}_j\right)}, \quad \text{for } j = 2, \ldots, J.$$

Fitted values, or "expected frequencies," for each covariate pattern can be calculated by multiplying the estimated probabilities $\widehat{\pi}_j$ by the total frequency of the covariate pattern.

The **Pearson chi-squared residuals** are given by

$$r_i = \frac{o_i - e_i}{\sqrt{e_i}}, \tag{8.5}$$

where o_i and e_i are the observed and expected frequencies for $i = 1, \ldots, N$, where N is J times the number of distinct covariate patterns. The residuals can be used to assess the adequacy of the model.

Summary statistics for goodness of fit are analogous to those for Binomial logistic regression:

(i) **Chi-squared statistic**

$$X^2 = \sum_{i=1}^{N} r_i^2; \tag{8.6}$$

(ii) **Deviance**, defined in terms of the maximum values of the log-likelihood function for the fitted model, $l(\mathbf{b})$, and for the maximal model, $l(\mathbf{b}_{\max})$,

$$D = 2\left[l(\mathbf{b}_{\max}) - l(\mathbf{b})\right]; \tag{8.7}$$

(iii) **Likelihood ratio chi-squared statistic**, defined in terms of the maximum value of the log likelihood function for the minimal model, $l(\mathbf{b}_{\min})$, and $l(\mathbf{b})$,

$$C = 2\left[l(\mathbf{b}) - l(\mathbf{b}_{\min})\right]; \tag{8.8}$$

(iv)

$$\text{Pseudo } R^2 = \frac{l(\mathbf{b}_{\min}) - l(\mathbf{b})}{l(\mathbf{b}_{\min})}; \tag{8.9}$$

(v) **Akaike information criterion**

$$AIC = -2l\left(\widehat{\boldsymbol{\pi}}; \mathbf{y}\right) + 2p. \tag{8.10}$$

If the model fits well, then both X^2 and D have, asymptotically, the distribution $\chi^2(N - p)$, where p is the number of parameters estimated. C has the asymptotic distribution $\chi^2[p - (J - 1)]$ because the minimal model will have one parameter for each logit defined in (8.4). AIC is used mainly for comparisons between models which are not nested.

Often it is easier to interpret the effects of explanatory factors in terms of odds ratios than the parameters $\boldsymbol{\beta}$. For simplicity, consider a response variable with J categories and a binary explanatory variable x which denotes whether an "exposure" factor is present ($x = 1$) or absent ($x = 0$). The odds ratio for exposure for response j $(j = 2, \ldots, J)$ relative to the reference category $j = 1$ is

$$OR_j = \frac{\pi_{jp}}{\pi_{ja}} \bigg/ \frac{\pi_{1p}}{\pi_{1a}},$$

where π_{jp} and π_{ja} denote the probabilities of response category j $(j = 1, \ldots, J)$ according to whether exposure is present or absent, respectively. For the model

$$\log\left(\frac{\pi_j}{\pi_1}\right) = \beta_{0j} + \beta_{1j}x, \quad j = 2, \ldots, J,$$

the log odds are

$$\log\left(\frac{\pi_{ja}}{\pi_{1a}}\right) = \beta_{0j} \quad \text{when } x = 0, \text{ indicating the exposure is absent, and}$$

$$\log\left(\frac{\pi_{jp}}{\pi_{1p}}\right) = \beta_{0j} + \beta_{1j} \text{ when } x = 1, \text{ indicating the exposure is present.}$$

Therefore, the logarithm of the odds ratio can be written as

$$\begin{aligned}
\log OR_j &= \log\left(\frac{\pi_{jp}}{\pi_{1p}}\right) - \log\left(\frac{\pi_{ja}}{\pi_{1a}}\right) \\
&= \beta_{1j}.
\end{aligned}$$

Hence, $OR_j = \exp(\beta_{1j})$ which is estimated by $\exp(b_{1j})$. If $\beta_{1j} = 0$, then $OR_j = 1$ which corresponds to the exposure factor having no effect. Also, for example, 95% confidence limits for OR_j are given by $\exp[b_{1j} \pm 1.96 \times \text{s.e.}(b_{1j})]$, where s.e.$(b_{1j})$ denotes the standard error of b_{1j}. Confidence intervals which do not include unity correspond to β values significantly different from zero.

For nominal logistic regression, the explanatory variables may be categorical or continuous. The choice of the reference category for the response variable will affect the parameter estimates **b** but not the estimated probabilities $\widehat{\pi}$ or the fitted values.

The following example illustrates the main characteristic of nominal logistic regression.

8.3.1 Example: Car preferences

In a study of motor vehicle safety, men and women driving small, medium and large cars were interviewed about vehicle safety and their preferences for cars, and various measurements were made of how close they sat to the steering wheel (McFadden et al. 2000). There were 50 subjects in each of the six categories (two sexes and three car sizes). They were asked to rate how important various features were to them when they were buying a car. Table 8.1 shows the ratings for air conditioning and power steering, according to the sex and age of the subject (the categories "not important" and "of little importance" have been combined).

Table 8.1 *Importance of air conditioning and power steering in cars (row percentages in brackets*).*

Sex	Age	Response			Total
		No or little importance	Important	Very important	
Women	18–23	26 (58%)	12 (27%)	7 (16%)	45
	24–40	9 (20%)	21 (47%)	15 (33%)	45
	> 40	5 (8%)	14 (23%)	41 (68%)	60
Men	18–23	40 (62%)	17 (26%)	8 (12%)	65
	24–40	17 (39%)	15 (34%)	12 (27%)	44
	> 40	8 (20%)	15 (37%)	18 (44%)	41
Total		105	94	101	300

* Row percentages may not add to 100 due to rounding.

The proportions of responses in each category by age and sex are shown in Figure 8.1. For these data the response, importance of air conditioning and power steering, is rated on an ordinal scale but for the purpose of this example the order is ignored and the 3-point scale is treated as nominal. The category "no or little" importance is chosen as the reference category. Age is also ordinal, but initially we will regard it as nominal.

Table 8.2 shows the results of fitting the nominal logistic regression model with reference categories of "Women" and "18–23 years," and

$$\log\left(\frac{\pi_j}{\pi_1}\right) = \beta_{0j} + \beta_{1j}x_1 + \beta_{2j}x_2 + \beta_{3j}x_3, \qquad j = 2, 3, \qquad (8.11)$$

where

$$x_1 = \begin{cases} 1 & \text{for men} \\ 0 & \text{for women} \end{cases}, \qquad x_2 = \begin{cases} 1 & \text{for age 24–40 years} \\ 0 & \text{otherwise} \end{cases}$$

$$\text{and } x_3 = \begin{cases} 1 & \text{for age > 40 years} \\ 0 & \text{otherwise} \end{cases}.$$

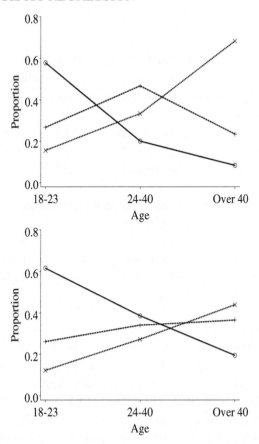

Figure 8.1 *Preferences for air conditioning and power steering: percentages of responses in each category by age and sex (solid lines denote "no/little importance," dashed lines denote "important" and dotted lines denote "very important." Top panel: women; bottom panel: men.*

For Stata the relevant command is

───────── Stata code (nominal logistic regression) ─────────
```
.mlogit c_resp c_sex _Ic_age_1 _Ic_age_2 [fweight=freq]
```

where _Ic_age_1 and _Ic_age_2 are indicator variables for the age groups 24–40 years and > 40 years, respectively. The corresponding command for R uses the function multinom which comes from the library "nnet" as follows (Venables and Ripley 2002):

───────── R code (nominal logistic regression) ─────────
```
>res.cars=multinom(response~factor(age)+factor(sex),
 weights=frequency,data=Cars)
```

Table 8.2 *Results of fitting the nominal logistic regression model (8.11) to the data in Table 8.1.*

Parameter β	Estimate b (std. error)	Odds ratio, $OR = e^b$ (95% confidence interval)	
$\log(\pi_2/\pi_1)$: important vs. no/little importance			
β_{02}: constant	-0.591 (0.284)		
β_{12}: men	-0.388 (0.301)	0.68	(0.38, 1.22)
β_{22}: 24–40	1.128 (0.342)	3.09	(1.58, 6.04)
β_{32}: > 40	1.588 (0.403)	4.89	(2.22, 10.78)
$\log(\pi_3/\pi_1)$: very important vs. no/little importance			
β_{03}: constant	-1.039 (0.331)		
β_{13}: men	-0.813 (0.321)	0.44	(0.24, 0.83)
β_{23}: 24–40	1.478 (0.401)	4.38	(2.00, 9.62)
β_{33}: > 40	2.917 (0.423)	18.48	(8.07, 42.34)

The maximum value of the log-likelihood function for the minimal model (with only two parameters, β_{02} and β_{03}) is -329.27 and for the fitted model (8.11) is -290.35, giving the likelihood ratio chi-squared statistic $C = 2\times (-290.35 + 329.27) = 77.84$, pseudo $R^2 = (-329.27 + 290.35)/(-329.27) = 0.118$ and $AIC = -2\times (-290.35) + 16 = 596.70$. The first statistic, which has 6 degrees of freedom (8 parameters in the fitted model minus 2 for the minimal model), is very significant compared with the $\chi^2(6)$ distribution, showing the overall importance of the explanatory variables. However, the second statistic suggests that only 11.8% of the "variation" is "explained" by these factors. From the Wald statistics $[b/\text{s.e.}(b)]$ and the odds ratios and the confidence intervals, it is clear that the importance of air-conditioning and power steering increased significantly with age. Also men considered these features less important than women did, although the statistical significance of this finding is dubious (especially considering the small frequencies in some cells).

To estimate the probabilities, first consider the preferences of women ($x_1 = 0$) aged 18–23 (so $x_2 = 0$ and $x_3 = 0$). For this group

$$\log\left(\frac{\widehat{\pi_2}}{\widehat{\pi_1}}\right) = -0.591, \quad \text{so} \quad \frac{\widehat{\pi_2}}{\widehat{\pi_1}} = e^{-0.591} = 0.5539,$$

$$\log\left(\frac{\widehat{\pi_3}}{\widehat{\pi_1}}\right) = -1.039, \quad \text{so} \quad \frac{\widehat{\pi_3}}{\widehat{\pi_1}} = e^{-1.039} = 0.3538$$

but $\widehat{\pi}_1 + \widehat{\pi}_2 + \widehat{\pi}_3 = 1$, so $\widehat{\pi}_1(1 + 0.5539 + 0.3538) = 1$; therefore, $\widehat{\pi}_1 = 1/1.9077 = 0.524$, and hence, $\widehat{\pi}_2 = 0.290$ and $\widehat{\pi}_3 = 0.186$. Now consider men ($x_1 = 1$) aged over 40 (so $x_2 = 0$, but $x_3 = 1$) so that $\log(\widehat{\pi}_2/\widehat{\pi}_1) = -0.591 - 0.388 + 1.588 = 0.609$, $\log(\widehat{\pi}_3/\widehat{\pi}_1) = 1.065$, and hence, $\widehat{\pi}_1 = 0.174$, $\widehat{\pi}_2 = 0.320$ and $\widehat{\pi}_3 = 0.505$ (correct to 3 decimal places). These estimated probabilities can be multiplied by the total frequency for each sex × age group to obtain the "expected" frequencies or fitted values. These are shown in Table 8.3, together with the Pearson residuals defined in (8.5). The sum of squares of the Pearson residuals, the chi-squared goodness of fit statistic (8.6), is $X^2 = 3.93$. (Note it is not usually necessary to calculate the estimated probabilities "by hand" like this but the calculations are presented here to illustrate the model in more detail).

Table 8.3 *Results from fitting the nominal logistic regression model (8.11) to the data in Table 8.1.*

Sex	Age	Importance Rating*	Obs. freq.	Estimated probability	Fitted value	Pearson residual
Women	18–23	1	26	0.524	23.59	0.496
		2	12	0.290	13.07	−0.295
		3	7	0.186	8.35	−0.466
	24–40	1	9	0.234	10.56	−0.479
		2	21	0.402	18.07	0.690
		3	15	0.364	16.37	−0.340
	> 40	1	5	0.098	5.85	−0.353
		2	14	0.264	15.87	−0.468
		3	41	0.638	38.28	0.440
Men	18–23	1	40	0.652	42.41	−0.370
		2	17	0.245	15.93	0.267
		3	8	0.102	6.65	0.522
	24–40	1	17	0.351	15.44	0.396
		2	15	0.408	17.93	−0.692
		3	12	0.241	10.63	0.422
	> 40	1	8	0.174	7.15	0.320
		2	15	0.320	13.13	0.515
		3	18	0.505	20.72	−0.600
Total			300		300	
Sum of squares						3.931

* 1 denotes "no/little" importance, 2 denotes "important," 3 denotes "very important."

The maximal model that can be fitted to these data involves terms for age, sex, and age $\times$ sex interactions. It has 6 parameters (a constant and coefficients for sex, two age categories and two age $\times$ sex interactions) for $j = 2$ and 6 parameters for $j = 3$, giving a total of 12 parameters. The maximum value of the log-likelihood function for the maximal model is -288.38. Therefore, the deviance for the fitted model (8.11) is $D = 2 \times (-288.38 + 290.35) = 3.94$. The degrees of freedom associated with this deviance are $12 - 8 = 4$ because the maximal model has 12 parameters and the fitted model has 8 parameters. As expected, the values of the goodness of fit statistics $D = 3.94$ and $X^2 = 3.93$ are very similar; when compared with the distribution $\chi^2(4)$, they suggest that model (8.11) provides a good description of the data.

An alternative model can be fitted with age group as a linear covariate, that is,

$$\log\left(\frac{\pi_j}{\pi_1}\right) = \beta_{0j} + \beta_{1j}x_1 + \beta_{2j}x_2; \quad j = 2, 3, \tag{8.12}$$

where

$$x_1 = \begin{cases} 1 & \text{for men} \\ 0 & \text{for women} \end{cases} \quad \text{and} \quad x_2 = \begin{cases} 0 & \text{for age group 18–23} \\ 1 & \text{for age group 24–40} \\ 2 & \text{for age group} > 40 \end{cases}.$$

This model fits the data almost as well as (8.11) but with two fewer parameters. The maximum value of the log likelihood function is -291.05, so the difference in deviance from model (8.11) is

$$\Delta D = 2 \times (-290.35 + 291.05) = 1.4,$$

which is not significant compared with the distribution $\chi^2(2)$. So on the grounds of parsimony model (8.12) is preferable.

8.4   Ordinal logistic regression

If there is an obvious natural order among the response categories, then this can be taken into account in the model specification. The example on car preferences (Section 8.3.1) provides an illustration as the study participants rated the importance of air conditioning and power steering in four categories from "not important" to "very important." Ordinal responses like this are common in areas such as market research, opinion polls and fields such as psychiatry where "soft" measures are common (Ashby et al. 1989).

In some situations there may, conceptually, be a continuous variable z, such as severity of disease, which is difficult to measure. It is assessed by

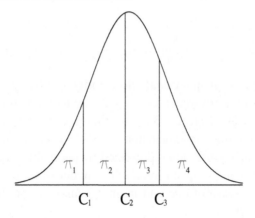

Figure 8.2 *Distribution of continuous latent variable and cutpoints that define an ordinal response variable with four categories.*

some crude method that amounts to identifying "cut points," C_j, for the **latent variable** so that, for example, patients with small values are classified as having "no disease," those with larger values of z are classified as having "mild disease" or "moderate disease" and those with high values are classified as having "severe disease" (see Figure 8.2). The cutpoints $C_1, \ldots, C_{J-1}$ define J ordinal categories with associated probabilities $\pi_1, \ldots, \pi_J$ (with $\sum_{j=1}^{J} \pi_j = 1$).

Not all ordinal variables can be thought of in this way because the underlying process may have many components, as in the car preference example. Nevertheless, the idea is helpful for interpreting the results from statistical models. For ordinal categories, there are several different commonly used models which are described in the next sections.

8.4.1   Cumulative logit model

The cumulative odds for the jth category are

$$\frac{P(z \leq C_j)}{P(z > C_j)} = \frac{\pi_1 + \pi_2 + \ldots + \pi_j}{\pi_{j+1} + \ldots + \pi_J};$$

see Figure 8.2. The cumulative logit model is

$$\log \frac{\pi_1 + \ldots + \pi_j}{\pi_{j+1} + \ldots + \pi_J} = \mathbf{x}_j^T \boldsymbol{\beta}_j. \tag{8.13}$$

8.4.2   Proportional odds model

If the linear predictor $\mathbf{x}_j^T \boldsymbol{\beta}_j$ in (8.13) has an intercept term β_{0j} which depends on the category j, but the other explanatory variables do not depend on j, then the model is

$$\log \frac{\pi_1 + \ldots + \pi_j}{\pi_{j+1} + \ldots + \pi_J} = \beta_{0j} + \beta_1 x_1 + \ldots + \beta_{p-1} x_{p-1}. \qquad (8.14)$$

This is called the **proportional odds model**. It is based on the assumption that the effects of the covariates $x_1, \ldots, x_{p-1}$ are the same for all categories on the logarithmic scale. Figure 8.3 shows the model for $J = 3$ response categories and one continuous explanatory variable x; on the log odds scale the probabilities for categories are represented by parallel lines.

As for the nominal logistic regression model (8.4), the odds ratio associated with an increase of one unit in an explanatory variable x_k is $\exp(\beta_k)$, where $k = 1, \ldots, p-1$.

If some of the categories are amalgamated, this does not change the parameter estimates $\beta_1, \ldots, \beta_{p-1}$ in (8.14)—although, of course, the terms β_{0j} will be affected (this is called the **collapsibility** property; see Ananth and Kleinbaum, 1997). This form of independence between the cutpoints C_j (in Figure 8.2) and the explanatory variables x_k is desirable for many applications, although it requires the strong assumption that wherever the cutpoints are, the odds ratio for a one unit change in x is the same for all response categories.

Another useful property of the proportional odds model is that it is not affected if the labelling of the categories is reversed – only the signs of the parameters will be changed.

The appropriateness of the proportional odds assumption can be tested by comparing models (8.13) and (8.14), if there is only one explanatory variable x. If there are several explanatory variables, the assumption can be tested separately for each variable by fitting (8.13) with the relevant parameter not depending on j.

The proportional odds model is the usual (or default) form of ordinal logistic regression provided by statistical software.

8.4.3   Adjacent categories logit model

One alternative to the cumulative odds model is to consider ratios of probabilities for successive categories, for example,

$$\frac{\pi_1}{\pi_2}, \frac{\pi_2}{\pi_3}, \ldots, \frac{\pi_{J-1}}{\pi_J}.$$

The adjacent category logit model is

$$\log \left(\frac{\pi_j}{\pi_{j+1}} \right) = \mathbf{x}_j^T \boldsymbol{\beta}_j. \qquad (8.15)$$

If this is simplified to

$$\log\left(\frac{\pi_j}{\pi_{j+1}}\right) = \beta_{0j} + \beta_1 x_1 + \ldots + \beta_{p-1} x_{p-1},$$

the effect of each explanatory variable is assumed to be the same for all adjacent pairs of categories. The parameters β_k are usually interpreted as odd ratios using $OR = \exp(\beta_k)$.

8.4.4 Continuation ratio logit model

Another alternative is to model the ratios of probabilities

$$\frac{\pi_1}{\pi_2}, \frac{\pi_1 + \pi_2}{\pi_3}, \ldots, \frac{\pi_1 + \ldots + \pi_{J-1}}{\pi_J}$$

or

$$\frac{\pi_1}{\pi_2 + \ldots + \pi_J}, \frac{\pi_2}{\pi_3 + \ldots + \pi_J}, \ldots, \frac{\pi_{J-1}}{\pi_J}.$$

The equation

$$\log\left(\frac{\pi_j}{\pi_{j+1} + \ldots + \pi_J}\right) = \mathbf{x}_j^T \boldsymbol{\beta}_j \tag{8.16}$$

models the odds of the response being in category j, that is, $C_{j-1} < z \leq C_j$ conditional upon $z > C_{j-1}$. For example, for the car preferences data (Section 8.3.1), one could estimate the odds of respondents regarding air conditioning and power steering as "unimportant" vs. "important" or "very important" and the odds of these features being "very important" given that they are

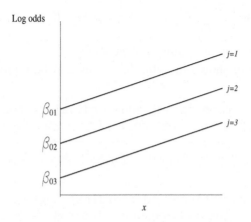

Figure 8.3 *Proportional odds model on log odds scale.*

"important" or "very important," using

$$\log\left(\frac{\pi_1}{\pi_2 + \pi_3}\right) \quad \text{and} \quad \log\left(\frac{\pi_2}{\pi_3}\right).$$

This model may be easier to interpret than the proportional odds model if the probabilities for individual categories π_j are of interest (Agresti 2007, Section 8.3.4).

8.4.5   Comments

Hypothesis tests for ordinal logistic regression models can be performed by comparing the fit of nested models or by using Wald statistics (or, less commonly, score statistics) based on the parameter estimates. Residuals and goodness of fit statistics are analogous to those for nominal logistic regression (Section 8.3).

The choice of model for ordinal data depends mainly on the practical problem being investigated. Comparisons of the models described in this chapter and some other models have been published by Holtbrugger and Schumacher (1991), Ananth and Kleinbaum (1997) and Agresti (2010), for example.

8.4.6   Example: Car preferences

The response variable for the car preference data is, of course, ordinal (Table 8.1). The following proportional odds model was fitted to these data:

$$\log\left(\frac{\pi_1}{\pi_2 + \pi_3}\right) = \beta_{01} + \beta_1 x_1 + \beta_2 x_2 + \beta_3 x_3$$

$$\log\left(\frac{\pi_1 + \pi_2}{\pi_3}\right) = \beta_{02} + \beta_1 x_1 + \beta_2 x_2 + \beta_3 x_3, \qquad (8.17)$$

where x_1, x_2 and x_3 are as defined for model (8.11).

The results are shown in Table 8.4. They can be reproduced using the Stata command

```
– Stata code (proportional odds ordinal regression model) –
.ologit c_resp c_sex _Ic_age_1 _Ic_age_2 [fweight=freq]
```

or the R command polr from the "MASS" library (Venables and Ripley 2002):

```
—— R code (proportional odds ordinal regression model) ——
>res.polr=polr(factor(response)~factor(age)+factor(sex),
 weights=frequency,data=Cars)
```

For model (8.17), the maximum value of the log-likelihood function is $l(\mathbf{b}) = -290.648$. For the minimal model, with only β_{01} and β_{02}, the maximum value is $l(\mathbf{b}_{\min}) = -329.272$, so from (8.8), $C = 2 \times (-290.648 + 329.272) = 77.248$, from (8.9), pseudo $R^2 = (-329.272 + 290.648)/(-329.272) = 0.117$ and from (8.10) $AIC = -2 \times (-290.648) + 2 \times 5 = 591.3$. These last two statistics show that there is very little difference in how well the proportional odds and the nominal logistic regression models describe the data.

The parameter estimates for the proportional odds model are all quite similar to those from the nominal logistic regression model (see Table 8.2). The estimated probabilities are also similar; for example, for females aged 18–23, $x_1 = 0$, $x_2 = 0$ and $x_3 = 0$, so from (8.17), $\log\frac{\widehat{\pi}_1}{\widehat{\pi}_2 + \widehat{\pi}_3} = 0.0435$ and $\log\frac{\widehat{\pi}_1 + \widehat{\pi}_2}{\widehat{\pi}_3} = 1.6550$. If these equations are solved with $\widehat{\pi}_1 + \widehat{\pi}_2 + \widehat{\pi}_3 = 1$, the estimates are $\widehat{\pi}_1 = 0.5109$, $\widehat{\pi}_2 = 0.3287$ and $\widehat{\pi}_3 = 0.1604$. The probabilities for other covariate patterns can be estimated similarly, or by using fitted(res.polr) in R. The probabilities can be used to calculate the expected frequencies, together with residuals and goodness of fit statistics. For the proportional odds model, $X^2 = 4.564$ which is consistent with distribution $\chi^2(7)$, indicating that the model described the data well (in this case $N = 18$, the maximal model has 12 parameters and model (8.14) has 5 parameters so degrees of freedom = 7).

For this example, the proportional odds logistic model for ordinal data and the nominal logistic model produce similar results. On the grounds of parsimony, model (8.17) would be preferred because it is simpler and takes into account the order of the response categories.

8.5 General comments

Although the models described in this chapter are developed from the logistic regression model for binary data, other link functions such as the probit or

Table 8.4 *Results of proportional odds ordinal regression model (8.17) for the data in Table 8.1.*

Parameter	Estimate b	Standard error, s.e.(b)	Odds ratio OR (95% confidence interval)	
β_{01}	0.044	0.232		
β_{02}	1.655	0.256		
β_1 : men	−0.576	0.226	0.56	(0.36, 0.88)
β_2 : 24–40	1.147	0.278	3.15	(1.83, 5.42)
β_3 : > 40	2.232	0.291	9.32	(5.28, 16.47)

complementary log-log functions can also be used. If the response categories are regarded as crude measures of some underlying latent variable, z (as in Figure 8.2), then the optimal choice of the link function can depend on the shape of the distribution of z (McCullagh 1980). Logits and probits are appropriate if the distribution is symmetric but the complementary log-log link may be better if the distribution is very skewed.

If there is doubt about the order of the categories, then nominal logistic regression will usually be a more appropriate model than any of the models based on assumptions that the response categories are ordinal. Although the resulting model will have more parameters and hence fewer degrees of freedom and less statistical power, it may give results very similar to the ordinal models (as in the car preference example).

The estimation methods and sampling distributions used for inference depend on asymptotic results. For small studies, or numerous covariate patterns, each with few observations, the asymptotic results may be poor approximations.

Multicategory logistic models have only been readily available in statistical software since the 1990s. Their use has grown because the results are relatively easy to interpret provided that one variable can clearly be regarded as a response and the others as explanatory variables. If this distinction is unclear, for example, if data from a cross-sectional study are cross-tabulated, then log-linear models may be more appropriate. These are discussed in Chapter 9.

8.6   Exercises

8.1 If there are only $J = 2$ response categories, show that models (8.4), (8.13), (8.15) and (8.16) all reduce to the logistic regression model for binary data.

8.2 The data in Table 8.5 are from an investigation into satisfaction with housing conditions in Copenhagen (derived from Example W in Cox and Snell, 1981, from original data from Madsen, 1971). Residents in selected areas living in rented homes built between 1960 and 1968 were questioned about their satisfaction and the degree of contact with other residents. The data were tabulated by type of housing.

   a. Summarize the data using appropriate tables of percentages to show the associations between levels of satisfaction and contact with other residents, levels of satisfaction and type of housing, and contact and type of housing.

   b. Use nominal logistic regression to model associations between level of

Table 8.5 *Satisfaction with housing conditions.*

Contact with other residents	Satisfaction					
	Low		Medium		High	
	Low	High	Low	High	Low	High
Tower block	65	34	54	47	100	100
Apartment	130	141	76	116	111	191
House	67	130	48	105	62	104

satisfaction and the other two variables. Obtain a parsimonious model that summarizes the patterns in the data.

c. Do you think an ordinal model would be appropriate for associations between the levels of satisfaction and the other variables? Justify your answer. If you consider such a model to be appropriate, fit a suitable one and compare the results with those from (b).

d. From the best model you obtained in (c), calculate the standardized residuals and use them to find where the largest discrepancies are between the observed frequencies and expected frequencies estimated from the model.

8.3 The data in Table 8.6 show tumor responses of male and female patients receiving treatment for small-cell lung cancer. There were two treatment regimes. For the sequential treatment, the same combination of chemotherapeutic agents was administered at each treatment cycle. For the alternating treatment, different combinations were alternated from cycle to cycle (data from Holtbrugger and Schumacher, 1991).

Table 8.6 *Tumor responses to two different treatments: numbers of patients in each category.*

Treatment	Sex	Progressive disease	No change	Partial remission	Complete remission
Sequential	Male	28	45	29	26
	Female	4	12	5	2
Alternating	Male	41	44	20	20
	Female	12	7	3	1

a. Fit a proportional odds model to estimate the probabilities for each response category taking treatment and sex effects into account.

b. Examine the adequacy of the model fitted in (a) using residuals and goodness of fit statistics.

c. Use a Wald statistic to test the hypothesis that there is no difference in responses for the two treatment regimes.

d. Fit two proportional odds models to test the hypothesis of no treatment difference. Compare the results with those for (c) above.

e. Fit adjacent category models and continuation ratio models using logit, probit and complementary log-log link functions. How do the different models affect the interpretation of the results?

8.4 Consider ordinal response categories which can be interpreted in terms of continuous latent variable as shown in Figure 8.2. Suppose the distribution of this underlying variable is Normal. Show that the probit is the natural link function in this situation (Hint: See Section 7.3).

Chapter 9

Poisson Regression and Log-Linear Models

9.1 Introduction

The number of times an event occurs is a common form of data. Examples of **count** or **frequency** data include the number of tropical cyclones crossing the North Queensland coast (Section 1.6.5) or the numbers of people in each cell of a contingency table summarizing survey responses (e.g., satisfaction ratings for housing conditions, Exercise 8.2).

The **Poisson distribution** $Po(\mu)$ is often used to model count data. If Y is the number of occurrences, its probability distribution can be written as

$$f(y) = \frac{\mu^y e^{-\mu}}{y!}, \quad y = 0, 1, 2, \ldots,$$

where μ is the average number of occurrences. It can be shown that $E(Y) = \mu$ and $\text{var}(Y) = \mu$ (see Exercise 3.4).

The parameter μ requires careful definition. Often it needs to be described as a rate; for example, the average number of customers who buy a particular product out of every 100 customers who enter the store. For motor vehicle crashes, the rate parameter may be defined in many different ways: crashes per 1,000 population, crashes per 1,000 licensed drivers, crashes per 1,000 motor vehicles, or crashes per 100,000 km travelled by motor vehicles. The time scale should be included in the definition; for example, the motor vehicle crash rate is usually specified as the rate per year (e.g., crashes per 100,000 km per year), while the rate of tropical cyclones refers to the cyclone season from November to April in Northeastern Australia. More generally, the rate is specified in terms of units of "exposure"; for instance, customers entering a store are "exposed" to the opportunity to buy the product of interest. For occupational injuries, each worker is exposed for the period he or she is at work, so the rate may be defined in terms of person-years "at risk."

The effect of explanatory variables on the response Y is modelled through the parameter μ. This chapter describes models for two situations.

In the first situation, the events relate to varying amounts of exposure which need to be taken into account when modelling the rate of events. **Poisson regression** is used in this case. The other explanatory variables (in addition to exposure) may be continuous or categorical.

In the second situation, exposure is constant (and therefore not relevant to the model) and the explanatory variables are usually categorical. If there are only a few explanatory variables the data are summarized in a cross-classified table. The response variable is the frequency or count in each cell of the table. The variables used to define the table are all treated as explanatory variables. The study design may mean that there are some constraints on the cell frequencies (for example, the totals for each row of the table may be equal), and these need to be taken into account in the modelling. The term **log-linear model**, which basically describes the role of the link function, is used for the generalized linear models appropriate for this situation.

The next section describes Poisson regression. A numerical example is used to illustrate the concepts and methods, including model checking and inference. Subsequent sections describe relationships between probability distributions for count data, constrained in various ways, and the log-linear models that can be used to analyze the data.

9.2   Poisson regression

Let $Y_1, \ldots, Y_N$ be independent random variables with Y_i denoting the number of events observed from exposure n_i for the ith covariate pattern. The expected value of Y_i can be written as

$$E(Y_i) = \mu_i = n_i \theta_i.$$

For example, suppose Y_i is the number of insurance claims for a particular make and model of car. This will depend on the number of cars of this type that are insured, n_i, and other variables that affect θ_i, such as the age of the cars and the location where they are used. The subscript i is used to denote the different combinations of make and model, age, location and so on.

The dependence of θ_i on the explanatory variables is usually modelled by

$$\theta_i = e^{\mathbf{x}_i^T \boldsymbol{\beta}}. \tag{9.1}$$

Therefore, the generalized linear model is

$$E(Y_i) = \mu_i = n_i e^{\mathbf{x}_i^T \boldsymbol{\beta}}; \quad Y_i \sim \mathrm{Po}(\mu_i). \tag{9.2}$$

The natural link function is the logarithmic function

$$\log \mu_i = \log n_i + \mathbf{x}_i^T \boldsymbol{\beta}. \tag{9.3}$$

Equation (9.3) differs from the usual specification of the linear component due to the inclusion of the term $\log n_i$. This term is called the **offset**. It is a known constant, which is readily incorporated into the estimation procedure. As usual, the terms $\mathbf{x}_i$ and $\boldsymbol{\beta}$ describe the covariate pattern and parameters, respectively.

For a binary explanatory variable denoted by an indictor variable, $x_j = 0$ if the factor is absent and $x_j = 1$ if it is present. The **rate ratio**, *RR*, for presence vs. absence is

$$RR = \frac{E(Y_i \mid present)}{E(Y_i \mid absent)} = e^{\beta_j}$$

from (9.1), provided all the other explanatory variables remain the same. Similarly, for a continuous explanatory variable x_k, a one-unit increase will result in a multiplicative effect of e^{β_k} on the rate μ. Therefore, parameter estimates are often interpreted on the exponential scale e^β in terms of ratios of rates.

Hypotheses about the parameters β_j can be tested using the Wald, score or likelihood ratio statistics. Confidence intervals can be estimated similarly. For example, for parameter β_j

$$\frac{b_j - \beta_j}{s.e.(b_j)} \sim N(0, 1) \tag{9.4}$$

approximately. Alternatively, hypothesis testing can be performed by comparing the goodness of fit of appropriately defined nested models (see Chapter 4).

The fitted values are given by

$$\widehat{Y}_i = \widehat{\mu}_i = n_i e^{\mathbf{x}_i^T \mathbf{b}}, \quad i = 1, \dots, N.$$

These are often denoted by e_i because they are estimates of the expected values $E(Y_i) = \mu_i$. As $\text{var}(Y_i) = E(Y_i)$ for the Poisson distribution, the standard error of Y_i is estimated by $\sqrt{e_i}$ so the **Pearson residuals** are

$$r_i = \frac{o_i - e_i}{\sqrt{e_i}}, \tag{9.5}$$

where o_i denotes the observed value of Y_i. As outlined in Section 6.2.6, these residuals may be further refined to

$$r_{pi} = \frac{o_i - e_i}{\sqrt{e_i}\sqrt{1 - h_i}},$$

where the leverage, h_i, is the ith element on the diagonal of the hat matrix.

For the Poisson distribution, the residuals given by (9.5) and the chi-squared goodness of fit statistic are related by

$$X^2 = \sum r_i^2 = \sum \frac{(o_i - e_i)^2}{e_i},$$

which is the usual definition of the chi-squared statistic for contingency tables.

The deviance for a Poisson model is given in Section 5.6.3. It can be written in the form

$$D = 2\sum [o_i \log(o_i/e_i) - (o_i - e_i)]. \qquad (9.6)$$

However, for most models $\sum o_i = \sum e_i$ (see Exercise 9.1), so the deviance simplifies to

$$D = 2\sum [o_i \log(o_i/e_i)]. \qquad (9.7)$$

The **deviance residuals** are the components of D in (9.6),

$$d_i = \text{sign}(o_i - e_i)\sqrt{2[o_i \log(o_i/e_i) - (o_i - e_i)]}, \quad i = 1,\dots,N \qquad (9.8)$$

so that $D = \sum d_i^2$.

The goodness of fit statistics X^2 and D are closely related. Using the Taylor series expansion given in Section 7.5,

$$o\log\left(\frac{o}{e}\right) = (o - e) + \tfrac{1}{2}\frac{(o - e)^2}{e} + \dots$$

so that, approximately, from (9.6)

$$\begin{aligned} D &= 2\sum_{i=1}^{N}\left[(o_i - e_i) + \tfrac{1}{2}\frac{(o_i - e_i)^2}{e_i} - (o_i - e_i)\right] \\ &= \sum_{i=1}^{N}\frac{(o_i - e_i)^2}{e_i} = X^2. \end{aligned}$$

The statistics D and X^2 can be used directly as measures of goodness of fit, as they can be calculated from the data and the fitted model (because they do not involve any nuisance parameters like σ^2 for the Normal distribution). They can be compared with the central chi-squared distribution with $N - p$ degrees of freedom, where p is the number of parameters that are estimated. The chi-squared distribution is likely to be a better approximation for the sampling distribution of X^2 than for the sampling distribution of D (see Section 7.5).

Two other summary statistics provided by some software are the likelihood ratio chi-squared statistic and pseudo R^2. These are based on comparisons between the maximum value of the log-likelihood function for a minimal model with the same rate parameter β_1 for all Y_i's and no covariates, $\log \mu_i = \log n_i + \beta_1$, and the maximum value of the log-likelihood function for Model (9.3) with p parameters. The likelihood ratio chi-squared statistic $C = 2[l(\mathbf{b}) - l(\mathbf{b}_{min})]$ provides an overall test of the hypotheses that $\beta_2 = \ldots = \beta_p = 0$, by comparison with the central chi-squared distribution with $p - 1$ degrees of freedom (see Exercise 7.4). Less formally, pseudo $R^2 = [l(\mathbf{b}_{min}) - l(\mathbf{b})]/l(\mathbf{b}_{min})$ provides an intuitive measure of fit.

Other diagnostics, such as delta-betas and related statistics, are also available for Poisson models.

9.2.1 Example of Poisson regression: British doctors' smoking and coronary death

The data in Table 9.1 are from a famous study conducted by Sir Richard Doll and colleagues. In 1951, all British doctors were sent a brief questionnaire about whether they smoked tobacco. Since then information about their deaths has been collected. Table 9.1 shows the numbers of deaths from coronary heart disease among male doctors 10 years after the survey. It also shows the total number of person-years of observation at the time of the analysis (Breslow and Day, 1987: Appendix 1A and page 112).

Table 9.1 *Deaths from coronary heart disease after 10 years among British male doctors categorized by age and smoking status in 1951.*

Age group	Smokers		Non-smokers	
	Deaths	Person-years	Deaths	Person-years
35–44	32	52407	2	18790
45–54	104	43248	12	10673
55–64	206	28612	28	5710
65–74	186	12663	28	2585
75–84	102	5317	31	1462

The questions of interest are

1. Is the death rate higher for smokers than non-smokers?
2. If so, by how much?

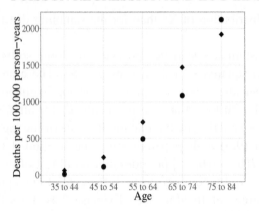

Figure 9.1 *Deaths rates from coronary heart disease per 100,000 person-years for smokers (diamonds) and non-smokers (dots).*

3. Is the differential effect related to age?

Figure 9.1 shows the death rates per 100,000 person-years from coronary heart disease for smokers and non-smokers. It is clear that the rates increase with age but more steeply than in a straight line. Death rates among smokers appear to be generally higher than among non-smokers but they do not rise as rapidly with age. Various models can be specified to describe these data well. One model, in the form of (9.3) is

$$\log(deaths_i) = \log(personyears_i) + \beta_1 + \beta_2 smoke_i + \beta_3 agecat_i$$
$$+ \beta_4 agesq_i + \beta_5 smkage_i \qquad (9.9)$$

where the subscript i denotes the ith subgroup defined by age group and smoking status ($i = 1, \ldots, 5$ for ages 35–44,...,75–84 for smokers and $i = 6, \ldots, 10$ for the corresponding age groups for non-smokers). The term $deaths_i$ denotes the expected number of deaths and $personyears_i$ denotes the number of doctors at risk and the observation periods in group i. For the other terms, $smoke_i$ is equal to 1 for smokers and 0 for non-smokers; $agecat_i$ takes the values $1, \ldots, 5$ for age groups 35–44,...,75–84; $agesq_i$ is the square of $agecat_i$ to take account of the non-linearity of the rate of increase; and $smkage_i$ is equal to $agecat_i$ for smokers and 0 for non-smokers; thus, describing a differential rate of increase with age.

In Stata, Model (9.9) can be fitted using either of the following commands

————————— Stata code (Poisson regression models) —————————
```
.poisson deaths agecat agesq smoke smkage, exposure(personyears)
```

```
.glm deaths agecat agesq smoke smkage, family(poisson) link(log)
lnoffset(personyears)
```

The option `irr` can be used to obtain the rate ratios and 95% confidence limits.

The corresponding command for R is

```
─────────── R code (Poisson regression model) ───────────
>data(doctors)
>res.doc<-glm(deaths~age + agesq + smoking + smoking:age +
offset(log(personyears)),family=poisson(),data=doctors)
```

Table 9.2 shows the parameter estimates in the form of rate ratios $e^{\hat{\beta}_j}$. The Wald statistics (9.4) to test $\beta_j = 0$ all have very small p-values and the 95% confidence intervals for e^{β_j} do not contain unity showing that all the terms are needed in the model. The estimates show that the risk of coronary deaths was, on average, about 4 times higher for smokers than non-smokers (based on the rate ratio for *smoke*), after the effect of age is taken into account. However, the effect is attenuated as age increases (coefficient for *smkage*). Table 9.3 shows that the model fits the data very well; the expected numbers of deaths estimated from (9.9) are quite similar to the observed numbers of deaths, and so the Pearson residuals calculated from (9.5) and deviance residuals from (9.8) are very small.

Table 9.2 *Parameter estimates obtained by fitting Model (9.9) to the data in Table 9.1.*

Term	agecat	agesq	smoke	smkage
$\hat{\beta}$	2.376	−0.198	1.441	−0.308
$s.e.(\hat{\beta})$	0.208	0.027	0.372	0.097
Wald statistic	11.43	−7.22	3.87	−3.17
p-value	< 0.001	< 0.001	< 0.001	0.002
Rate ratio	10.77	0.82	4.22	0.74
95% confidence interval	7.2, 16.2	0.78, 0.87	2.04, 8.76	0.61, 0.89

To obtain these results from Stata after the command `poisson` use `predict fit` and then calculate the residuals, or use `poisgof` to obtain the deviance statistic D or `poisgof, pearson` to obtain the statistic X^2. Alternatively, after `glm` use `predict fit`; `predict d, deviance`; and `predict c, pearson`.

The R commands are as follows

─────────── R code (Poisson model residuals) ───────────

Table 9.3 *Observed and estimated expected numbers of deaths and residuals for the model described in Table 9.2.*

Age category	Smoking category	Observed deaths	Expected deaths	Pearson residual	Deviance residual
1	1	32	29.58	0.444	0.438
2	1	104	106.81	−0.272	−0.273
3	1	206	208.20	−0.152	−0.153
4	1	186	182.83	0.235	0.234
5	1	102	102.58	−0.057	−0.057
1	0	2	3.41	−0.766	−0.830
2	0	12	11.54	0.135	0.134
3	0	28	27.74	0.655	0.641
4	0	28	30.23	−0.405	−0.411
5	0	31	31.07	−0.013	−0.013
Sum of squares*				1.550	1.635

* Calculated from residuals correct to more significant figures than shown here.

```
>fit_p=c(fitted(res.doc))
>pearsonresid<-(doctors$deaths-fit_p)/sqrt(fit_p)
>chisq<-sum(pearsonresid*pearsonresid)
>devres<-sign(doctors$deaths-fit_p)*(sqrt(2*(doctors$deaths*
 log(doctors$deaths/fit_p)-(doctors$deaths-fit_p))))
>deviance<-sum(devres*devres)
```

For the minimal model, with only the parameter β_1, the maximum value for the log-likelihood function is $l(b_{min}) = -495.067$. The corresponding value for Model (9.9) is $l(\mathbf{b}) = -28.352$. Therefore, an overall test of the model (testing $\beta_j = 0$ for $j = 2,\ldots,5$) is $C = 2[l(\mathbf{b}) - l(b_{min})] = 933.43$ which is highly statistically significant compared with the chi-squared distribution with 4 degrees of freedom. The *pseudo* R^2 value is 0.94, or 94%, which suggests a good fit. More formal tests of the goodness of fit are provided by the statistics $X^2 = 1.550$ and $D = 1.635$ which are small compared with the chi-squared distribution with $N - p = 10 - 5 = 5$ degree of freedom.

9.3   Examples of contingency tables

Before specifying log-linear models for frequency data summarized in contingency tables, it is important to consider how the design of the study may determine constraints on the data. The study design also affects the choice of probability models to describe the data. These issues are illustrated in the following three examples.

9.3.1   Example: Cross-sectional study of malignant melanoma

These data are from a cross-sectional study of patients with a form of skin cancer called malignant melanoma (Roberts et al. 1981). For a sample of $n = 400$ patients, the site of the tumor and its histological type were recorded. The data, numbers of patients with each combination of tumor type and site, are given in Table 9.4.

Table 9.4 *Malignant melanoma: frequencies for tumor type and site (Roberts et al. 1981).*

	Site			
Tumor type	Head & neck	Trunk	Extrem -ities	Total
Hutchinson's melanotic freckle	22	2	10	34
Superficial spreading melanoma	16	54	115	185
Nodular	19	33	73	125
Indeterminate	11	17	28	56
Total	68	106	226	400

The question of interest is whether there is any association between tumor type and site. Table 9.5 shows the data displayed as percentages of row and column totals. It appears that Hutchinson's melanotic freckle is more common on the head and neck but there is little evidence of association between other tumor types and sites.

Let Y_{jk} denote the frequency for the (j,k)th cell with $j = 1,\ldots,J$ and $k = 1,\ldots,K$. In this example, there are $J = 4$ rows, $K = 3$ columns and the constraint that $\sum_{j=1}^{J}\sum_{k=1}^{K} Y_{jk} = n$, where $n = 400$ is fixed by the design of the study. If the Y_{jk}'s are independent random variables with Poisson distributions with parameters $E(Y_{jk}) = \mu_{jk}$, then their sum has the Poisson distribution with parameter $E(n) = \mu = \sum\sum \mu_{jk}$. Hence, the joint probability distribution of the Y_{jk}'s, conditional on their sum n, is the Multinomial distribution

$$f(\mathbf{y}|n) = n! \prod_{j=1}^{J} \prod_{k=1}^{K} \theta_{jk}^{y_{jk}} / y_{jk}! \, ,$$

where $\theta_{jk} = \mu_{jk}/\mu$. This result is derived in Section 8.2. The sum of the terms θ_{jk} is unity because $\sum\sum \mu_{jk} = \mu$; also $0 < \theta_k < 1$. Thus, θ_{jk} can be interpreted

Table 9.5 *Malignant melanoma: row and column percentages for tumor type and site.*

	Site			
Tumor type	Head & neck	Trunk	Extrem -ities	Total
Row percentages				
Hutchinson's melanotic freckle	64.7	5.9	29.4	100
Superficial spreading melanoma	8.6	29.2	62.2	100
Nodular	15.2	26.4	58.4	100
Indeterminate	19.6	30.4	50.0	100
All types	17.0	26.5	56.5	100
Column percentages				
Hutchinson's melanotic freckle	32.4	1.9	4.4	8.50
Superficial spreading melanoma	23.5	50.9	50.9	46.25
Nodular	27.9	31.1	32.3	31.25
Indeterminate	16.2	16.0	12.4	14.00
All types	100.0	99.9	100.0	100.0

as the probability of an observation in the (j,k)th cell of the table. Also the expected value of Y_{jk} is

$$E(Y_{jk}) = \mu_{jk} = n\theta_{jk}.$$

The usual link function for a Poisson model gives

$$\log \mu_{jk} = \log n + \log \theta_{jk},$$

which is like Equation (9.3), except that the term $\log n$ is the same for all the Y_{jk}'s.

9.3.2 Example: Randomized controlled trial of influenza vaccine

In a prospective study of a new living attenuated recombinant vaccine for influenza, patients were randomly allocated to two groups, one of which was given the new vaccine and the other a saline placebo. The responses were titre levels of hemagglutinin inhibiting antibody found in the blood six weeks after vaccination; they were categorized as "small," "medium" or "large." The cell frequencies in the rows of Table 9.6 are constrained to add to the number of subjects in each treatment group (35 and 38, respectively). We want to know if the pattern of responses is the same for each treatment group.

Table 9.6 *Flu vaccine trial.*

	Response			
	Small	Moderate	Large	Total
Placebo	25	8	5	38
Vaccine	6	18	11	35

(Data from R.S. Gillett, personal communication, 1992)

In this example the row totals are fixed. Thus, the joint probability distribution for each row is Multinomial

$$f(y_{j1}, y_{j2}, \ldots, y_{jK} \,|\, y_{j.}) = y_{j.}! \prod_{k=1}^{K} \theta_{jk}^{y_{jk}} / y_{jk}!,$$

where $y_{j.} = \sum_{k=1}^{K} y_{jk}$ is the row total and $\sum_{k=1}^{K} \theta_{jk} = 1$. So the joint probability distribution for all the cells in the table is the **product multinomial distribution**

$$f(\mathbf{y} \,|\, y_{1.}, y_{2.}, \ldots, y_{J.}) = \prod_{j=1}^{J} y_{j.}! \prod_{k=1}^{K} \theta_{jk}^{y_{jk}} / y_{jk}!,$$

where $\sum_{k=1}^{K} \theta_{jk} = 1$ for each row. In this case $E(Y_{jk}) = y_{j.} \theta_{jk}$ so that

$$\log E(Y_{jk}) = \log \mu_{jk} = \log y_{j.} + \log \theta_{jk}.$$

If the response pattern was the same for both groups, then $\theta_{jk} = \theta_{.k}$ for $k = 1, \ldots, K$.

9.3.3 Example: Case–control study of gastric and duodenal ulcers and aspirin use

In this retrospective case–control study, a group of ulcer patients was compared with a group of control patients not known to have peptic ulcer, but who were similar to the ulcer patients with respect to age, sex and socioeconomic status (Duggan et al. 1986). The ulcer patients were classified according to the site of the ulcer: gastric or duodenal. Aspirin use was ascertained for all subjects. The results are shown in Table 9.7.

This is a $2 \times 2 \times 2$ contingency table. Some questions of interest are

1. Is gastric ulcer associated with aspirin use?
2. Is duodenal ulcer associated with aspirin use?

Table 9.7 *Gastric and duodenal ulcers and aspirin use: frequencies (Duggan et al. 1986).*

| | Aspirin use | | |
	Non-user	User	Total
Gastric ulcer			
Control	62	6	68
Cases	39	25	64
Duodenal ulcer			
Control	53	8	61
Cases	49	8	57

Table 9.8 *Gastric and duodenal ulcers and aspirin use: row percentages for the data in Table 9.7.*

| | Aspirin use | | |
	Non-user	User	Total
Gastric ulcer			
Control	91	9	100
Cases	61	39	100
Duodenal ulcer			
Control	87	13	100
Cases	86	14	100

3. Is any association with aspirin use the same for both ulcer sites?

When the data are presented as percentages of row totals (Table 9.8), it appears that aspirin use is more common among ulcer patients than among controls for gastric ulcer but not for duodenal ulcer.

In this example, the numbers of patients with each type of ulcer and the numbers in each of the groups of controls, that is, the four row totals in Table 9.7, were all fixed.

Let $j = 1$ or 2 denote the controls or cases, respectively; $k = 1$ or 2 denote gastric ulcers or duodenal ulcers, respectively; and $l = 1$ for patients who did not use aspirin and $l = 2$ for those who did. In general, let Y_{jkl} denote the frequency of observations in category (j,k,l) with $j = 1,\ldots,J$, $k = 1,\ldots,K$ and $l = 1,\ldots,L$. If the marginal totals $y_{jk.}$ are fixed, the joint probability distribution for the Y_{jkl}'s is

$$f(\mathbf{y} \,|\, y_{11.},\ldots,y_{JK.}) = \prod_{j=1}^{J}\prod_{k=1}^{K} y_{jk.}! \prod_{l=1}^{L} \theta_{jkl}^{y_{jkl}} / y_{jkl}!,$$

where $\mathbf{y}$ is the vector of Y_{jkl}'s and $\sum_l \theta_{jkl} = 1$ for $j = 1, \ldots, J$ and $k = 1, \ldots, K$. This is another form of **product multinomial distribution**. In this case, $E(Y_{jkl}) = \mu_{jkl} = y_{jk.} \theta_{jkl}$, so that

$$\log \mu_{jkl} = \log y_{jk.} + \log \theta_{jkl}.$$

9.4 Probability models for contingency tables

The examples in Section 9.3 illustrate the main probability models for contingency table data. In general, let the vector $\mathbf{y}$ denote the frequencies Y_i in N cells of a cross-classified table.

9.4.1 Poisson model

If there were no constraints on the Y_i's, they could be modelled as independent random variables with the parameters $E(Y_i) = \mu_i$ and joint probability distribution

$$f(\mathbf{y}; \boldsymbol{\mu}) = \prod_{i=1}^{N} \mu_i^{y_i} e^{-\mu_i} / y_i!,$$

where $\boldsymbol{\mu}$ is a vector of μ_i's.

9.4.2 Multinomial model

If the only constraint is that the sum of the Y_i's is n, then the following Multinomial distribution may be used

$$f(\mathbf{y}; \boldsymbol{\mu} \mid n) = n! \prod_{i=1}^{N} \theta_i^{y_i} / y_i!,$$

where $\sum_{i=1}^{N} \theta_i = 1$ and $\sum_{i=1}^{N} y_i = n$. In this case, $E(Y_i) = n\theta_i$.

For a two-dimensional contingency table (such as Table 9.4 for the melanoma data), if j and k denote the rows and columns, then the most commonly considered hypothesis is that the row and column variables are independent so that

$$\theta_{jk} = \theta_{j.} \theta_{.k},$$

where $\theta_{j.}$ and $\theta_{.k}$ denote the marginal probabilities with $\sum_j \theta_{j.} = 1$ and $\sum_k \theta_{.k} = 1$. This hypothesis can be tested by comparing the fit of two linear models for the logarithm of $\mu_{jk} = E(Y_{jk})$; namely

$$\log \mu_{jk} = \log n + \log \theta_{jk}$$

and

$$\log \mu_{jk} = \log n + \log \theta_{j.} + \log \theta_{.k}.$$

9.4.3 Product multinomial models

If there are more fixed marginal totals than just the overall total n, then appropriate products of multinomial distributions can be used to model the data.

For example, for a three-dimensional table with J rows, K columns and L layers, if the row totals are fixed in each layer, the joint probability for the Y_{jkl}'s is

$$f(\mathbf{y}|y_{j.l}, j = 1,\dots,J, l = 1,\dots,L) = \prod_{j=1}^{J}\prod_{l=1}^{L} y_{j.l}! \prod_{k=1}^{K} \theta_{jkl}^{y_{jkl}}/y_{jkl}!,$$

where $\sum_k \theta_{jkl} = 1$ for each combination of j and l. In this case, $E(Y_{jkl}) = y_{j.l}\theta_{jkl}$.

If only the layer totals are fixed, then

$$f(\mathbf{y}|y_{..l}, l = 1,\dots,L) = \prod_{l=1}^{L} y_{..l}! \prod_{j=1}^{J}\prod_{k=1}^{K} \theta_{jkl}^{y_{jkl}}/y_{jkl}!$$

with $\sum_j\sum_k \theta_{jkl} = 1$ for $l = 1,\dots,L$. Also $E(Y_{jkl}) = y_{..l}\theta_{jkl}$.

9.5 Log-linear models

All the probability models given in Section 9.4 are based on the Poisson distribution and in all cases $E(Y_i)$ can be written as a product of parameters and other terms. Thus, the natural link function for the Poisson distribution, the logarithmic function, yields a linear component

$$\log E(Y_i) = \text{constant} + \mathbf{x}_i^T \boldsymbol{\beta}.$$

The term **log-linear model** is used to describe all these generalized linear models.

For the melanoma Example 9.3.1, if there are no associations between site and type of tumor so that these two variables are independent, their joint probability θ_{jk} is the product of the marginal probabilities

$$\theta_{jk} = \theta_{j.}\theta_{.k}, \quad j = 1,\dots,J \text{ and } k = 1,\dots,K.$$

The hypothesis of independence can be tested by comparing the additive model (on the logarithmic scale)

$$\log E(Y_{jk}) = \log n + \log \theta_{j.} + \log \theta_{.k} \tag{9.10}$$

with the model

$$\log E(Y_{jk}) = \log n + \log \theta_{jk}. \tag{9.11}$$

This is analogous to analysis of variance for a two-factor experiment without replication (see Section 6.4.2). Equation (9.11) can be written as the saturated model

$$\log E(Y_{jk}) = \mu + \alpha_j + \beta_k + (\alpha\beta)_{jk},$$

and Equation (9.10) can be written as the additive model

$$\log E(Y_{jk}) = \mu + \alpha_j + \beta_k.$$

Since the term $\log n$ has to be in all models, the minimal model is

$$\log E(Y_{jk}) = \mu.$$

For the flu vaccine trial, Example 9.3.2, $E(Y_{jk}) = y_{j.}\theta_{jk}$ if the distribution of responses described by the θ_{jk}'s differs for the j groups, or $E(Y_{jk}) = y_{j.}\theta_{.k}$ if it is the same for all groups. So the hypothesis of **homogeneity** of the response distributions can be tested by comparing the model

$$\log E(Y_{jk}) = \mu + \alpha_j + \beta_k + (\alpha\beta)_{jk},$$

corresponding to $E(Y_{jk}) = y_{j.}\theta_{jk}$, and the model

$$\log E(Y_{jk}) = \mu + \alpha_j + \beta_k,$$

corresponding to $E(Y_{jk}) = y_{j.}\theta_{.k}$. The minimal model for these data is

$$\log E(Y_{jk}) = \mu + \alpha_j$$

because the row totals, corresponding to the subscript j, are fixed by the design of the study.

More generally, the specification of the linear components for log-linear models bears many resemblances to the specification for ANOVA models. The models are **hierarchical**, meaning that if a higher-order (interaction) term is included in the model, then all the related lower-order terms are also included. Thus, if the two-way (first-order) interaction $(\alpha\beta)_{jk}$ is included, then so are the main effects α_j and β_k and the constant μ. Similarly, if second-order interactions $(\alpha\beta\gamma)_{jkl}$ are included, then so are the first-order interactions $(\alpha\beta)_{jk}$, $(\alpha\gamma)_{jl}$ and $(\beta\gamma)_{kl}$.

If log-linear models are specified analogously to ANOVA models, they include too many parameters, so sum-to-zero or corner-point constraints are needed. Interpretation of the parameters is usually simpler if reference or corner-point categories are identified so that parameter estimates describe effects for other categories relative to the reference categories.

For contingency tables the main questions almost always relate to associations between variables. Therefore, in log-linear models, the terms of primary interest are the interactions involving two or more variables.

9.6 Inference for log-linear models

Although three types of probability distributions are used to describe contingency table data (see Section 9.4), Birch (1963) showed that for any log-linear model the maximum likelihood estimators are the same for all these distributions provided that the parameters which correspond to the fixed marginal totals are always included in the model. This means that for the purpose of estimation, the Poisson distribution can always be assumed. As the Poisson distribution belongs to the exponential family and the parameter constraints can be incorporated into the linear component, all the standard methods for generalized linear models can be used.

The adequacy of a model can be assessed using the goodness of fit statistics X^2 or D (and sometimes C and pseudo R^2) summarized in Section 9.2 for Poisson regression. More insight into model adequacy can often be obtained by examining the Pearson or deviance residuals given by Equations (9.5) and (9.8), respectively. Hypothesis tests can be conducted by comparing the difference in goodness of fit statistics between a general model corresponding to an alternative hypothesis and a nested, simpler model corresponding to a null hypothesis.

These methods are illustrated in the following examples.

9.7 Numerical examples

9.7.1 Cross-sectional study of malignant melanoma

For the data in Table 9.4 the question of interest is whether there is an association between tumor type and site. This can be examined by testing the null hypothesis that the variables are independent.

The conventional chi-squared test of independence for a two-dimensional table is performed by calculating expected frequencies for each cell based on the marginal totals, $e_{jk} = y_{j.}y_{.k}/n$, calculating the chi-squared statistic $X^2 = \sum_j \sum_k (y_{jk} - e_{jk})^2 / e_{jk}$ and comparing this with the central chi-squared distribution with $(J-1)(K-1)$ degrees of freedom. The observed and expected frequencies are shown in Table 9.9. These give

$$X^2 = \frac{(22-5.78)^2}{5.78} + \ldots + \frac{(28-31.64)^2}{31.64} = 65.8.$$

The value $X^2 = 65.8$ is very significant compared with the $\chi^2(6)$ distribution. Examination of the observed frequencies y_{jk} and expected frequencies e_{jk} shows that Hutchinson's melanotic freckle is more common on the head and neck than would be expected if site and type were independent.

Table 9.9 *Conventional chi-squared test of independence for melanoma data in Table 9.4; expected frequencies are shown in brackets.*

Tumor type	Site			
	Head & Neck	Trunk	Extrem -ities	Total
Hutchinson's melanotic freckle	22 (5.78)	2 (9.01)	10 (19.21)	34
Superficial spreading melanoma	16 (31.45)	54 (49.03)	115 (104.52)	185
Nodular	19 (21.25)	33 (33.13)	73 (70.62)	125
Indeterminate	11 (9.52)	17 (14.84)	28 (31.64)	56
Total	68	106	226	400

The corresponding analysis using log-linear models involves fitting the additive Model (9.10) corresponding to the hypothesis of independence. The saturated Model (9.11) and the minimal model with only a term for the mean effect are also fitted for illustrative purposes. In Stata the commands for the three models are

```
──────────────── Stata code (log-linear models) ────────────
.xi:glm frequency i.tumor i.site i.tumor*i.site, family(poisson)
  link(log)
.xi:glm frequency i.tumor i.site, family(poisson) link(log)

.glm frequency, family(poisson) link(log)
```

The corresponding commands in R (site and tumor should be text variables)

```
──────────────── R code (log-linear models) ────────────
>ressat.melanoma<-glm(frequency~site*tumor,family=poisson(),
  data=melanoma)
```

Table 9.10 *Log-linear models for the melanoma data in Table 9.4; coefficients, b, with standard errors in brackets.*

Term*	Saturated Model (9.10)	Additive Model (9.9)	Minimal model
Constant	3.091 (0.213)	1.754 (0.204)	3.507 (0.05)
SSM	−0.318 (0.329)	1.694 (0.187)	
NOD	−0.147 (0.313)	1.302 (0.193)	
IND	−0.693 (0.369)	0.499 (0.217)	
TNK	−2.398 (0.739)	0.444 (0.155)	
EXT	−0.788 (0.381)	1.201 (0.138)	
SSM * TNK	3.614 (0.792)		
SSM * EXT	2.761 (0.465)		
NOD * TNK	2.950 (0.793)		
NOD * EXT	2.134 (0.460)		
IND * TNK	2.833 (0.834)		
IND * EXT	1.723 (0.522)		
log-likelihood	−29.556	−55.453	−177.16
X^2	0.0	65.813	
D	0.0	51.795	

*Reference categories are Hutchinson's melanotic freckle (*HMF*) and head and neck (*HNK*). Other categories are for type, superficial spreading melanoma (*SSM*), nodular (*NOD*) and indeterminate (*IND*), and for site, trunk (*TNK*) and extremities (*EXT*).

```
>resadd.melanoma<-glm(frequency~site + tumor,family=poisson(),
 data=melanoma)
>resmin.melanoma<-glm(frequency~1, family=poisson(),
data=melanoma)
```

The results for all three models are shown in Table 9.10. For the reference category of Hutchinson's melanotic freckle (*HMF*) on the head or neck (*HNK*), the expected frequencies are as follows:

minimal model: $e^{3.507} = 33.35$;

additive model: $e^{1.754} = 5.78$, as in Table 9.9;

saturated model: $e^{3.091} = 22$, equal to observed frequency.

For indeterminate tumors (*IND*) in the extremities (*EXT*), the expected frequencies are minimal model: $e^{3.507} = 33.35$;

additive model: $e^{1.754+0.499+1.201} = 31.64$, as in Table 9.9;

saturated model: $e^{3.091-0.693-0.788+1.723} = 28$, equal to observed frequency.

The saturated model with 12 parameters fits the 12 data points exactly. The additive model corresponds to the conventional analysis. The deviance for the additive model can be calculated from the sum of squares of the deviance residuals given by (9.8), or from twice the difference between the maximum values of the log-likelihood function for this model and the saturated model, $\triangle D = 2[-29.556 - (-55.453)] = 51.79$.

For this example, the conventional chi-squared test for independence and log-linear modelling produce exactly the same results. The advantage of log-linear modelling is that it provides a method for analyzing more complicated cross-tabulated data as illustrated by the next example.

9.7.2 Case–control study of gastric and duodenal ulcer and aspirin use

Preliminary analysis of the 2×2 tables for gastric ulcer and duodenal ulcer separately suggests that aspirin use may be a risk factor for gastric ulcer but not for duodenal ulcer. For analysis of the full data set, Table 9.7, the main effects for case–control status (CC), ulcer site (GD) and the interaction between these terms $(CC \times GD)$ have to be included in all models (as these correspond to the fixed marginal totals). Table 9.11 shows the results of fitting this and several more complex models involving aspirin use (AP).

Table 9.11 *Results of log-linear modelling of data in Table 9.7.*

Terms in model	d.f.*	Deviance
$GD + CC + GD \times CC$	4	126.708
$GD + CC + GD \times CC + AP$	3	21.789
$GD + CC + GD \times CC + AP + AP \times CC$	2	10.538
$GD + CC + GD \times CC + AP + AP \times CC$		
$\qquad\qquad\qquad\qquad + AP \times GD$	1	6.283

*d.f. denotes degrees of freedom = number of observations (8) minus number of parameters

In Stata the commands are

```
────────────── Stata code (log-linear models) ──────────────
.xi:glm frequency i.GD i.CC i.GD*i.CC, family(poisson) link(log)
.xi:glm frequency i.GD i.CC i.GD*i.CC i.AP, family(poisson)
link(log)
```

```
.xi:glm frequency i.GD i.CC i.GD*i.CC i.AP i.AP*i.CC,
family(poisson) link(log)
.xi:glm frequency i.GD i.CC i.GD*i.CC i.AP i.AP*i.CC i.AP*i.GD,
family(poisson) link(log)
```

In R the corresponding commands are

──────────────────── R code (log-linear models) ────────────────────
```
>data(ulcer)
>res1.aspirin<-glm(frequency~GD + CC + GD*CC, family=poisson(),
 data=ulcer)
>res2.aspirin<-glm(frequency~GD + CC + GD*CC + AP,
 family=poisson(), data=ulcer)
>res3.aspirin<-glm(frequency~GD + CC + GD*CC + AP + AP*CC,
 family=poisson(), data=ulcer)
>res4.aspirin<-glm(frequency~GD + CC + GD*CC + AP + AP*CC +
 AP*GD, family=poisson(), data=ulcer)
```

The comparison of aspirin use between cases and controls can be summarized by the deviance difference for the second and third rows of Table 9.11, $\triangle D = 11.25$. This value is statistically significant compared with the $\chi^2(1)$ distribution, suggesting that aspirin is a risk factor for ulcers. Comparison between the third and fourth rows of the table, $\triangle D = 4.26$, provides only weak evidence of a difference between ulcer sites, possibly due to the lack of statistical power (p-value = 0.04 from the distribution $\chi^2(1)$).

The fit of the model with all three two-way interactions is shown in Table 9.12. The goodness of fit statistics for this table are $X^2 = 6.49$ and $D = 6.28$, which suggest that the model is not particularly good (compared with the $\chi^2(1)$ distribution) even though $p = 7$ parameters have been used to describe $N = 8$ data points.

9.8 Remarks

Two issues relevant to the analysis of a count data have not yet been discussed in this chapter.

First, **overdispersion** occurs when $\text{var}(Y_i)$ is greater than $\text{E}(Y_i)$, although $\text{var}(Y_i) = \text{E}(Y_i)$ for the Poisson distribution. The **negative binomial distribution** provides an alternative model with $\text{var}(Y_i) = \phi\, \text{E}(Y_i)$, where $\phi > 1$ is a parameter that can be estimated. Overdispersion can be due to lack of independence between the observations, in which case the methods described in Chapter 11 for correlated data can be used.

Second, contingency tables may include cells which cannot have any observations (e.g., male hysterectomy cases). This phenomenon, termed **structural zeros** may not be easily incorporated in Poisson regression unless the parameters can be specified to accommodate the situation. Alternative approaches are discussed by Agresti (2013) and Hilbe (2014).

9.9 Exercises

9.1 Let $Y_1, \ldots, Y_N$ be independent random variables with $Y_i \sim Po(\mu_i)$ and $\log \mu_i = \beta_1 + \sum_{j=2}^{J} x_{ij}\beta_j$, $i = 1, \ldots, N$.

a. Show that the score statistic for β_1 is $U_1 = \sum_{i=1}^{N}(Y_i - \mu_i)$.

b. Hence, show that for maximum likelihood estimates $\widehat{\mu}_i$, $\sum \widehat{\mu}_i = \sum y_i$.

c. Deduce that the expression for the deviance in (9.6) simplifies to (9.7) in this case.

9.2 The data in Table 9.13 are numbers of insurance policies, n, and numbers of claims, y, for cars in various insurance categories, CAR, tabulated by age of policy holder, AGE, and district where the policy holder lived ($DIST = 1$, for London and other major cities, and $DIST = 0$, otherwise). The table is derived from the $CLAIMS$ data set in Aitkin et al. (2005) obtained from a paper by Baxter et al. (1980).

a. Calculate the rate of claims y/n for each category and plot the rates by AGE, CAR and $DIST$ to get an idea of the main effects of these factors.

b. Use Poisson regression to estimate the main effects (each treated as categorical and modelled using indicator variables) and interaction terms.

c. Based on the modelling in (b), Aitkin et al. (2005) determined that all

Table 9.12 *Comparison of observed frequencies and expected frequencies obtained from the log-linear model with all two-way interaction terms for the data in Table 9.7; expected frequencies in brackets.*

	Aspirin use		
	Non-user	User	Total
Gastric ulcer			
Controls	62 (58.53)	6 (9.47)	68
Cases	39 (42.47)	25 (21.53)	64
Duodenal ulcer			
Controls	53 (56.47)	8 (4.53)	61
Cases	49 (45.53)	8 (11.47)	57

the interactions were unimportant and decided that *AGE* and *CAR* could be treated as though they were continuous variables. Fit a model incorporating these features and compare it with the best model obtained in (b). What conclusions do you reach?

Table 9.13 *Car insurance claims: based on the CLAIMS data set reported by Aitkin et al. (2005).*

		DIST = 0		DIST = 1	
CAR	AGE	y	n	y	n
1	1	65	317	2	20
1	2	65	476	5	33
1	3	52	486	4	40
1	4	310	3259	36	316
2	1	98	486	7	31
2	2	159	1004	10	81
2	3	175	1355	22	122
2	4	877	7660	102	724
3	1	41	223	5	18
3	2	117	539	7	39
3	3	137	697	16	68
3	4	477	3442	63	344
4	1	11	40	0	3
4	2	35	148	6	16
4	3	39	214	8	25
4	4	167	1019	33	114

9.3 This question relates to the flu vaccine trial data in Table 9.6.

   a. Using a conventional chi-squared test and an appropriate log-linear model, test the hypothesis that the distribution of responses is the same for the placebo and vaccine groups.

   b. For the model corresponding to the hypothesis of homogeneity of response distributions, calculate the fitted values, the Pearson and deviance residuals, and the goodness of fit statistics X^2 and D. Which of the cells of the table contribute most to X^2 (or D)? Explain and interpret these results.

   c. Re-analyze these data using ordinal logistic regression to estimate cutpoints for a latent continuous response variable and to estimate a location shift between the two treatment groups. Sketch a rough diagram to illustrate the model which forms the conceptual base for this analysis (see Exercise 8.4).

9.4 For a 2×2 contingency table, the maximal log-linear model can be written as

$$
\begin{aligned}
\eta_{11} &= \mu + \alpha + \beta + (\alpha\beta), & \eta_{12} &= \mu + \alpha - \beta - (\alpha\beta), \\
\eta_{21} &= \mu - \alpha + \beta - (\alpha\beta), & \eta_{22} &= \mu - \alpha - \beta + (\alpha\beta),
\end{aligned}
$$

where $\eta_{jk} = \log E(Y_{jk}) = \log(n\theta_{jk})$ and $n = \sum\sum Y_{jk}$.
Show that the interaction term $(\alpha\beta)$ is given by

$$
(\alpha\beta) = \tfrac{1}{4}\log\phi,
$$

where ϕ is the **odds ratio** $(\theta_{11}\theta_{22})/(\theta_{12}\theta_{21})$, and hence that $\phi = 1$ corresponds to no interaction.

9.5 Use log-linear models to examine the housing satisfaction data in Table 8.5. The numbers of people surveyed in each type of housing can be regarded as fixed.

a. First, analyze the associations between level of satisfaction (treated as a nominal categorical variable) and contact with other residents, separately for each type of housing.

b. Next, conduct the analyses in (a) simultaneously for all types of housing.

c. Compare the results from log-linear modelling with those obtained using nominal or ordinal logistic regression (see Exercise 8.2).

9.6 Consider a $2 \times K$ contingency table (Table 9.14) in which the column totals $y_{.k}$ are fixed for $k = 1, \ldots, K$.

Table 9.14 *Contingency table with 2 rows and K columns.*

	1	$\cdots$	k	$\cdots$	K
Success	y_{11}		y_{1k}		y_{1K}
Failure	y_{21}		y_{2k}		y_{2K}
Total	$y_{.1}$		$y_{.k}$		$y_{.K}$

a. Show that the product multinomial distribution for this table reduces to

$$
f(z_1, \ldots, z_K \mid n_1, \ldots, n_K) = \sum_{k=1}^{K} \binom{n_k}{z_k} \pi_k^{z_k} (1 - \pi_k)^{n_k - z_k},
$$

where $n_k = y_{.k}, z_k = y_{1k}, n_k - z_k = y_{2k}, \pi_k = \theta_{1k}$ and $1 - \pi_k = \theta_{2k}$ for $k = 1, \ldots, K$. This is the **product binomial distribution** and is the joint distribution for Table 7.1 (with appropriate changes in notation).

b. Show that the log-linear model with

$$\eta_{1k} = \log E(Z_k) = \mathbf{x}_{1k}^T \boldsymbol{\beta}$$

and

$$\eta_{2k} = \log E(n_k - Z_k) = \mathbf{x}_{2k}^T \boldsymbol{\beta}$$

is equivalent to the logistic model

$$\log \left(\frac{\pi_k}{1 - \pi_k} \right) = \mathbf{x}_k^T \boldsymbol{\beta},$$

where $\mathbf{x}_k = \mathbf{x}_{1k} - \mathbf{x}_{2k}, k = 1, \ldots, K$.

c. Based on (b), analyze the case–control study data on aspirin use and ulcers using logistic regression and compare the results with those obtained using log-linear models.

9.7 Mittlbock and Heinzl (2001) compare Poisson and logistic regression models for data in which the event rate is small so that the Poisson distribution provides a reasonable approximation to the Binomial distribution. An example is the number of deaths from coronary heart disease among British doctors (Table 9.1). In Section 9.2.1 we fitted the model $Y_i \sim Po(deaths_i)$ with Equation (9.9)

$$\log(deaths_i) = \log(personyears_i) + \beta_1 + \beta_2 smoke_i + \beta_3 agecat_i + \beta_4 agesq_i + \beta_5 smkage_i.$$

An alternative is $Y_i \sim Bin(personyears_i, \pi_i)$ with

$$\text{logit}(\pi_i) = \beta_1 + \beta_2 smoke_i + \beta_3 agecat_i + \beta_4 agesq_i + \beta_5 smkage_i.$$

Another version is based on a Bernoulli distribution $Z_j \sim B(\pi_i)$ for each doctor in group i with

$$Z_j = \begin{cases} 1, & j = 1, \ldots, deaths_i \\ 0, & j = deaths_i + 1, \ldots, personyears_i \end{cases}$$

and

$$\text{logit}(\pi_i) = \beta_1 + \beta_2 smoke_i + \beta_3 agecat_i + \beta_4 agesq_i + \beta_5 smkage_i.$$

a. Fit all three models (in Stata the Bernoulli model cannot be fitted with glm; use blogit instead). Verify that the β estimates are very similar.

b. Calculate the statistics D, X^2 and pseudo R^2 for all three models. Notice that the pseudo R^2 is much smaller for the Bernoulli model. As Mittlbock and Heinzl (2001) point out this is because the Poisson and Binomial models are estimating the probability of death for each group (which is relatively easy) whereas the Bernoulli model is estimating the probability of death for an individual (which is much more difficult).

Chapter 10

Survival Analysis

10.1 Introduction

An important type of data is the time from a well-defined starting point until some event, called "failure," occurs. In engineering, this may be the time from initial use of a component until it fails to operate properly. In medicine, it may be the time from when a patient is diagnosed with a disease until he or she dies. Analysis of these data focuses on summarizing the main features of the distribution, such as median or other percentiles of time to failure, and on examining the effects of explanatory variables. Data on times until failure, or more optimistically, duration of survival or **survival times**, have two important features:

(a) the times are non-negative and typically have skewed distributions with long tails;

(b) some of the subjects may survive beyond the study period so that their actual failure times may not be known; in this case, and other cases where the failure times are not known completely, the data are said to be **censored**.

Examples of various forms of censoring are shown in Figure 10.1. The horizontal lines represent the survival times of subjects. T_O and T_C are the beginning and end of the study period, respectively. D denotes "death" or "failure" and A denotes "alive at the end of the study." L indicates that the subject was known to be alive at the time shown but then became lost to the study and so the subsequent life course is unknown.

For subjects 1 and 2, the entire survival period (e.g., from diagnosis until death, or from installation of a machine until failure) occurred within the study period. For subject 3, "death" occurred after the end of the study so that only the solid part of the line is recorded and the time is said to be **right censored** at time T_C. For subject 4, the observed survival time was right censored due to loss to follow up at time T_L. For subject 5, the survival time commenced before the study began so the period before T_O (i.e., the dotted

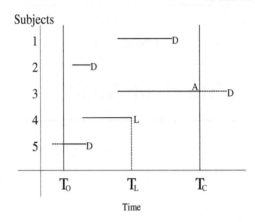

Figure 10.1 *Diagram of types of censoring for survival times.*

line) is not recorded and the recorded survival time is said to be **left censored** at time T_O.

The analysis of survival time data is the topic of numerous books and papers. Procedures to implement the calculations are available in most statistical programs. In this book, only continuous scale survival time data are considered. Furthermore only parametric models are considered; that is, models requiring the specification of a probability distribution for the survival times. In particular, this means that one of the best known forms of survival analysis, the **Cox proportional hazards model** (Cox 1972), is not considered because it is a **semi-parametric model** in which dependence on the explanatory variables is modelled explicitly but no specific probability distribution is assumed for the survival times. An advantage of parametric models, compared with the Cox proportional hazards model, is that inferences are usually more precise and there is a wider range of models with which to describe the data, including **accelerated failure time models** (Wei 1992). The parametric models considered here are ones that can be fitted using generalized linear models.

Important topics not considered here include time-dependent explanatory variables (Kalbfleisch and Prentice 2002) and discrete survival time models (Fleming and Harrington 2005). Some books that describe the analysis of survival data in detail include Collett (2014), Hosmer and Lemeshow (2008), Klein et al. (2013), Kleinbaum and Klein (2012), Lawless (2002) and Lee and Wang (2003).

The next section explains various functions of the probability distribution of survival times which are useful for model specification. This is followed by descriptions of the two distributions most commonly used for survival data—the exponential and Weibull distributions.

Estimation and inference for survival data are complicated by the presence of censored survival times. The likelihood function contains two components, one involving the uncensored survival times and the other making as much use as possible of information about the survival times which are censored. In some cases this formulation makes it possible to fit models using Poisson regression, as illustrated in Section 10.4. In other cases the requirements for generalized linear models are not fully met. Nevertheless, estimation based on the Newton–Raphson method for maximizing the likelihood function, described in Chapter 4, and the inference methods described in Chapter 5 all apply quite well, at least for large sample sizes.

The methods discussed in this chapter are illustrated using a small data set so that the calculations are relatively easy, even though the asymptotic properties of the methods apply only approximately.

10.2   Survivor functions and hazard functions

Let the random variable Y denote the survival time and let $f(y)$ denote its probability density function. Then the probability of failure before a specific time y is given by the cumulative probability distribution

$$F(y) = \Pr(Y < y) = \int_0^y f(t)dt.$$

The **survivor function** is the probability of survival beyond time y. It is given by

$$S(y) = \Pr(Y \geq y) = 1 - F(y). \tag{10.1}$$

The **hazard function** is the probability of death in an infinitesimally small time between y and $(y + \delta y)$, given survival up to time y,

$$
\begin{aligned}
h(y) &= \lim_{\delta y \to 0} \frac{\Pr(y \leqslant Y < y + \delta y \mid Y > y)}{\delta y} \\
&= \lim_{\delta y \to 0} \frac{F(y + \delta y) - F(y)}{\delta y} \times \frac{1}{S(y)}.
\end{aligned}
$$

But

$$\lim_{\delta y \to 0} \frac{F(y + \delta y) - F(y)}{\delta y} = f(y)$$

by the definition of a derivative. Therefore,

$$h(y) = \frac{f(y)}{S(y)}, \tag{10.2}$$

which can also be written as

$$h(y) = -\frac{d}{dy}\{\log[S(y)]\}. \tag{10.3}$$

Hence,

$$S(y) = \exp[-H(y)] \quad \text{where} \quad H(y) = \int_0^y h(t)dt,$$

or

$$H(y) = -\log[S(y)]. \tag{10.4}$$

$H(y)$ is called the **cumulative hazard function** or **integrated hazard function**.

The "average" survival time is usually estimated by the median of the distribution. This is preferable to the expected value because of the skewness of the distribution. The **median survival time**, $y(50)$, is given by the solution of the equation $F(y) = \frac{1}{2}$. Other percentiles can be obtained similarly; for example, the pth percentile $y(p)$ is the solution of $F[y(p)] = p/100$ or $S[y(p)] = 1 - (p/100)$. For some distributions these percentiles may be obtained explicitly; for others, the percentiles may need to be calculated from the estimated survivor function (see Section 10.6).

10.2.1 Exponential distribution

The simplest model for a survival time Y is the **exponential distribution** with probability density function

$$f(y; \theta) = \theta e^{-\theta y}, \qquad y \geq 0, \, \theta > 0. \tag{10.5}$$

This is a member of the exponential family of distributions (see Exercise 3.3(b)) and has $E(Y) = 1/\theta$ and $\text{var}(Y) = 1/\theta^2$ (see Exercise 4.2). The cumulative distribution is

$$F(y; \theta) = \int_0^y \theta e^{-\theta t} dt = 1 - e^{-\theta y}.$$

So the survivor function is

$$S(y; \theta) = e^{-\theta y}, \tag{10.6}$$

the hazard function is

$$h(y; \theta) = \theta$$

and the cumulative hazard function is

$$H(y; \theta) = \theta y.$$

The hazard function does not depend on y, so the probability of failure in the time interval $[y, y + \delta y]$ is not related to how long the subject has already survived. This "**lack of memory**" property may be a limitation because, in practice, the probability of failure often increases with time. In such situations an accelerated failure time model, such as the Weibull distribution, may be more appropriate. One way to examine whether data satisfy the constant hazard property is to estimate the cumulative hazard function $H(y)$ (see Section 10.3) and plot it against survival time y. If the plot is nearly linear, then the exponential distribution may provide a useful model for the data.

The **median survival time** is given by the solution of the equation

$$F(y; \theta) = \frac{1}{2} \quad \text{which is} \quad y(50) = \frac{1}{\theta} \log 2.$$

This is a more appropriate description of the "average" survival time than $E(Y) = 1/\theta$ because of the skewness of the exponential distribution.

10.2.2 Proportional hazards models

For an exponential distribution, the dependence of Y on explanatory variables could be modelled as $E(Y) = \mathbf{x}^T \boldsymbol{\beta}$. In this case the identity link function would be used. To ensure that $\theta > 0$, however, it is more common to use

$$\theta = e^{\mathbf{x}^T \boldsymbol{\beta}}.$$

In this case the hazard function has the multiplicative form

$$h(y; \boldsymbol{\beta}) = \theta = e^{\mathbf{x}^T \boldsymbol{\beta}} = \exp(\sum_{i=1}^{p} x_i \beta_i).$$

For a binary explanatory variable with values $x_k = 0$ if the exposure is absent and $x_k = 1$ if the exposure is present, the **hazard ratio** or **relative hazard** for presence vs. absence of exposure is

$$\frac{h_1(y; \boldsymbol{\beta})}{h_0(y; \boldsymbol{\beta})} = e^{\beta_k}, \tag{10.7}$$

provided that $\sum_{i \neq k} x_i \beta_i$ is constant. A one-unit change in a continuous explanatory variable x_k will also result in the hazard ratio given in (10.7).

More generally, models of the form

$$h_1(y) = h_0(y)e^{\mathbf{x}^T \boldsymbol{\beta}} \tag{10.8}$$

are called **proportional hazards models** and $h_0(y)$, which is the hazard function corresponding to the reference levels for all the explanatory variables, is called the **baseline hazard**.

For proportional hazards models, the cumulative hazard function is given by

$$H_1(y) = \int_0^y h_1(t)dt = \int_0^y h_0(t)e^{\mathbf{x}^T \boldsymbol{\beta}}dt = H_0(y)e^{\mathbf{x}^T \boldsymbol{\beta}},$$

so

$$\log H_1(y) = \log H_0(y) + \sum_{i=1}^{p} x_i \beta_i.$$

Therefore, for two groups of subjects which differ only with respect to the presence (denoted by P) or absence (denoted by A) of some exposure, from (10.7)

$$\log H_P(y) = \log H_A(y) + \beta_k, \tag{10.9}$$

so the **log cumulative hazard functions** differ by a constant.

10.2.3 Weibull distribution

Another commonly used model for survival times is the Weibull distribution which has the probability density function

$$f(y; \lambda, \theta) = \frac{\lambda y^{\lambda-1}}{\theta^\lambda} \exp\left[-\left(\frac{y}{\theta}\right)^\lambda\right], \qquad y \geq 0, \quad \lambda > 0, \quad \theta > 0$$

(see Example 4.2). The parameters λ and θ determine the shape of the distribution and the scale, respectively. To simplify some of the notation, it is convenient to reparameterize the distribution using $\theta^{-\lambda} = \phi$. Then the probability density function is

$$f(y; \lambda, \phi) = \lambda \phi y^{\lambda-1} \exp\left(-\phi y^\lambda\right). \tag{10.10}$$

The exponential distribution is a special case of the Weibull distribution with $\lambda = 1$.

The survivor function for the Weibull distribution is

$$S(y; \lambda, \phi) = \int_y^\infty \lambda \phi u^{\lambda-1} \exp\left(-\phi u^\lambda\right) du$$

$$= \exp\left(-\phi y^\lambda\right), \tag{10.11}$$

the hazard function is

$$h(y; \lambda, \phi) = \lambda \phi y^{\lambda-1} \tag{10.12}$$

and the cumulative hazard function is

$$H(y; \lambda, \phi) = \phi y^\lambda.$$

The hazard function depends on y, and with suitable values of λ it can increase or decrease with increasing survival time. Thus, the Weibull distribution yields **accelerated failure time** models. The appropriateness of this feature for modelling a particular data set can be assessed using

$$\log H(y) = \log \phi + \lambda \log y \tag{10.13}$$

$$= \log[-\log S(y)].$$

The empirical survivor function $\widehat{S}(y)$ can be used to plot $\log[-\log \widehat{S}(y)]$ (or $\widehat{S}(y)$ can be plotted on the complementary log-log scale) against the logarithm of the survival times. For the Weibull (or exponential) distribution the points should lie approximately on a straight line. This technique is illustrated in Section 10.3.

It can be shown that the expected value of the survival time Y is

$$E(Y) = \int_0^\infty \lambda \phi y^\lambda \exp\left(-\phi y^\lambda\right) dy$$

$$= \phi^{-1/\lambda} \Gamma(1 + 1/\lambda),$$

where $\Gamma(u) = \int_0^\infty s^{u-1} e^{-s} ds$. Also the median, given by the solution of

$$S(y; \lambda, \phi) = \frac{1}{2},$$

is

$$y(50) = \phi^{-1/\lambda} (\log 2)^{1/\lambda}.$$

These statistics suggest that the relationship between Y and explanatory variables should be modelled in terms of ϕ and it should be multiplicative. In particular, if

$$\phi = \alpha e^{\mathbf{x}^T \boldsymbol{\beta}},$$

then the hazard function (10.12) becomes

$$h(y; \lambda, \phi) = \lambda \alpha y^{\lambda-1} e^{\mathbf{x}^T \boldsymbol{\beta}}. \tag{10.14}$$

If $h_0(y)$ is the baseline hazard function corresponding to reference levels of all the explanatory variables, then

$$h(y) = h_0(y) e^{\mathbf{x}^T \boldsymbol{\beta}},$$

which is a proportional hazards model.

In fact, the Weibull distribution is the only distribution for survival time data that has the properties of accelerated failure times and proportional hazards; see Exercises 10.3 and 10.4.

10.3   Empirical survivor function

The cumulative hazard function $H(y)$ is an important tool for examining how well a particular distribution describes a set of survival time data. For example, for the exponential distribution, $H(y) = \theta y$ is a linear function of time (see Section 10.2.1) and this can be assessed from the data.

The empirical survivor function, an estimate of the probability of survival beyond time y, is given by

$$\tilde{S}(y) = \frac{\text{number of subjects with survival times} \geqslant y}{\text{total number of subjects}}.$$

The most common way to calculate this function is to use the **Kaplan–Meier estimate**, which is also called the **product limit estimate**. It is calculated by first arranging the observed survival times in order of increasing magnitude $y_{(1)} \leqslant y_{(2)} \leqslant \ldots \leqslant y_{(k)}$. Let n_j denote the number of subjects who are alive just before time $y_{(j)}$ and let d_j denote the number of deaths that occur at time $y_{(j)}$ (or, strictly within a small time interval from $y_{(j)} - \delta$ to $y_{(j)}$). Then the estimated probability of survival past $y_{(j)}$ is $(n_j - d_j)/n_j$. Assuming that the times $y_{(j)}$ are independent, the Kaplan–Meier estimate of the survivor function at time y is

$$\hat{S}(y) = \prod_{j=1}^{k} \left(\frac{n_j - d_j}{n_j} \right)$$

for y between times $y_{(j)}$ and $y_{(j+1)}$.

Table 10.1 *Remission times (in weeks) of leukemia patients; data from Gehan (1965).*

Controls										
1	1	2	2	3	4	4	5	5	8	8
8	8	11	11	12	12	15	17	22	23	
Treatment										
6	6	6	6*	7	9*	10	10*	11*	13	16
17*	19*	20*	22	23	25*	32*	32*	34*	35*	
* indicates censoring										

Table 10.2 *Calculation of Kaplan–Meier estimate of the survivor function for the treatment group for the data in Table 10.1.*

Time y_j	No. n_j alive just before time y_j	No. d_j deaths at time y_j	$\widehat{S}(y) = \prod \left(\frac{n_j - d_j}{n_j} \right)$
0– <6	21	0	1
6– <7	21	3	0.857
7– <10	17	1	0.807
10– <13	15	1	0.753
13– <16	12	1	0.690
16– <22	11	1	0.627
22– <23	7	1	0.538
≥23	6	1	0.448

10.3.1   Example: Remission times

The calculation of $\widehat{S}(y)$ is illustrated using an old data set of times to remission of leukemia patients (Gehan 1965). There are two groups each of $n = 21$ patients. In the control group who were treated with a placebo there was no censoring, whereas in the active treatment group, who were given 6 mercaptopurine, more than half of the observations were censored. The data for both groups are given in Table 10.1. Details of the calculation of $\widehat{S}(y)$ for the treatment group are shown in Table 10.2.

Figure 10.2 shows dot plots of the uncensored times (dots) and censored times (squares) for each group. Due to the high level of censoring in the treatment group, the distributions are not really comparable. Nevertheless, the plots show the skewed distributions and suggest that survival times were longer in the treatment group. Figure 10.3 shows the Kaplan–Meier estimates of the survivor functions for the two groups. The solid line represents the

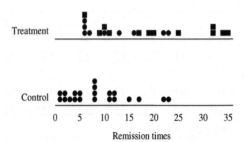

Figure 10.2 *Dot plots of remission time data in Table 10.1: dots represent uncensored times and squares represent censored times.*

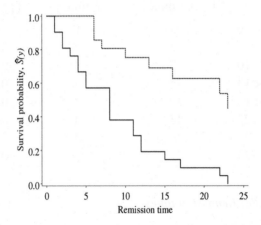

Figure 10.3 *Empirical survivor functions (Kaplan–Meier estimates) for data in Table 10.1: the solid line represents the control group and the dotted line represents the treatment group.*

control group and the dotted line represents the treatment group. Survival was obviously better in the treatment group. Figure 10.4 shows the logarithm of the cumulative hazard function plotted against log y. The two lines are fairly straight which suggests that the Weibull distribution is appropriate, from (10.13). Furthermore, the lines are parallel which suggests that the proportional hazards model is appropriate, from (10.9). The slopes of the lines are near unity which suggests that the simpler exponential distribution may provide as good a model as the Weibull distribution. The distance between the

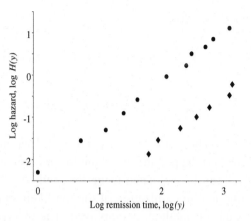

Figure 10.4 *Log cumulative hazard function plotted against log of remission time for data in Table 10.1; dots represent the control group and diamonds represent the treatment group.*

lines is about 1.4 which indicates that the hazard ratio is about $\exp(1.4) \simeq 4$, from (10.9).

10.4   Estimation

For the jth subject, the data recorded are y_j the survival time, δ_j a censoring indicator with $\delta_j = 1$ if the survival time is uncensored and $\delta_j = 0$ if it is censored, and $\mathbf{x}_j$ a vector of explanatory variables. Let $y_1, \ldots, y_r$ denote the uncensored observations and $y_{r+1}, \ldots, y_n$ denote the censored ones. The contribution of the uncensored variables to the likelihood function is

$$\prod_{j=1}^{r} f(y_j).$$

For a censored variable the survival time Y is at least y_j $(r+1 \leqslant j \leqslant n)$ and the probability of this is $\Pr(Y \geqslant y_j) = S(y_j)$, so the contribution of the censored variables to the likelihood function is

$$\prod_{j=r+1}^{n} S(y_j).$$

The full likelihood is

$$L = \prod_{j=1}^{n} f(y_j)^{\delta_j} S(y_j)^{1-\delta_j}, \tag{10.15}$$

so the log-likelihood function is

$$l = \sum_{j=1}^{n} [\delta_j \log f(y_j) + (1 - \delta_j) \log S(y_j)]$$

$$= \sum_{j=1}^{n} [\delta_j \log h(y_j) + \log S(y_j)] \tag{10.16}$$

from Equation (10.2). These functions depend on the parameters of the probability distributions and the parameters in the linear component $\mathbf{x}^T \boldsymbol{\beta}$.

The parameters can be estimated using the methods described in Chapter 4. Usually numerical maximization of the log-likelihood function, based on the Newton–Raphson method, is employed. The inverse of the information matrix which is used in the iterative procedure provides an asymptotic estimate of the variance–covariance matrix of the parameter estimates.

The main difference between the parametric models for survival data described in this book and the commonly used Cox proportional hazards regression model is in the function (10.15). For the Cox model, the functions f and S are not fully specified; for more details, see Collett (2014), for example.

10.4.1   Example: Exponential model

Suppose we have survival time data with censoring and we believe that the exponential distribution is a suitable model. Then the likelihood function is $L(\boldsymbol{\theta}; \mathbf{y}) = \prod_{j=1}^{n} (\theta_j e^{-\theta_j y_j})^{\delta_j} (e^{-\theta_j y_j})^{1-\delta_j}$ from Equations (10.5), (10.6) and (10.15). The log-likelihood function is

$$l(\boldsymbol{\theta}; \mathbf{y}) = \sum_{j=1}^{n} \delta_j \log \theta_j + \sum_{j=1}^{n} [\delta_j (-\theta_j y_j) + (1 - \delta_j)(-\theta_j y_j)]$$

$$= \sum_{j=1}^{n} (\delta_j \log \theta_j - \theta_j y_j). \tag{10.17}$$

Now we can consider the censoring data δ_j, $j = 1, \ldots, n$ as observations of random variables D_j. The expression on the right-hand side of Equation (10.17) is proportional to the log-likelihood function of n independent Poisson variables, $D_j \sim \text{Po}(\theta_j y_j)$ (see Section 3.2.1). A consequence of this result is that we can use generalized linear modelling methods for the censoring variables to model censored survival time data (Aitkin and Clayton 1980). Since $\text{E}(D_j) = \mu_j = \theta_j y_j$, the usual link function is

$$\log \mu_j = \log \theta_j + \log y_j,$$

so the survival times y_j are included in the model as offset terms $\log y_j$. Proportional hazards can be modelled using $\theta_j = e^{\mathbf{x}_j^T \boldsymbol{\beta}}$. This approach is illustrated in the example in Section 10.7. There are many extensions of this approach, including using Binomial distributions with the complementary log-log link for grouped survival data; for example, see Dickman et al. (2004).

10.4.2 Example: Weibull model

If the data for subject j are $\{y_j, \delta_j$ and $\mathbf{x}_j\}$ and the Weibull distribution is thought to provide a suitable model (for example, based on initial exploratory analysis), then the log-likelihood function is

$$l = \sum_{j=1}^{n} \left[\delta_j \log(\lambda \alpha y_j^{\lambda-1} e^{\mathbf{x}_j^T \boldsymbol{\beta}}) - (\alpha y_j^{\lambda} e^{\mathbf{x}_j^T \boldsymbol{\beta}}) \right] \qquad (10.18)$$

from Equations (10.14) and (10.16). The right-hand side of Equation (10.18) is proportional to the log-likelihood function of n independent random variables D_j, the censoring variables, with distributions $\text{Po}(\mu_j)$ with $\mu_j = \alpha y_j^{\lambda} e^{\mathbf{x}_j^T \boldsymbol{\beta}}$. The log link function gives

$$\log \mu_j = \log \alpha + \lambda \log y_j + \mathbf{x}_j^T \boldsymbol{\beta}.$$

So the survival times are included via the offset terms which also include the unknown Weibull shape parameter (and the intercept $\log \alpha + \beta_0$ includes the Weibull scale parameter). This model can be fitted iteratively estimating λ using maximum likelihood as follows

$$\begin{aligned}
\frac{\partial l}{\partial \lambda} &= \sum_{j=1}^{n} \left[\frac{\delta_j}{\lambda} + \delta_j \log y_j - \log y_j (\alpha y_j^{\lambda} e^{\mathbf{x}_j^T \boldsymbol{\beta}}) \right] \\
&= \frac{m}{\lambda} + \sum_{j=1}^{n} [(\delta_j - \mu_j) \log y_j] = 0,
\end{aligned}$$

where m is the number of δ_j's that are non-zero, that is, the number of uncensored observations. Hence,

$$\widehat{\lambda} = \frac{m}{\sum_{j=1}^{n} [(\mu_j - \delta_j) \log y_j]}.$$

Starting with the exponential distribution so that $\lambda^{(0)} = 1$, the model can be fitted using Poisson regression to produce fitted values $\boldsymbol{\mu}$ which in turn are used to estimate $\widehat{\lambda}$. Aitkin and Clayton (1980) recommend damping successive estimates of λ using $\lambda^{(k)} = (\lambda^{(k-1)} + \widehat{\lambda})/2$ to improve convergence.

10.5　Inference

The Newton–Raphson iteration procedure used to obtain maximum likelihood estimates also produces the information matrix $\mathfrak{J}$ which can be inverted to give the approximate variance–covariance matrix for the estimators. Hence, inferences for any parameter θ can be based on the maximum likelihood estimator $\widehat{\theta}$ and the standard error, $s.e.(\widehat{\theta})$, obtained by taking the square root of the relevant element of the diagonal of $\mathfrak{J}^{-1}$. Then the Wald statistic $(\widehat{\theta} - \theta)/s.e.(\widehat{\theta})$ can be used to test hypotheses about θ or to calculate approximate confidence limits for θ assuming that the statistic has the standard Normal distribution $N(0, 1)$ (see Section 5.4).

For the Weibull and exponential distributions, the maximum value of the log-likelihood function can be calculated by substituting the maximum likelihood estimates of the parameters, denoted by the vector $\widehat{\boldsymbol{\theta}}$, into the expression in (10.16) to obtain $l(\widehat{\boldsymbol{\theta}}; \mathbf{y})$. For censored data, the statistic $-2l(\widehat{\boldsymbol{\theta}}; \mathbf{y})$ may not have a chi-squared distribution, even approximately. For nested models M_1, with p parameters and maximum value $\widehat{l_1}$ of the log-likelihood function, and M_0, with $q < p$ parameters and maximum value $\widehat{l_0}$ of the log-likelihood function, the difference

$$D = 2(\widehat{l_1} - \widehat{l_0})$$

will approximately have a chi-squared distribution with $p - q$ degrees of freedom if both models fit well. The statistic D, which is analogous to the **deviance**, provides another method for testing hypotheses (see Section 5.7).

10.6　Model checking

To assess the adequacy of a model it is necessary to check assumptions, such as the proportional hazards and accelerated failure time properties, in addition to looking for patterns in the residuals (see Section 2.3.4) and examining influential observations using statistics analogous to those for multiple linear regression (see Section 6.2.7).

The empirical survivor function $\widehat{S}(y)$ described in Section 10.3 can be used to examine the appropriateness of the probability model. For example, for the exponential distribution, the plot of $-\log[\widehat{S}(y)]$ against y should be approximately linear from (10.6). More generally, for the Weibull distribution, the plot of the log cumulative hazard function $\log[-\log \widehat{S}(y)]$ against $\log y$ should be linear, from (10.13). If the plot shows curvature, then some alternative model such as the log-logistic distribution may be better (see Exercise 10.2).

The general proportional hazards model given in (10.8) is

$$h(y) = h_0(y)e^{x^T \beta},$$

where h_0 is the baseline hazard. Consider a binary explanatory variable x_k with values $x_k = 0$ if a characteristic is absent and $x_k = 1$ if it is present. The log-cumulative hazard functions are related by

$$\log H_P = \log H_A + \beta_k;$$

see (10.9). Therefore, if the empirical hazard functions $\widehat{S}(y)$ are calculated separately for subjects with and without the characteristic and the log-cumulative hazard functions $\log[-\log \widehat{S}(y)]$ are plotted against $\log y$, the lines should have the same slope but be separated by a distance β_k.

More generally, parallel lines for the plots of the log cumulative hazard functions provide support for the proportional hazards assumption. For a fairly small number of categorical explanatory variables, the proportional hazards assumption can be examined in this way. If the lines are not parallel, this may suggest that there are interaction effects among the explanatory variables. If the lines are curved but still parallel this supports the proportional hazards assumption but suggests that the accelerated failure time model is inadequate. For more complex situations it may be necessary to rely on general diagnostics based on residuals and specific tests for checking the proportional hazards property.

The simplest residuals for survival time data are the **Cox–Snell residuals**. If the survival time for subject j is uncensored, then the Cox–Snell residual is

$$r_{Cj} = \widehat{H}_j(y_j) = -\log[\widehat{S}_j(y_j)], \tag{10.19}$$

where $\widehat{H}_j$ and $\widehat{S}_j$ are the estimated cumulative hazard and survivor functions for subject j at time y_j. For proportional hazards models, (10.19) can be written as

$$r_{Cj} = \exp(x_j^T \widehat{\beta})\widehat{H}_0(y_j)$$

where $\widehat{H}_0(y_j)$ is the baseline hazard function evaluated at y_j.

It can be shown that if the model fits the data well, then these residuals have an exponential distribution with a parameter of one. In particular, their mean and variance should be approximately equal to one.

For censored observations, r_{Cj} will be too small and various modifications have been proposed of the form

$$r'_{Cj} = \begin{cases} r_{Cj} \text{ for uncensored observations} \\ r_{Cj} + \Delta \text{ for censored observations} \end{cases},$$

where $\Delta = 1$ or $\Delta = \log 2$ (Crowley and Hu 1977). The distribution of the r'_{Cj}'s can be compared with the exponential distribution with unit mean using exponential probability plots (analogous Normal probability plots) which are available in various statistical software. An exponential probability plot of the residuals r'_{Cj} may be used to identify outliers and systematic departures from the assumed distribution.

Martingale residuals provide an alternative approach. For the jth subject the Martingale residual is

$$r_{Mj} = \delta_j - r_{Cj},$$

where $\delta_j = 1$ if the survival time is uncensored and $\delta_j = 0$ if it is censored. These residuals have an expected value of zero but a negatively skewed distribution.

Deviance residuals (which are somewhat misnamed because the sum of their squares is not, in fact, equal to the deviance mentioned in Section 10.5) are defined by

$$r_{Dj} = \text{sign}(r_{Mj})\{-2[r_{Mj} + \delta_j \log(r_{Cj})]\}^{\frac{1}{2}}.$$

The r_{Dj}'s are approximately symmetrically distributed about zero and large values may indicate outlying observations.

In principle, any of the residuals r'_{Cj}, r_{Mj} or r_{Dj} are suitable for sequence plots against the order in which the survival times were measured, or any other relevant order (to detect lack of independence among the observations) and for plots against explanatory variables that have been included in the model (and those that have not) to detect any systematic patterns which would indicate that the model is not correctly specified. However, the skewness of the distributions of r'_{Cj} and r_{Mj} makes them less useful than r_{Dj}, in practice.

Diagnostics to identify influential observations can be defined for survival time data, by analogy with similar statistics for multiple linear regression and other generalized linear models. For example, for any parameter β_k delta-betas $\Delta_j \beta_k$, one for each subject j, show the effect on the estimate of β_k caused by omitting the data for subject j from the calculations. Plotting the $\Delta_j \beta_k$'s against the order of the observations or against the survival times y_j may indicate systematic effects or particularly influential observations.

10.7   Example: Remission times

Figure 10.4 suggests that a proportional hazards model with a Weibull, or even an exponential, distribution should provide a good model for the remis-

Table 10.3 *Results of fitting proportional hazards models based on the exponential and Weibull distributions to the data in Table 10.1.*

	Weibull shape parameter λ	Group difference $\beta_1(s.e.)$	Constant $\beta_0(s.e.)$
Exponential	1.00*	-1.527 (0.390)	-2.159 (0.218)
Weibull iterations			
1	1.316	-1.704	-2.944
2	1.355	-1.725	-3.042
3	1.363	-1.729	-3.064
4	1.365	-1.731	-3.069
5	1.366	-1.731	-3.071

* Shape parameter is unity for the exponential distribution.

sion time data in Table 10.1. The models are

$$h(y) = \exp(\beta_0 + \beta_1 x), \qquad Y \sim \text{Exp};$$
$$h(y) = \lambda y^{\lambda-1} \exp(\beta_0 + \beta_1 x), \qquad Y \sim \text{Wei},$$

where $x = 0$ for the control group, $x = 1$ for the treatment group and λ is the shape parameter. Using Poisson regression as described in Section 10.4 the exponential model can be fitted using the Stata command

```
──────────────── Stata code (exponential model) ────────────────
.glm censor grpcat, family(poisson) link(log) offset(logtime)
```

The corresponding R command is

```
──────────────── R code (exponential model) ────────────────
>res.gehanexp<-glm(censored==0~group + offset(log(time)),
 family=poisson(), data=remission)
```

The Weibull model has to be fitted iteratively using fitted values $\widehat{\mu}_j$ for Poisson regression to estimate λ by maximum likelihood and then using this estimate in the offset term and refitting the Poisson regression model (see Section 10.4.2). The results of fitting these models are shown in Table 10.3. The parameter estimates are identical to those using conventional survival analysis methods instead of Poisson regression. These can be obtained using the Stata commands stset (to identify the survival times and censoring data) and

```
──────────── Stata code (exponential/Weibull models) ────────────
.streg grpcat, dist(exponential) nohr
.streg grpcat, dist(weibull) nohr
```

or using the R library **survival** and the commands

```
_____ R code (exponential/Weibull models) _____
>res.gehan<-survreg(Surv(time,censored==0)~group,
 dist="exponential", data=remission)
>res.gehan<-survreg(Surv(time,censored==0)~group,
 dist="weibull", data=remission)
```

(The parameterization of the Weibull model fitted by R differs from the one used above.)

The standard errors for $\boldsymbol{\beta}$ for the Weibull model fitted using Poisson regression are too small as they do not take into account the variability of $\hat{\lambda}$.

Model fit can be assessed using the AIC. For the exponential model AIC $= 2.429$ and for the Weibull model AIC $= 2.782$. Therefore, the exponential distribution is about as good as the Weibull distribution for modelling the data but the exponential model would be preferred on the grounds of parsimony. The exponential model suggests that the parameter β_1 is non-zero and provides the estimate $\exp(1.53) = 4.62$ for the relative hazard.

Figure 10.5 shows box plots of Cox–Snell and deviance residuals for the exponential model. The skewness of the Cox–Snell residuals and the more symmetric distribution of the deviance residuals is apparent. Additionally, the difference in location between the distributions of the treatment and control groups suggests the model has failed to describe fully the patterns of remission times for the two groups of patients.

10.8   Exercises

10.1 The data in Table 10.4 are survival times, in weeks, for leukemia patients. There is no censoring. There are two covariates, white blood cell count (WBC) and the results of a test (AG positive and AG negative). The data set is from Feigl and Zelen (1965) and the data for the 17 patients with AG positive test results are described in Exercise 4.2.

   (a) Obtain the empirical survivor functions $\widehat{S}(y)$ for each group (AG positive and AG negative), ignoring WBC.

   (b) Use suitable plots of the estimates $\widehat{S}(y)$ to select an appropriate probability distribution to model the data.

   (c) Use a parametric model to compare the survival times for the two groups, after adjustment for the covariate WBC, which is best transformed to log(WBC).

   (d) Check the adequacy of the model using residuals and other diagnostic tests.

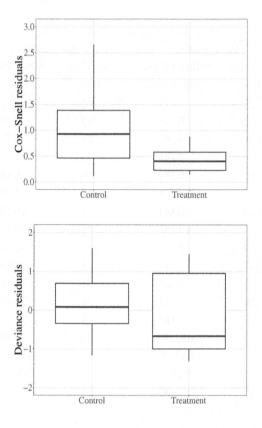

Figure 10.5 *Boxplots of Cox–Snell and deviance residuals from the exponential Model (10.19) for the data in Table 10.1.*

(e) Based on this analysis, is AG a useful prognostic indicator?

10.2 The **log-logistic distribution** with the probability density function

$$f(y) = \frac{e^\theta \lambda y^{\lambda-1}}{(1 + e^\theta y^\lambda)^2}$$

is sometimes used for modelling survival times.

(a) Find the survivor function $S(y)$, the hazard function $h(y)$ and the cumulative hazard function $H(y)$.

(b) Show that the median survival time is $\exp(-\theta/\lambda)$.

(c) Plot the hazard function for $\lambda = 1$ and $\lambda = 5$ with $\theta = -5$, $\theta = -2$ and $\theta = \frac{1}{2}$.

10.3 For **accelerated failure time models** the explanatory variables for subject

Table 10.4 *Leukemia survival times.*

AG positive		AG negative	
Survival time	White blood cell count	Survival time	White blood cell count
65	2.30	56	4.40
156	0.75	65	3.00
100	4.30	17	4.00
134	2.60	7	1.50
16	6.00	16	9.00
108	10.50	22	5.30
121	10.00	3	10.00
4	17.00	4	19.00
39	5.40	2	27.00
143	7.00	3	28.00
56	9.40	8	31.00
26	32.00	4	26.00
22	35.00	3	21.00
1	100.00	30	79.00
1	100.00	4	100.00
5	52.00	43	100.00
65	100.00		

i, η_i, act multiplicatively on the time variable so that the hazard function for subject i is

$$h_i(y) = \eta_i h_0(\eta_i y),$$

where $h_0(y)$ is the baseline hazard function. Show that the Weibull and log-logistic distributions both have this property but the exponential distribution does not. (Hint: Obtain the hazard function for the random variable $T = \eta_i Y$.)

10.4 For **proportional hazards models** the explanatory variables for subject i, η_i, act multiplicatively on the hazard function. If $\eta_i = e^{x_i^T \beta}$, then the hazard function for subject i is

$$h_i(y) = e^{x_i^T \beta} h_0(y), \tag{10.20}$$

where $h_0(y)$ is the baseline hazard function.

(a) For the exponential distribution if $h_0 = \theta$, show that if $\theta_i = e^{x_i^T \beta} \theta$ for the ith subject, then (10.20) is satisfied.

(b) For the Weibull distribution if $h_0 = \lambda \phi y^{\lambda-1}$, show that if $\phi_i = e^{x_i^T \beta} \phi$ for the ith subject, then (10.20) is satisfied.

(c) For the log-logistic distribution if $h_0 = e^{\theta} \lambda y^{\lambda-1}/(1 + e^{\theta} y^{\lambda})$, show that if $e^{\theta_i} = e^{\theta + x_i^T \beta}$ for the ith subject, then (10.20) is not satisfied. Hence, or otherwise, deduce that the log-logistic distribution does not have the proportional hazards property.

10.5 As the survivor function $S(y)$ is the probability of surviving beyond time y, the odds of survival past time y are

$$O(y) = \frac{S(y)}{1 - S(y)}.$$

For **proportional odds models** the explanatory variables for subject i, η_i, act multiplicatively on the odds of survival beyond time y

$$O_i = \eta_i O_0,$$

where O_0 is the baseline odds.

(a) Find the odds of survival beyond time y for the exponential, Weibull and log-logistic distributions.

(b) Show that only the log-logistic distribution has the proportional odds property.

(c) For the log-logistic distribution show that the log odds of survival beyond time y are

$$\log O(y) = \log \left[\frac{S(y)}{1 - S(y)} \right] = -\theta - \lambda \log y.$$

Therefore, if $\log \widehat{O}_i$ (estimated from the empirical survivor function) plotted against $\log y$ is approximately linear, then the log-logistic distribution may provide a suitable model.

(d) From (b) and (c) deduce that for two groups of subjects with explanatory variables η_1 and η_2 plots of $\log \widehat{O}_1$ and $\log \widehat{O}_2$ against $\log y$ should produce approximately parallel straight lines.

10.6 The data in Table 10.5 are survival times, in months, of 44 patients with chronic active hepatitis. They participated in a randomized controlled trial of prednisolone compared with no treatment. There were 22 patients in each group. One patient was lost to follow-up and several in each group were still alive at the end of the trial. The data are from Altman and Bland (1998).

(a) Calculate the empirical survivor functions for each group.

(b) Use suitable plots to investigate the properties of accelerated failure times, proportional hazards and proportional odds, using the results from Exercises 10.3, 10.4 and 10.5, respectively.

(c) Based on the results from (b) fit an appropriate model to the data in Table 10.5 to estimate the relative effect of prednisolone.

Table 10.5 *Survival times in months of patients with chronic active hepatitis in a randomized controlled trial of prednisolone versus no treatment; data from Altman and Bland (1998).*

Prednisolone							
2	6	12	54	56**	68	89	96
96	125*	128*	131*	140*	141*	143	145*
146	148*	162*	168	173*	181*		
No treatment							
2	3	4	7	10	22	28	29
32	37	40	41	54	61	63	71
127*	140*	146*	158*	167*	182*		
* indicates censoring, ** indicates loss to follow-up.							

Chapter 11

Clustered and Longitudinal Data

11.1 Introduction

In all the models considered so far the outcomes Y_i, $i = 1, \ldots, n$ are assumed to be independent. There are two common situations where this assumption is implausible. In one situation the outcomes are repeated measurements over time on the same subjects; for example, the weights of the same people when they are 30, 40, 50 and 60 years old. This is an example of **longitudinal data**. Measurements on the same person at different times may be more alike than measurements on different people because they are affected by persistent characteristics as well as potentially more variable factors; for instance, weight is likely to be related to an adult's genetic makeup and height as well as their eating habits and level of physical activity. For this reason longitudinal data for the same individuals are likely to exhibit correlation between successive measurements.

The other situation in which data are likely to be correlated is where they are measurements on related subjects; for example, the weights of samples of women aged 40 years selected from specific locations. In this case the locations are the **primary sampling units** or **clusters** and the women are sub samples within each primary sampling unit. Women from the same geographic area are likely to be more similar to one another, due to shared socioeconomic and environmental conditions, than they are to women from other locations. Any comparison of women's weights between areas that failed to take this within-area correlation into account could produce misleading results. For example, the standard deviation of the mean difference in weights between two areas will be too small if the observations which are correlated are assumed to be independent.

The term **repeated measures** is used to describe both longitudinal and clustered data. In both cases, models that include correlation are needed in order to make valid statistical inferences. There are two approaches to modelling such data.

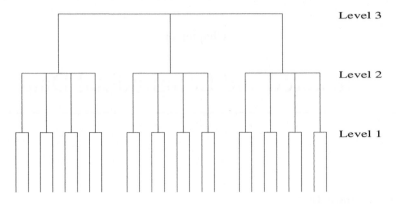

Figure 11.1 *Multilevel study.*

One approach involves dropping the usual assumption of independence between the outcomes Y_i and modelling the correlation structure explicitly. This method goes under various names such as **repeated measures** (for example, **repeated measures analysis of variance**) and the **generalized estimating equation** approach. The estimation and inference procedures for these models are, in principle, analogous to those for generalized linear models for independent outcomes.

The alternative approach for modelling repeated measures is based on considering the hierarchical structure of the study design. This is called **multilevel modelling**. For example, suppose there are repeated, longitudinal measurements, level 1, on different subjects, level 2, who were randomized to experimental groups, level 3. This nested structure is illustrated in Figure 11.1 which shows three groups, each of four subjects, on whom measurements are made at two times (for example, before and after some intervention). On each branch, outcomes at the same level are assumed to be independent and the correlation is a result of the multilevel structure (see Section 11.4).

In the next section an example is presented of an experiment with longitudinal outcome measures. Descriptive data analyses are used to explore the study hypothesis and the assumptions that are made in various models which might be used to test the hypothesis.

Repeated measures models for Normal data are described in Section 11.3. In Section 11.4, repeated measures models are described for non-Normal data such as counts and proportions which might be analyzed using Poisson, Binomial and other distributions (usually from the exponential family). These sections include details of the relevant estimation and inferential procedures. For repeated measures models, it is necessary to choose a correlation structure likely to reflect the relationships between the observations. Usually the correlation parameters are not of particular interest (i.e., they are nuisance parameters), but they need to be included in the model in order to obtain consistent estimates of those parameters that are of interest and to correctly calculate the standard errors of these estimates.

For multilevel models described in Section 11.5, the effects of levels may be described either by fixed parameters (e.g., for group effects) or random variables (e.g., for subjects randomly allocated to groups). If the linear predictor of the model has both fixed and random effects, the term **mixed model** is used. The correlation between observations is due to the random effects. This may make the correlation easier to interpret in multilevel models than in repeated measures models. Also the correlation parameters may be of direct interest. For Normally distributed data, multilevel models are well established and estimation and model checking procedures are available in most general purpose statistical software.

In Section 11.6, both repeated measures and multilevel models are fitted to data from the stroke example in Section 11.2. The results are used to illustrate the connections between the various models.

Finally, in Section 11.7, a number of issues that arise in the modelling of clustered and longitudinal data are mentioned. These include methods of exploratory analysis, consequences of using inappropriate models and problems that arise from missing data.

11.2 Example: Recovery from stroke

The data in Table 11.1 are from an experiment to promote the recovery of stroke patients. There were three experimental groups:

- A was a new occupational therapy intervention;
- B was the existing stroke rehabilitation program conducted in the same hospital where A was conducted;
- C was the usual care regime for stroke patients provided in a different hospital.

There were eight patients in each experimental group. The response variable was a measure of functional ability, the Bartel index; higher scores correspond to better outcomes and the maximum score is 100. Each patient was assessed weekly over the eight weeks of the study. The study was conducted by C. Cropper, at the University of Queensland, and the data were obtained from the Oz-

Table 11.1 *Functional ability scores measuring recovery from stroke for patients in three experimental groups over 8 weeks of the study.*

		Week							
Subject	Group	1	2	3	4	5	6	7	8
1	A	45	45	45	45	80	80	80	90
2	A	20	25	25	25	30	35	30	50
3	A	50	50	55	70	70	75	90	90
4	A	25	25	35	40	60	60	70	80
5	A	100	100	100	100	100	100	100	100
6	A	20	20	30	50	50	60	85	95
7	A	30	35	35	40	50	60	75	85
8	A	30	35	45	50	55	65	65	70
9	B	40	55	60	70	80	85	90	90
10	B	65	65	70	70	80	80	80	80
11	B	30	30	40	45	65	85	85	85
12	B	25	35	35	35	40	45	45	45
13	B	45	45	80	80	80	80	80	80
14	B	15	15	10	10	10	20	20	20
15	B	35	35	35	45	45	45	50	50
16	B	40	40	40	55	55	55	60	65
17	C	20	20	30	30	30	30	30	30
18	C	35	35	35	40	40	40	40	40
19	C	35	35	35	40	40	40	45	45
20	C	45	65	65	65	80	85	95	100
21	C	45	65	70	90	90	95	95	100
22	C	25	30	30	35	40	40	40	40
23	C	25	25	30	30	30	30	35	40
24	C	15	35	35	35	40	50	65	65

Dasl website developed by Gordon Smyth—http://www.statsci.org/data/oz/stroke.html.

The hypothesis was that the patients in group A would do better than those in group B or C. Figure 11.2 shows the time course of scores for every patient. Figure 11.3 shows the time course of the average scores for each experimental group. Clearly most patients improved. Also it appears that those in group A recovered best and those in group C did worst (however, people in group C may have started at a lower level).

The scatter plot matrix in Figure 11.4 shows data for all 24 patients at different times. The corresponding Pearson correlation coefficients are given

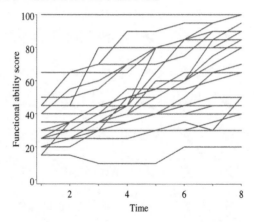

Figure 11.2 *Stroke recovery scores of individual patients.*

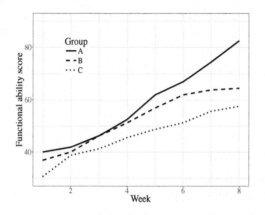

Figure 11.3 *Average stroke recovery scores for groups of patients.*

in Table 11.2. These show high positive correlation between measurements made one week apart and decreasing correlation between observations further apart in time.

A **naive analysis**, sometimes called a **pooled analysis**, of these data is to fit an analysis of covariance model in which all 192 observations (for 3 groups × 8 subjects × 8 times) are assumed to be independent with

$$E(Y_{ijk}) = \alpha_i + \beta t_k + e_{ijk} \tag{11.1}$$

where Y_{ijk} is the score at time t_k ($k = 1, \dots, 8$) for patient j ($j = 1, \dots, 8$) in group i (where $i = 1$ for group A, $i = 2$ for group B and $i = 3$ for group C); α_i is the mean score for group i; β is a common slope parameter; t_k denotes time ($t_k = k$ for week k, $k = 1, \dots, 8$); and the random error terms e_{ijk} are

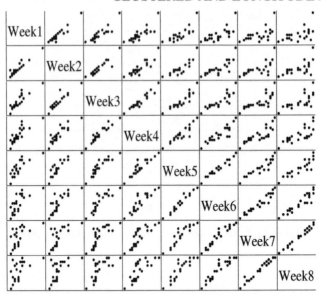

Figure 11.4 *Scatter plot matrix for stroke recovery scores in Table 11.2.*

Table 11.2 *Correlation coefficients for the stroke recovery scores in Table 11.1.*

| | Week | | | | | | |
	1	2	3	4	5	6	7
Week 2	0.93						
Week 3	0.88	0.92					
Week 4	0.83	0.88	0.95				
Week 5	0.79	0.85	0.91	0.92			
Week 6	0.71	0.79	0.85	0.88	0.97		
Week 7	0.62	0.70	0.77	0.83	0.92	0.96	
Week 8	0.55	0.64	0.70	0.77	0.88	0.93	0.98

all assumed to be independent. The null hypothesis H_0: $\alpha_1 = \alpha_2 = \alpha_3$ can be compared with an alternative hypothesis such as H_1: $\alpha_1 > \alpha_2 > \alpha_3$ by fitting models with different group parameters α_i. Figure 11.3 suggests that the slopes may differ between the three groups so the following model was also fitted

$$E(Y_{ijk}) = \alpha_i + \beta_i t_k + e_{ijk}, \tag{11.2}$$

where the slope parameter β_i denotes the rate of recovery for group i. Models (11.1) and (11.2) can be compared to test the hypothesis H_0: $\beta_1 = \beta_2 = \beta_3$ against an alternative hypothesis that the β's differ. Neither of these naive models takes account of the fact that measurements of the same patient at

Table 11.3 *Results of naive analyses of stroke recovery scores in Table 11.1, assuming all the data are independent and using models (11.1) and (11.2).*

Parameter	Estimate	Standard error
Model (11.1)		
α_1	36.842	3.971
$\alpha_2 - \alpha_1$	−5.625	3.715
$\alpha_3 - \alpha_1$	−12.109	3.715
β	4.764	0.662
Model (11.2)		
α_1	29.821	5.774
$\alpha_2 - \alpha_1$	3.348	8.166
$\alpha_3 - \alpha_1$	−0.022	8.166
β_1	6.324	1.143
$\beta_2 - \beta_1$	−1.994	1.617
$\beta_3 - \beta_1$	−2.686	1.617

different times are likely to be more similar than measurements of different patients. This is analogous to using an unpaired t-test for paired data (see Exercise 2.2).

Table 11.3 shows the results of fitting these models, which will be compared later with results from more appropriate analyses. Note, however, that for model (11.2), the Wald statistics for $\alpha_2 - \alpha_1$ $(3.348/8.166 = 0.41)$ and for $\alpha_3 - \alpha_1$ $(-0.022/8.166 = -0.003)$ are very small compared with the standard Normal distribution which suggests that the intercepts are not different (i.e., on average the groups started with the same level of functional ability).

A preferable form of **exploratory analysis**, sometimes called **data reduction** or **data summary**, consists of summarizing the response profiles for each subject by a small number of descriptive statistics based on assuming that measurements on the same subject are independent. For the stroke data, appropriate summary statistics are the intercept and slope of the individual regression lines. Other examples of summary statistics that may be appropriate in particular situations include peak values, areas under curves, or coefficients of quadratic or exponential terms in non-linear growth curves. These subject-specific statistics are used as the data for subsequent analyses.

The intercept and slope estimates and their standard errors for each of the 24 stroke patients are shown in Table 11.4. These results show considerable variability between subjects which should, in principle, be taken into account in any further analyses. Tables 11.5 and 11.6 show analyses comparing intercepts and slopes between the experimental groups, assuming independence

Table 11.4 *Estimates of intercepts and slopes (and their standard errors) for each subject in Table 11.1.*

Subject	Intercept (std. error)		Slope (std. error)	
1	30.000	(7.289)	7.500	(1.443)
2	15.536	(4.099)	3.214	(0.812)
3	39.821	(3.209)	6.429	(0.636)
4	11.607	(3.387)	8.393	(0.671)
5	100.000	(0.000)	0.000	(0.000)
6	0.893	(5.304)	11.190	(1.050)
7	15.357	(4.669)	7.976	(0.925)
8	25.357	(1.971)	5.893	(0.390)
9	38.571	(3.522)	7.262	(0.698)
10	61.964	(2.236)	2.619	(0.443)
11	14.464	(5.893)	9.702	(1.167)
12	26.071	(2.147)	2.679	(0.425)
13	48.750	(8.927)	5.000	(1.768)
14	10.179	(3.209)	1.071	(0.636)
15	31.250	(1.948)	2.500	(0.386)
16	34.107	(2.809)	3.810	(0.556)
17	21.071	(2.551)	1.429	(0.505)
18	34.107	(1.164)	0.893	(0.231)
19	32.143	(1.164)	1.607	(0.231)
20	42.321	(3.698)	7.262	(0.732)
21	48.571	(6.140)	7.262	(1.216)
22	24.821	(1.885)	2.262	(0.373)
23	22.321	(1.709)	1.845	(0.339)
24	13.036	(4.492)	6.548	(0.890)

between the subjects but ignoring the differences in precision (standard errors) between the estimates. Notice that although the estimates are the same as those for model (11.2) in Table 11.3, the standard errors are (correctly) much larger and the data do not provide much evidence of differences in either the intercepts or the slopes.

Although the analysis of subject specific summary statistics does not require the implausible assumption of independence between observations within subjects, it ignores the random error in the estimates. Ignoring this information can lead to underestimation of effect sizes and underestimation of the overall variation. To avoid these biases, models are needed that better

Table 11.5 *Analysis of variance of intercept estimates in Table 11.4.*

Source	d.f.	Mean square	F	p-value
Groups	2	30	0.07	0.94
Error	21	459		

Parameter	Estimate	Std. error
α_1	29.821	7.572
$\alpha_2 - \alpha_1$	3.348	10.709
$\alpha_3 - \alpha_1$	−0.018	10.709

Table 11.6 *Analysis of variance of slope estimates in Table 11.4.*

Source	d.f.	Mean square	F	p-value
Groups	2	15.56	1.67	0.21
Error	21	9.34		

Parameter	Estimate	Std. error
β_1	6.324	1.080
$\beta_2 - \beta_1$	−1.994	1.528
$\beta_3 - \beta_1$	−2.686	1.528

describe the data structure that arises from the study design. Such models are described in the next three sections.

11.3   Repeated measures models for Normal data

Suppose there are N study units or subjects with n_i measurements for subject i (e.g., n_i longitudinal observations for person i or n_i observations for cluster i). Let $\mathbf{y}_i$ denote the vector of responses for subject i and let $\mathbf{y}$ denote the vector of responses for all subjects

$$\mathbf{y} = \begin{bmatrix} \mathbf{y}_1 \\ \vdots \\ \mathbf{y}_N \end{bmatrix}, \qquad \text{so } \mathbf{y} \text{ has length } \sum_{i=1}^{N} n_i.$$

A Normal linear model for $\mathbf{y}$ is

$$E(\mathbf{y}) = \mathbf{X}\boldsymbol{\beta} = \boldsymbol{\mu}; \qquad \mathbf{y} \sim \text{MVN}(\boldsymbol{\mu}, \mathbf{V}), \qquad (11.3)$$

where

$$
\mathbf{X} = \begin{bmatrix} \mathbf{X}_1 \\ \mathbf{X}_2 \\ \vdots \\ \mathbf{X}_N \end{bmatrix}, \quad \boldsymbol{\beta} = \begin{bmatrix} \beta_1 \\ \vdots \\ \beta_p \end{bmatrix},
$$

$\mathbf{X}_i$ is the $n_i \times p$ design matrix for subject i and $\boldsymbol{\beta}$ is a parameter vector of length p. The variance–covariance matrix for measurements for subject i is

$$
\mathbf{V}_i = \begin{bmatrix} \sigma_{i11} & \sigma_{i12} & \cdots & \sigma_{i1n_i} \\ \sigma_{i21} & \ddots & & \vdots \\ \vdots & & \ddots & \\ \sigma_{in1} & & & \sigma_{in_i n_i} \end{bmatrix},
$$

and the overall variance–covariance matrix has the block diagonal form

$$
\mathbf{V} = \begin{bmatrix} \mathbf{V}_1 & \mathbf{O} & & \mathbf{O} \\ \mathbf{O} & \mathbf{V}_2 & & \mathbf{O} \\ & & \ddots & \\ \mathbf{O} & \mathbf{O} & & \mathbf{V}_N \end{bmatrix},
$$

assuming that responses for different subjects are independent (where $\mathbf{O}$ denotes a matrix of zeros). Usually the matrices $\mathbf{V}_i$ are assumed to have the same form for all subjects.

If the elements of $\mathbf{V}$ are known constants, then $\boldsymbol{\beta}$ can be estimated from the likelihood function for model (11.3) or by the method of least squares. The maximum likelihood estimator is obtained by solving the score equations

$$
\mathbf{U}(\boldsymbol{\beta}) = \frac{\partial l}{\partial \boldsymbol{\beta}} = \mathbf{X}^T \mathbf{V}^{-1} (\mathbf{y} - \mathbf{X}\boldsymbol{\beta}) = \sum_{i=1}^{N} \mathbf{X}_i^T \mathbf{V}_i^{-1} (\mathbf{y}_i - \mathbf{X}_i \boldsymbol{\beta}) = \mathbf{0} \qquad (11.4)
$$

where l is the log-likelihood function. The solution is

$$
\widehat{\boldsymbol{\beta}} = (\mathbf{X}^T \mathbf{V}^{-1} \mathbf{X})^{-1} \mathbf{X}^T \mathbf{V}^{-1} \mathbf{y} = \left(\sum_{i=1}^{N} \mathbf{X}_i^T \mathbf{V}_i^{-1} \mathbf{X}_i \right)^{-1} \left(\sum_{i=1}^{N} \mathbf{X}_i^T \mathbf{V}_i^{-1} \mathbf{y}_i \right) \qquad (11.5)
$$

with

$$
\mathrm{var}(\widehat{\boldsymbol{\beta}}) = (\mathbf{X}^T \mathbf{V}^{-1} \mathbf{X})^{-1} = \left(\sum_{i=1}^{N} \mathbf{X}_i^T \mathbf{V}_i^{-1} \mathbf{X}_i \right)^{-1} \qquad (11.6)
$$

and $\widehat{\boldsymbol{\beta}}$ is asymptotically Normal (see Chapter 6).

In practice, $\mathbf{V}$ is usually not known and has to be estimated from the data by an iterative process. This involves starting with an initial $\mathbf{V}$ (for instance, the identity matrix), calculating an estimate $\widehat{\boldsymbol{\beta}}$ and hence the linear predictors $\widehat{\boldsymbol{\mu}} = \mathbf{X}\widehat{\boldsymbol{\beta}}$ and the residuals $\mathbf{r} = \mathbf{y} - \widehat{\boldsymbol{\mu}}$. The variances and covariances of the residuals are used to calculate $\widehat{\mathbf{V}}$, which in turn is used in (11.5) to obtain a new estimate $\widehat{\boldsymbol{\beta}}$. The process alternates between estimating $\widehat{\boldsymbol{\beta}}$ and estimating $\widehat{\mathbf{V}}$ until convergence is achieved.

If the estimate $\widehat{\mathbf{V}}$ is substituted for $\mathbf{V}$ in Equation (11.6), the variance of $\widehat{\boldsymbol{\beta}}$ is likely to be underestimated. Therefore, a preferable alternative is

$$\mathbf{V}_s(\widehat{\boldsymbol{\beta}}) = \mathfrak{I}^{-1}\mathbf{C}\mathfrak{I}^{-1},$$

where

$$\mathfrak{I} = \mathbf{X}^T\widehat{\mathbf{V}}^{-1}\mathbf{X} = \sum_{i=1}^{N}\mathbf{X}_i^T\widehat{\mathbf{V}}_i^{-1}\mathbf{X}_i$$

and

$$\mathbf{C} = \sum_{i=1}^{N}\mathbf{X}_i^T\widehat{\mathbf{V}}_i^{-1}(\mathbf{y}_i - \mathbf{X}_i\widehat{\boldsymbol{\beta}})(\mathbf{y}_i - \mathbf{X}_i\widehat{\boldsymbol{\beta}})^T\widehat{\mathbf{V}}_i^{-1}\mathbf{X}_i,$$

where $\widehat{\mathbf{V}}_i$ denotes the ith sub matrix of $\widehat{\mathbf{V}}$. $\mathbf{V}_s(\widehat{\boldsymbol{\beta}})$ is called the **information sandwich estimator**, because $\mathfrak{I}$ is the information matrix (see Chapter 5). It is also sometimes called the **Huber estimator**. It is a consistent estimator of $\text{var}(\widehat{\boldsymbol{\beta}})$ when $\mathbf{V}$ is not known, and it is robust to mis-specification of $\mathbf{V}$.

There are several commonly used forms for the matrix $\mathbf{V}_i$.

1. All the off-diagonal elements are equal so that

$$\mathbf{V}_i = \sigma^2 \begin{bmatrix} 1 & \rho & \cdots & \rho \\ \rho & 1 & & \rho \\ \vdots & & \ddots & \vdots \\ \rho & \rho & \cdots & 1 \end{bmatrix}. \tag{11.7}$$

This is appropriate for clustered data where it is plausible that all measurements are equally correlated, for example, for elements within the same primary sampling unit such as people living in the same area. The term ρ is called the **intra-class correlation coefficient**. The **exchangeable** matrix in (11.7) is called **equicorrelation** or **spherical**. If the off-diagonal term ρ can be written in the form $\sigma_a^2/(\sigma_a^2 + \sigma_b^2)$, the matrix is said to have **compound symmetry**. The number of parameters needed for this variance–covariance matrix is $P = 2$, one for the variance (σ^2) and one for the correlation (ρ).

2. The off-diagonal terms decrease with "distance" between observations; for example, if all the vectors $\mathbf{y}_i$ have the same length n and

$$
\mathbf{V}_i = \sigma^2 \begin{bmatrix} 1 & \rho_{12} & \cdots & \rho_{1n} \\ \rho_{21} & 1 & & \rho_{2n} \\ \vdots & & \ddots & \vdots \\ \rho_{n1} & \rho_{n2} & \cdots & 1 \end{bmatrix}, \tag{11.8}
$$

where ρ_{jk} depends on the "distance" between observations j and k. Examples include $\rho_{jk} = |t_j - t_k|$ for measurements at times t_j and t_k (provided these are defined so that $-1 \le \rho_{jk} \le 1$), or $\rho_{jk} = \exp(-|j - k|)$. One commonly used form is the first-order **autoregressive model** with $\rho^{|j-k|}$, where $|\rho| < 1$ so that

$$
\mathbf{V}_i = \sigma^2 \begin{bmatrix} 1 & \rho & \rho^2 & \cdots & \rho^{n-1} \\ \rho & 1 & \rho & & \rho^{n-2} \\ \rho^2 & \rho & 1 & & \vdots \\ \vdots & & & \ddots & \\ \rho^{n-1} & \cdots & & \rho & 1 \end{bmatrix}. \tag{11.9}
$$

The number of parameters needed for this variance–covariance matrix is $P = 2$.

3. All the correlation terms may be different

$$
\mathbf{V}_i = \sigma^2 \begin{bmatrix} 1 & \rho_{12} & \cdots & \rho_{1n} \\ \rho_{21} & 1 & & \rho_{2n} \\ \vdots & & \ddots & \vdots \\ \rho_{n1} & \rho_{n2} & \cdots & 1 \end{bmatrix}. \tag{11.10}
$$

This **unstructured correlation matrix** involves no assumptions about correlations between measurements but all the vectors $\mathbf{y}_i$ must be the same length n. It is only practical to use this form when the matrix $\mathbf{V}_i$ is not large relative to the number of subjects because the number of nuisance parameters ρ_{jk} is $P = n(n-1)/2$, which increases quadratically with n and may lead to convergence problems in the iterative estimation process. Sometimes it may be useful to fit a model with an unstructured correlation matrix and examine the estimates $\widehat{\rho}_{jk}$ for patterns that may suggest a simpler model.

The Akaike Information Criterion can be used to choose the best variance–covariance matrix (Barnett et al. 2010). To do this, a range of different variance–covariance matrices are tried, from the simplest naive independent matrix to the most complex unstructured matrix. These models must use the same design matrix so that the number of regression parameters p stays fixed, but the number of nuisance parameters P varies.

The term **repeated measures analysis of variance** is often used when the data are assumed to be Normally distributed. The calculations can be performed using most general purpose statistical software although, sometimes, the correlation structure is assumed to be either exchangeable or unstructured and correlations which are functions of the times between measurements cannot be modelled. Sometimes repeated measures are treated as a special case of multivariate data—for example, by not distinguishing between heights of children in the same class (i.e., clustered data), heights of children when they are measured at different ages (i.e., longitudinal data), and heights, weights and girths of children (multivariate data). This is inappropriate for longitudinal data in which the time order of the observations matters. The **multivariate approach** to analyzing Normally distributed repeated measures data is explained in detail by Verbeke and Molenberghs (2000), while the inappropriateness of these methods for longitudinal data is illustrated by Senn et al. (2000).

11.4   Repeated measures models for non-Normal data

The score equations for Normal models (11.4) can be generalized to other distributions using ideas from Chapter 4. For the generalized linear model

$$E(Y_i) = \mu_i, \quad g(\mu_i) = \mathbf{x}_i^T \boldsymbol{\beta} = \eta_i$$

for independent random variables $Y_1, Y_2, \ldots, Y_N$ with a distribution from the exponential family, the scores given by Equation (4.18) are

$$U_j = \sum_{i=1}^{N} \frac{(y_i - \mu_i)}{\text{var}(Y_i)} x_{ij} \left(\frac{\partial \mu_i}{\partial \eta_i} \right)$$

for parameters β_j, $j = 1, \ldots, p$. The last two terms come from

$$\frac{\partial \mu_i}{\partial \beta_j} = \frac{\partial \mu_i}{\partial \eta_i} \cdot \frac{\partial \eta_i}{\partial \beta_j} = \frac{\partial \mu_i}{\partial \eta_i} x_{ij}.$$

Therefore, the score equations for the generalized model (with independent responses $Y_i, i = 1, \ldots, N$) can be written as

$$U_j = \sum_{i=1}^{N} \frac{(y_i - \mu_i)}{\text{var}(Y_i)} \frac{\partial \mu_i}{\partial \beta_j} = 0, \qquad j = 1, \ldots, p. \tag{11.11}$$

For repeated measures, let $\mathbf{y}_i$ denote the vector of responses for subject i with $E(\mathbf{y}_i) = \boldsymbol{\mu}_i$, $g(\boldsymbol{\mu}_i) = \mathbf{X}_i^T \boldsymbol{\beta}$ and let $\mathbf{D}_i$ be the matrix of derivatives $\partial \boldsymbol{\mu}_i / \partial \beta_j$. To simplify the notation, assume that all the subjects have the same number of measurements n.

The **generalized estimating equations** (GEEs) analogous to equations (11.11) are

$$\mathbf{U} = \sum_{i=1}^{N} \mathbf{D}_i^T \mathbf{V}_i^{-1}(\mathbf{y}_i - \boldsymbol{\mu}_i) = \mathbf{0}. \tag{11.12}$$

These are also called the **quasi-score equations**. The matrix $\mathbf{V}_i$ can be written as

$$\mathbf{V}_i = \mathbf{A}_i^{\frac{1}{2}} \mathbf{R}_i \mathbf{A}_i^{\frac{1}{2}} \phi,$$

where $\mathbf{A}_i$ is the diagonal matrix with elements $\text{var}(y_{ik})$, $\mathbf{R}_i$ is the correlation matrix for $\mathbf{y}_i$ and ϕ is a constant to allow for overdispersion.

Liang and Zeger (1986) showed that if the correlation matrices $\mathbf{R}_i$ are correctly specified, the estimator $\widehat{\boldsymbol{\beta}}$ is consistent and asymptotically Normal. Furthermore, $\widehat{\boldsymbol{\beta}}$ is fairly robust against mis-specification of $\mathbf{R}_i$. They used the term **working correlation matrix** for $\mathbf{R}_i$ and suggested that knowledge of the study design and results from exploratory analyses should be used to select a plausible form. Preferably, $\mathbf{R}_i$ should depend on only a small number of parameters, using assumptions such as an exchangeable or autoregressive structure (see Section 11.3 above).

The GEEs given by Equation (11.12) are used iteratively. Starting with $\mathbf{R}_i$ as the identity matrix and $\phi = 1$, the parameters $\boldsymbol{\beta}$ are estimated by solving Equations (11.12). The estimates are used to calculate fitted values $\widehat{\boldsymbol{\mu}}_i = g^{-1}(\mathbf{X}_i^T \widehat{\boldsymbol{\beta}})$ and hence the residuals $\mathbf{y}_i - \widehat{\boldsymbol{\mu}}_i$. These are used to estimate the parameters of $\mathbf{A}_i$, $\mathbf{R}_i$ and ϕ. Then (11.12) is solved again to obtain improved estimates $\widehat{\boldsymbol{\beta}}$, and so on, until convergence is achieved.

Software for solving GEEs is available in most commercially available software and free-ware programs. While the concepts underlying GEEs are relatively simple, there are a number of complications that occur in practice. For example, for binary data, correlation is not a natural measure of association and alternative measures using odds ratios have been proposed (Lipsitz et al. 1991).

For GEEs it is even more important to use a sandwich estimator for $\text{var}(\widehat{\boldsymbol{\beta}})$ than for the Normal case (see Section 11.3). This is given by

$$\mathbf{V}_s(\widehat{\boldsymbol{\beta}}) = \mathfrak{I}^{-1}\mathbf{C}\mathfrak{I}^{-1},$$

where

$$\mathfrak{I} = \sum_{i=1}^{N} \mathbf{D}_i^T \widehat{\mathbf{V}}_i^{-1} \mathbf{D}_i$$

is the information matrix and

$$\mathbf{C} = \sum_{i=1}^{N} \mathbf{D}_i^T \widehat{\mathbf{V}}_i^{-1} (\mathbf{y}_i - \widehat{\boldsymbol{\mu}}_i)(\mathbf{y}_i - \widehat{\boldsymbol{\mu}}_i)^T \widehat{\mathbf{V}}_i^{-1} \mathbf{D}_i.$$

Then asymptotically, $\widehat{\boldsymbol{\beta}}$ has the distribution $\mathrm{MVN}\left(\boldsymbol{\beta}, \mathbf{V}_s(\widehat{\boldsymbol{\beta}})\right)$, and inferences can be made using Wald statistics.

11.5 Multilevel models

An alternative approach for analyzing repeated measures data is to use hierarchical models based on the study design. Consider a survey conducted using cluster randomized sampling. Let Y_{jk} denote the response of the kth subject in the jth cluster. For example, suppose Y_{jk} is the income of the kth randomly selected household in council area j, where council areas, the primary sampling units, are chosen randomly from all councils within a country or state. If the goal is to estimate the average household income μ, then a suitable model might be

$$Y_{jk} = \mu + a_j + e_{jk}, \tag{11.13}$$

where a_j is the effect of area j and e_{jk} is the random error term. As areas were randomly selected and the area effects are not of primary interest, the terms a_j can be defined as independent, identically distributed random variables with $a_j \sim \mathrm{N}(0, \sigma_a^2)$. Similarly, the terms e_{jk} are independently, identically distributed random variables $e_{jk} \sim \mathrm{N}(0, \sigma_e^2)$ and the a_j's and e_{jk}'s are independent. In this case

$$\mathrm{E}(Y_{jk}) = \mu,$$

$$\mathrm{var}(Y_{jk}) = \mathrm{E}\left[(Y_{jk} - \mu)^2\right] = \mathrm{E}\left[(a_j + e_{jk})^2\right] = \sigma_a^2 + \sigma_e^2.$$

For households in the same area,

$$\mathrm{cov}(Y_{jk}, Y_{jm}) = \mathrm{E}\left[(a_j + e_{jk})(a_j + e_{jm})\right] = \sigma_a^2,$$

and for households in different areas,

$$\mathrm{cov}(Y_{jk}, Y_{lm}) = \mathrm{E}\left[(a_j + e_{jk})(a_l + e_{lm})\right] = 0.$$

If $\mathbf{y}_j$ is the vector of responses for households in area j, then the variance–covariance matrix for $\mathbf{y}_j$ is

$$
\mathbf{V}_j = \begin{bmatrix}
\sigma_a^2 + \sigma_e^2 & \sigma_a^2 & \sigma_a^2 & \cdots & \sigma_a^2 \\
\sigma_a^2 & \sigma_a^2 + \sigma_e^2 & \sigma_a^2 & & \sigma_a^2 \\
\sigma_a^2 & \sigma_a^2 & \sigma_a^2 + \sigma_e^2 & & \\
\vdots & & & \ddots & \\
\sigma_a^2 & & & \sigma_a^2 & \sigma_a^2 + \sigma_e^2
\end{bmatrix}
$$

$$
= \sigma_a^2 + \sigma_e^2 \begin{bmatrix}
1 & \rho & \rho & \cdots & \rho \\
\rho & 1 & \rho & & \rho \\
\rho & \rho & 1 & & \\
\vdots & & & \ddots & \\
\rho & & & \rho & 1
\end{bmatrix},
$$

where $\rho = \sigma_a^2/(\sigma_a^2 + \sigma_e^2)$ is the intra-class correlation coefficient which describes the proportion of the total variance due to within-cluster variance. If the responses within a cluster are much more alike than responses from different clusters, then σ_e^2 is much smaller than σ_a^2, so ρ will be near unity; thus, ρ is a relative measure of the within-cluster similarity. The matrix $\mathbf{V}_j$ is the same as (11.7), the exchangeable matrix.

In model (11.13), the parameter μ is a **fixed effect** and the term a_j is a **random effect**. This is an example of a **mixed model** with both fixed and random effects. The parameters of interest are μ, σ_a^2 and σ_e^2 (and hence ρ).

As another example, consider longitudinal data in which Y_{jk} is the measurement at time t_k on subject j who was selected at random from the population of interest. A linear model for this situation is

$$
Y_{jk} = \beta_0 + a_j + (\beta_1 + b_j)t_k + e_{jk}, \tag{11.14}
$$

where β_0 and β_1 are the intercept and slope parameters for the population, a_j and b_j are the differences from these parameters specific to subject j, t_k denotes the time of the kth measurement, and e_{jk} is the random error term. The terms a_j, b_j and e_{jk} may be considered as random variables with $a_j \sim N(0, \sigma_a^2)$, $b_j \sim N(0, \sigma_b^2)$, $e_{jk} \sim N(0, \sigma_e^2)$, and they are all assumed to be independent. For this model

$$
E(Y_{jk}) = \beta_0 + \beta_1 t_k,
$$

$$
\text{var}(Y_{jk}) = \text{var}(a_j) + t_k^2 \text{var}(b_j) + \text{var}(e_{jk}) = \sigma_a^2 + t_k^2 \sigma_b^2 + \sigma_e^2,
$$

$$
\text{cov}(Y_{jk}, Y_{jm}) = \sigma_a^2 + t_k t_m \sigma_b^2
$$

for measurements on the same subject, and

$$\text{cov}(Y_{jk}, Y_{lm}) = 0$$

for measurements on different subjects. Therefore, the variance–covariance matrix for subject j is of the form shown in (11.8) with terms dependent on t_k and t_m. In model (11.14), β_0 and β_1 are fixed effects, a_j and b_j are random effects and the aim is to estimate β_0, β_1, σ_a^2, σ_b^2 and σ_e^2.

In general, mixed models for Normal responses can be written in the form

$$\mathbf{y} = \mathbf{X}\boldsymbol{\beta} + \mathbf{Z}\mathbf{u} + \mathbf{e}, \tag{11.15}$$

where $\boldsymbol{\beta}$ are the fixed effects, and $\mathbf{u}$ and $\mathbf{e}$ are random effects. The matrices $\mathbf{X}$ and $\mathbf{Z}$ are design matrices. Both $\mathbf{u}$ and $\mathbf{e}$ are assumed to be Normally distributed. $\text{E}(\mathbf{y}) = \mathbf{X}\boldsymbol{\beta}$ summarizes the non-random component of the model. $\mathbf{Z}\mathbf{u}$ describes the between-subjects random effects and $\mathbf{e}$ the within-subjects random effects. If $\mathbf{G}$ and $\mathbf{R}$ denote the variance–covariance matrices for $\mathbf{u}$ and $\mathbf{e}$, respectively, then the variance–covariance matrix for $\mathbf{y}$ is

$$\mathbf{V}(\mathbf{y}) = \mathbf{Z}\mathbf{G}^T\mathbf{Z} + \mathbf{R}. \tag{11.16}$$

The parameters of interest are the elements of $\boldsymbol{\beta}$ and the variance and covariance elements in $\mathbf{G}$ and $\mathbf{R}$. For Normal models these can be estimated using the methods of maximum likelihood or residual maximum likelihood (REML). Computational procedures are available in many general purpose statistical programs.

Mixed models for non-Normal data are less readily implemented although they were first described by Zeger et al. (1988) and have been the subject of many publications since then; see, for example, Twisk (2006), Stroup (2012), and Molenberghs and Verbeke (2005). The models can be specified as follows

$$\text{E}(\mathbf{y}|\mathbf{u}) = \boldsymbol{\mu}, \qquad \text{var}(\mathbf{y}|\mathbf{u}) = \phi \mathbf{V}(\mu), \qquad g(\boldsymbol{\mu}) = \mathbf{X}\boldsymbol{\beta} + \mathbf{Z}\mathbf{u},$$

where the random coefficients $\mathbf{u}$ have some distribution $f(\mathbf{u})$ and the conditional distribution of $\mathbf{y}$ given $\mathbf{u}$, written as $\mathbf{y}|\mathbf{u}$, follows the usual properties for a generalized linear model with link function g. The unconditional mean and variance–covariance for $\mathbf{y}$ can, in principle, be obtained by integrating over the distribution of $\mathbf{u}$. To make the calculations more tractable, it is common to use particular pairs of distributions, called conjugate distributions; for example, Normal for $\mathbf{y}|\mathbf{u}$ and Normal for $\mathbf{u}$; Poisson for $\mathbf{y}|\mathbf{u}$ and Gamma for $\mathbf{u}$; Binomial for $\mathbf{y}|\mathbf{u}$ and Beta for $\mathbf{u}$; or Binomial for $\mathbf{y}|\mathbf{u}$ and Normal for $\mathbf{u}$. This approach is similar to Bayesian analysis which is discussed in Chapters 12–14.

11.6 Stroke example continued

The results of the exploratory analyses and fitting GEEs and mixed models with different intercepts and slopes to the stroke recovery data are shown in Table 11.7. The data need to be formatted so that the records for the same subject for each time are listed one below the next (in Stata this is called a "long" format) and the variable defining the treatment groups needs to be declared to be a factor.

Table 11.7 *Comparison of analyses of the stroke recovery data using various different models.*

	Intercept estimates		
	$\widehat{\alpha}_1$ (s.e.)	$\widehat{\alpha}_2 - \widehat{\alpha}_1$ (s.e.)	$\widehat{\alpha}_3 - \widehat{\alpha}_1$ (s.e.)
Pooled	29.821 (5.774)	3.348 (8.166)	−0.022 (8.166)
Data reduction	29.821 (5.772)	3.348 (10.709)	−0.018 (10.709)
GEE, independent	29.821 (10.395)	3.348 (11.884)	−0.022 (11.130)
GEE, exchangeable	29.821 (10.395)	3.348 (11.884)	−0.022 (11.130)
GEE, AR(1)	33.492 (9.924)	−0.270 (11.139)	−6.396 (10.551)
GEE, unstructured	30.703 (10.297)	2.058 (11.564)	−1.403 (10.906)
Random effects	29.821 (7.047)	3.348 (9.966)	−0.022 (9.966)
	Slope estimates		
	$\widehat{\beta}_1$ (s.e.)	$\widehat{\beta}_2 - \widehat{\beta}_1$ (s.e.)	$\widehat{\beta}_3 - \widehat{\beta}_1$ (s.e.)
Pooled	6.324 (1.143)	−1.994 (1.617)	−2.686 (1.617)
Data reduction	6.324 (1.080)	−1.994 (1.528)	−2.686 (1.528)
GEE, independent	6.324 (1.156)	−1.994 (1.509)	−2.686 (1.502)
GEE, exchangeable	6.324 (1.156)	−1.994 (1.509)	−2.686 (1.502)
GEE, AR(1)	6.074 (1.057)	−2.142 (1.360)	−2.236 (1.504)
GEE, unstructured	7.126 (1.272)	−3.559 (1.563)	−4.012 (1.598)
Random effects	6.324 (0.463)	−1.994 (0.655)	−2.686 (0.655)

For Stata the GEE models are as follows:

```
──────────────────── Stata code (GEE models) ────────────────────
.xtgee ability _Igroup_2 _Igroup_3 time _IgroXtime_2 _IgroXtime_3
 ,family(gaussian) link(identity) corr(independent) robust
.xtgee ability _Igroup_2 _Igroup_3 time _IgroXtime_2 _IgroXtime_3
 ,family(gaussian) link(identity) corr(exchangeable) robust
.xtgee ability _Igroup_2 _Igroup_3 time _IgroXtime_2 _IgroXtime_3
 ,family(gaussian) link(identity) corr(ar 1) robust
.xtgee ability _Igroup_2 _Igroup_3 time _IgroXtime_2 _IgroXtime_3
 ,family(gaussian) link(identity) corr(unstructured) robust
```

Sandwich estimates of the standard errors were calculated for all the GEE models.

For random effects model the Stata commands are

```
──────────── Stata code (Random effects model) ────────────
.tsset subject time
.xtreg ability _Igroup_2 _Igroup_3 _IgroXtime_2 _IgroXtime_3
time, mle
```

In R, relevant programs for GEEs can be found in the "geepack" library (Højsgaard, Halekoh, and Yan 2005) and the commands are

```
──────────────── R code (GEE models) ────────────────
>gee.ind<-geeglm(ability~Group+time+Group*time,family=gaussian,
 data=stroke,id=Subject,wave=time,corst="independence")
>gee.exch<-geeglm(ability~Group+time+Group*time,family=gaussian,
 data=stroke,id=Subject,wave=time,corst="exchangeable")
>gee.ar1<-geeglm(ability~Group+time+Group*time,family=gaussian,
 data=stroke,id=Subject,wave=time,corst="ar1")
>gee.un<-geeglm(ability~Group+time+Group*time,family=gaussian,
 data=stroke,id=Subject,wave=time,corst="unstructured")
```

For random effects models the "nlme" library (Pinheiro, Bates, DebRoy, Sarkar, and R Core Team 2017) can be used and the following command produces comparable results to the Stata analysis:

```
──────── R code (Random intercepts and slopes model) ────────
>rndeff<-lme(ability~group + time + group*time,data=stroke,
random=~1|subject)
```

The results shown in Table 11.7 are from models that were fitted using Stata. Fitting a GEE, assuming independence between observations for the same subject, is the same as the naive or pooled analysis in Table 11.3. The estimate of σ_e is 20.96 (this is the square root of the deviance divided by the degrees of freedom $192 - 6 = 186$). These results suggest that neither intercepts nor slopes differ between groups as the estimates of differences from $\widehat{\alpha}_1$ and $\widehat{\beta}_1$ are small relative to their standard errors.

The data reduction approach, which uses the estimated intercepts and slopes for every subject as the data for comparisons of group effects, produces the same point estimates but different standard errors. From Tables 11.5 and 11.6, the standard deviations are 21.42 for the intercepts and 3.056 for the slopes and the data do not support hypotheses of differences between the groups.

The GEE analysis, assuming equal correlation (exchangeable correlation)

among the observations in different weeks, produced the same estimates for the intercept and slope parameters. The estimate of the common correlation coefficient, $\widehat{\rho} = 0.831$, is about the average of the values in Table 11.2, but the assumption of equal correlation is not really plausible. The estimate of σ_e is 20.96, the same as for the models based on independence.

In view of the pattern of correlation coefficients in Table 11.2, an autoregressive model of order 1, AR(1), shown in Equation (11.9) seems plausible. In this case, the estimates for ρ and σ_e are 0.960 and 21.08, respectively. The estimates of intercepts and slopes, and their standard errors, differ from the previous models. Wald statistics for the differences in slope support the hypothesis that the patients in group A improved significantly faster than patients in the other two groups.

The GEE model with an unstructured correlation matrix involved estimating $P = 28$ $((8 \times 7)/2)$ correlation parameters. The estimate of σ_e was 21.21. While the point estimates differ from those for the other GEE models with correlation structures, the conclusion that the slopes differ significantly is the same.

The final model fitted was the mixed model (11.14) estimated by the method of maximum likelihood. The point estimates and standard errors for the fixed parameters were similar to those from the GEE model with the exchangeable matrix. This is not surprising as the estimated intra-class correlation coefficient is $\widehat{\rho} = 0.831$.

A generalized least squares procedure from the "nlme" library in R can be used to estimate the AIC for the different correlation matrices.

```
——— R code (AIC to compare correlation structures) ———
>ind<-corIdent(form = ~ 1 | Subject)
>gls.ind<-gls(ability~Group+time+Group*time, data=stroke,
correlation=ind)
>exch<-corCompSymm(form = ~ 1 | Subject)
>gls.exch<-gls(ability~Group+time+Group*time, data=stroke,
correlation=exch)
>ar1<-corAR1(form = ~ 1 | Subject)
>gls.ar1<-gls(ability~Group+time+Group*time, data=stroke,
correlation=ar1)
>un<-corSymm(form = ~ 1 | Subject)
>gls.un<-gls(ability~Group+time+Group*time, data=stroke,
correlation=un)
>AIC(gls.ind,gls.exch,gls.ar1,gls.un)
```

Table 11.8 *Akaike information criteria for the four correlation matrices used to model the stroke recovery data. P is the number of "nuisance" variance–covariance parameters. Every model has p = 6 regression parameters.*

Correlation matrix	P	AIC	Difference in AIC from the autoregressive model
Autoregressive	2	1320.3	0.0
Unstructured	29	1338.1	17.8
Exchangeable	2	1452.7	132.4
Independent	1	1703.6	383.3

As shown by the AIC in Table 11.8, the independent correlation matrix is clearly not a good choice. An exchangeable correlation is a big improvement, but it is far from the best indicating that all observations are not equally correlated, which we saw earlier in Table 11.2. An unstructured correlation model does much better than the exchangeable one, but the improvement in fit is achieved using many parameters. The much simpler autoregressive structure gives the best trade-off between fit and complexity according to the AIC. The single correlation parameter has captured the decay in correlation with increasing time gap between observations.

This example illustrates both the importance of taking into account the correlation between repeated measures and the robustness of the results regardless of how the correlation is modelled. Without considering the correlation, it was not possible to detect the statistically significantly better outcomes for patients in group A. It is also important to note that different software may produce slightly different parameter estimates and standard error estimates, even for the same model. This is because the correlation parameters may be estimated slightly differently. Also GEEs depend on large sample sizes to achieve optimal performance and this data set is too small for the necessary robustness. Nevertheless, the inferences are essentially the same— in this example, a common intercept may provide an adequate model and the first-order autoregressive model is appropriate for describing the correlation parsimoniously.

11.7 Comments

The models described in this chapter provide the means to analyse longitudinal data, which are becoming increasingly important, especially in the health field because they can provide strong evidence for the temporal order of cause and effect. There are now many books that describe these methods in detail;

for example, Verbeke and Molenberghs (2000), (Rabe-Hesketh and Skrondal 2012), Diggle et al. (2002), Twisk (2013), and Fitzmaurice et al. (2012).

Exploratory analyses for repeated measures data should follow the main steps outlined in Section 11.2. For longitudinal data these include plotting the time course for individual subjects or groups of subjects, and using an appropriate form of data reduction to produce summary statistics that can be examined to identify patterns for the population overall or for sub samples. For clustered data it is worthwhile to calculate summary statistics at each level of a multilevel model to examine both the main effects and the variability.

Missing data can present problems. With suitable software it may be possible to perform calculations on unbalanced data (e.g., different numbers of observations per subject) but this is dangerous without careful consideration of why data are missing. Diggle et al. (2002) discuss the problem in more detail and provide some suggestions about how adjustments may be made in some situations.

Unbalanced data and longitudinal data in which the observations are not equally spaced or do not all occur at the planned times can be accommodated in mixed models and generalized estimating equations; for example, see Cnaan et al. (1997), Burton et al. (1998), and Carlin et al. (1999).

Inference for models fitted by GEEs is best undertaken using Wald statistics with a robust sandwich estimator for the variance. The optimal choice of the correlation matrix is not critical because the estimator is robust with respect to the choice of working correlation matrix, but a poor choice can reduce the efficiency of the estimator. In practice, the choice may be affected by the number of correlation parameters to be estimated; for example, use of a large unstructured correlation matrix may produce unstable estimates or the calculations may not converge. Model checking can be carried out with the usual range of residual plots.

For multilevel data, nested models can be compared using likelihood ratio statistics. Residuals used for checking the model assumptions need to be standardized or "shrunk," to apportion the variance appropriately at each level of the model (Goldstein 2011). If the primary interest is in the random effects, then Bayesian methods described in Chapters 12–14 may be more appropriate than the frequentist approach.

11.8   Exercises

11.1 The measurement of left ventricular volume of the heart is important for studies of cardiac physiology and clinical management of patients with heart disease. An indirect way of measuring the volume, y, involves a mea-

Table 11.9 *Measurements of left ventricular volume and parallel conductance volume on five dogs under eight different load conditions: data from Boltwood et al. (1989).*

Dog		1	2	3	4	5	6	7	8
					Conditions				
1	y	81.7	84.3	72.8	71.7	76.7	75.8	77.3	86.3
	x	54.3	62.0	62.3	47.3	53.6	38.0	54.2	54.0
2	y	105.0	113.6	108.7	83.9	89.0	86.1	88.7	117.6
	x	81.5	80.8	74.5	71.9	79.5	73.0	74.7	88.6
3	y	95.5	95.7	84.0	85.8	98.8	106.2	106.4	115.0
	x	65.0	68.3	67.9	61.0	66.0	81.8	71.4	96.0
4	y	113.1	116.5	100.8	101.5	120.8	95.0	91.9	94.0
	x	87.5	93.6	70.4	66.1	101.4	57.0	82.5	80.9
5	y	99.5	99.2	106.1	85.2	106.3	84.6	92.1	101.2
	x	79.4	82.5	87.9	66.4	68.4	59.5	58.5	69.2

surement called *parallel conductance volume*, x. Boltwood et al. (1989) found an approximately linear association between y and x in a study of dogs under various "load" conditions. The results, reported by Glantz and Slinker (1990), are shown in Table 11.9.

(a) Conduct an exploratory analysis of these data.

(b) Let (Y_{jk}, x_{jk}) denote the kth measurement on dog j, $(j = 1, \ldots, 5; k = 1, \ldots, 8)$. Fit the linear model

$$E(Y_{jk}) = \mu = \alpha + \beta x_{jk}, \qquad Y \sim N(\mu, \sigma^2),$$

assuming the random variables Y_{jk} are independent (i.e., ignoring the repeated measures on the same dogs). Compare the estimates of the intercept α and slope β and their standard errors from this pooled analysis with the results you obtain using a data reduction approach.

(c) Fit a suitable random effects model.

(d) Fit a clustered model using a GEE.

(e) Compare the results you obtain from each approach. Which method(s) do you think are most appropriate? Why?

11.2 Suppose that (Y_{jk}, x_{jk}) are observations on the kth subject in cluster k (with $j = 1, \ldots, J; k = 1, \ldots, K$) and the goal is to fit a "regression through the origin" model

$$E(Y_{jk}) = \beta x_{jk},$$

where the variance–covariance matrix for Y's in the same cluster is

$$V_j = \sigma^2 \begin{bmatrix} 1 & \rho & \cdots & \rho \\ \rho & 1 & & \rho \\ \vdots & & \ddots & \vdots \\ \rho & \rho & \cdots & 1 \end{bmatrix}$$

and Y's in different clusters are independent.

(a) From Section 11.3, if the Y's are Normally distributed, then

$$\widehat{\boldsymbol{\beta}} = \left(\sum_{j=1}^{J} \mathbf{x}_j^T \mathbf{V}_j^{-1} \mathbf{x}_j \right)^{-1} \left(\sum_{j=1}^{J} \mathbf{x}_j^T \mathbf{V}_j^{-1} \mathbf{y}_j \right) \text{ with } \text{var}(\widehat{\boldsymbol{\beta}}) = \left(\sum_{j=1}^{J} \mathbf{x}_j^T \mathbf{V}_j^{-1} \mathbf{x}_j \right)^{-1},$$

where $\mathbf{x}_j^T = [x_{j1}, \ldots, x_{jK}]$. Deduce that the estimate b of β is unbiased.

(b) As

$$\mathbf{V}_j^{-1} = c \begin{bmatrix} 1 & \phi & \cdots & \phi \\ \phi & 1 & & \phi \\ \vdots & & \ddots & \vdots \\ \phi & \phi & \cdots & 1 \end{bmatrix},$$

where

$$\frac{1}{\sigma^2[1+(K-1)\phi\rho]} \quad \text{and} \quad \phi = \frac{-\rho}{1+(K-2)\rho},$$

show that

$$\text{var}(b) = \frac{\sigma^2[1+(K-1)\phi\rho]}{\sum_j \{\sum_k x_{jk}^2 + \phi[(\sum_k x_{jk})^2 - \sum_k x_{jk}^2]\}}.$$

(c) If the clustering is ignored, show that the estimate b^* of β has $\text{var}(b^*) = \sigma^2 / \sum_j \sum_k x_{jk}^2$.

(d) If $\rho = 0$, show that $\text{var}(b) = \text{var}(b^*)$ as expected if there is no correlation within clusters.

(e) If $\rho = 1$, $\mathbf{V}_j / \sigma^2$ is a matrix of ones, so the inverse does not exist. But the case of maximum correlation is equivalent to having just one element per cluster. If $K = 1$, show that $\text{var}(b) = \text{var}(b^*)$, in this situation.

(f) If the study is designed so that $\sum_k x_{jk} = 0$ and $\sum_k x_{jk}^2$ is the same for all clusters, let $W = \sum_j \sum_k x_{jk}^2$ and show that

$$\text{var}(b) = \frac{\sigma^2[1+(K-1)\phi\rho]}{W(1-\phi)}.$$

Table 11.10 *Numbers of ears clear of acute otitis media at 14 days, tabulated by antibiotic treatment and age of the child. Data from Rosner (1989).*

| | CEF | | | | AMO | | | |
| | Number clear | | | | Number clear | | | |
Age	0	1	2	Total	0	1	2	Total
< 2	8	2	8	18	11	2	2	15
2–5	6	6	10	22	3	1	5	9
≥ 6	0	1	3	4	1	0	6	7
Total	14	9	21	44	15	3	13	31

(g) With this notation $\text{var}(b^*) = \sigma^2/W$; hence, show that

$$\frac{\text{var}(b)}{\text{var}(b^*)} = \frac{[1 + (K-1)\phi\rho]}{1 - \phi} = 1 - \rho.$$

Deduce the effect on the estimated standard error of the slope estimate for this model if the clustering is ignored.

11.3 Data on the ears or eyes of subjects are a classical example of clustering— the ears or eyes of the same subject are unlikely to be independent. The data in Table 11.10 are the responses to two treatments coded CEF and AMO of children who had acute otitis media in both ears (data from Rosner, 1989).

(a) Conduct an exploratory analysis to compare the effects of treatment and age of the child on the success of the treatments, ignoring the clustering within each child.

(b) Let Y_{ijkl} denote the response of the lth ear of the kth child in the treatment group j and age group i. The Y_{ijkl}'s are binary variables with possible values of 1 denoting cured and 0 denoting not cured. A possible model is

$$\text{logit}\left(\frac{\pi_{ijkl}}{1 - \pi_{ijkl}}\right) = \beta_0 + \beta_1\text{age} + \beta_2\text{treatment} + b_k,$$

where b_k denotes the random effect for the kth child and β_0, β_1 and β_2 are fixed parameters. Fit this model (and possibly other related models) to compare the two treatments. How well do the models fit? What do you conclude about the treatments?

(c) An alternative approach, similar to the one proposed by Rosner, is to use nominal logistic regression with response categories 0, 1 or 2 cured ears for each child. Fit a model of this type and compare the results with those obtained in (b). Which approach is preferable considering the assumptions made, ease of computation and ease of interpretation?

Chapter 12

Bayesian Analysis

12.1 Frequentist and Bayesian paradigms

The methods presented in this book so far have been from the classical or **frequentist** statistical paradigm. The frequentist paradigm requires thinking about the random process that produces the observed data; for example, a single toss of a coin can result in one of two possible outcomes "head" or "tail." The frequentist imagines the random process being repeated a large number of times so that for the coin example the data comprise the number of heads in n tosses. The proportion of heads is y/n. The frequentist formulates the concept of the probability of a head π as the value of y/n as n becomes infinitely large. Consequently the parameter π can be estimated from the observed data, $\hat{\pi} = y/n$, for example, by maximising the likelihood function $L(\pi; y, n)$ as described in Section 1.6.1.

This idea of repeating the random process that produces the data is fundamental to the frequentist approach, and it underlies the definition of two commonly used statistics, namely p-values and confidence intervals.

12.1.1 Alternative definitions of p-values and confidence intervals

The frequentist definitions of a p-value and 95% confidence interval are

a. A p-value is the probability of observing more extreme data (if the random process were repeated) given that the null hypothesis is correct.

b. A 95% confidence interval is an interval that contains the true value on 95% of occasions, if the random process could be repeated many times. By definition, a 95% confidence interval does not include the true value on 5% of occasions.

These definitions depend on the concept of multiple repetitions of the random process that yields the data. Alternative, arguably more natural, definitions rely on just the actual observed data.

Ideal definitions of a p-value and 95% confidence interval are

a. A p-value is the estimated probability that the null hypothesis is true, given the observed data.

b. A 95% confidence interval is an interval that contains the true value with a probability of 0.95.

In practice frequentist p-values and confidence intervals are often misinterpreted to fit these ideals (Goodman 2008). These ideal definitions are possible under an alternative statistical paradigm: **Bayesian analysis.**

12.1.2   Bayes' equation

The difference between frequentist and Bayesian paradigms is best explained using Bayes' equation,

$$P(\theta|\mathbf{y}) = \frac{P(\mathbf{y}|\theta)P(\theta)}{P(\mathbf{y})}, \tag{12.1}$$

where θ is an unknown parameter (assuming a single parameter) and $\mathbf{y}$ is the observed data. Because the denominator on the right side of this equation is the same for every value of the numerator, Equation (12.1) can be simplified to

$$P(\theta|\mathbf{y}) \propto P(\mathbf{y}|\theta)P(\theta), \tag{12.2}$$

where $\propto$ means "proportional to." This equation gives the **posterior** probability of θ, $P(\theta|\mathbf{y})$, as a function of the **likelihood**, $P(\mathbf{y}|\theta)$, and **prior**, $P(\theta)$ (*posterior* and *prior* are hereafter used as nouns). The likelihood is the key function in frequentist methodology and is the conditional probability of the data dependent on the parameter. The frequentist p-value is based on the same conditioning as it is the probability of observing more extreme data conditioned on a specified range for the parameter. Under frequentist methodology the data are random (i.e., subject to change) and the parameter is fixed.

Bayes' equation reverses this conditioning, as the value of the parameter θ is dependent on the data $\mathbf{y}$. Under Bayesian methodology the parameter is a random variable and the data are fixed. Most people find this view more intuitive, as once data are created they cannot be recreated (even under exactly the same conditions).

Because the parameter is a random variable under Bayesian methodology, this allows us to make the ideal statements concerning the probability of θ and intervals of likely values (Section 12.1.1). In Bayesian methodology the 95% **posterior interval** (or 95% **credible interval**) for θ has a 0.95 probability of containing the true value.

12.1.3 Parameter space

The denominator of Equation (12.1) is the probability of the data. This probability can be evaluated by conditioning on θ to give

$$P(\mathbf{y}) = \sum P(\mathbf{y}|\theta)P(\theta),$$

where the summation is over all possible values of θ. This range of possible values is called the **parameter space** and is denoted by Θ.

Using the expanded version of the denominator in Equation (12.1) gives

$$P(\theta|\mathbf{y}) = \frac{P(\mathbf{y}|\theta)P(\theta)}{\sum P(\mathbf{y}|\theta)P(\theta)}. \tag{12.3}$$

This equation shows that Bayes' equation evaluates the combination of likelihood and prior of a specific value for θ relative to every other value. Also if the terms in (12.3) are summed over all possible values of θ, then $\sum P(\theta|\mathbf{y}) = 1$, and hence, $P(\mathbf{y})$ is known as the normalising constant.

In this case the parameter space is assumed to be discrete so it can be summed. For a continuous probability density, an integral is used in Equation (12.3) rather than a sum.

The size of the parameter space is often very large and can become extremely large for problems involving multiple parameters $(\theta_1, \ldots, \theta_m)$. The number of calculations needed to evaluate Equation (12.3) can hence be very large. For problems with large parameter spaces or complex likelihoods, it may even be infeasible to calculate Equation (12.3) exactly. This is the reason that Bayesian analysis has become popular only in the last 20 years; an increase in computing power was needed to allow previously intractable problems to be estimated using simulation techniques. These estimation techniques are discussed in the next chapter.

12.1.4 Example: Schistosoma japonicum

Here is a simple numerical example of Bayes' equation and the effect of prior probabilities. *Schistosoma japonicum* is a zoonotic parasitic worm, and humans become infected when they are exposed to fresh water infested with the infective larval stage (*cercariae*) of the parasite, which is released by the intermediate host, an amphibious snail. Infection is considered endemic in a village if over half of the population are infected. Two mutually exclusive hypotheses for a new village are as follows:

H_0: Infection is not endemic ($\theta \leq 0.5$),
H_1: Infection is endemic ($\theta > 0.5$).

Table 12.1 *Calculations for updating the investigator's priors for H_0 and H_1 to posteriors using the observed data on positive samples for* Schistosoma japonicum.

θ	Hypothesis	$P(\theta)$ Prior	$P(y\|\theta)$ Likelihood	$P(y\|\theta) \times P(\theta)$ Likelihood×Prior	$P(\theta\|y)$ Posterior
0.0	H_0	0.0333	0.0000	0.0000	0.0000
0.1	H_0	0.0333	0.0000	0.0000	0.0000
0.2	H_0	0.0333	0.0008	0.0000	0.0002
0.3	H_0	0.0333	0.0090	0.0003	0.0024
0.4	H_0	0.0333	0.0425	0.0014	0.0114
0.5	H_0	0.0333	0.1172	0.0039	0.0315
Sum		0.2000			0.0455
0.6	H_1	0.1600	0.2150	0.0344	0.2771
0.7	H_1	0.1600	0.2668	0.0427	0.3439
0.8	H_1	0.1600	0.2013	0.0322	0.2595
0.9	H_1	0.1600	0.0574	0.0092	0.0740
1.0	H_1	0.1600	0.0000	0.0000	0.0000
Sum		0.8000		0.1241	0.9545

An investigator arrives at a village and after observing the villagers' contact with water, she is 80% sure that infection will be endemic in this village. The investigator takes stool samples from 10 villagers and 7 are positive for the parasite.

The calculations for updating the investigator's prior estimates of H_0 and H_1 to posterior estimates are given in Table 12.1. Assuming independence between stool samples, the data y have the Binomial distribution, $\text{Bin}(10, \theta)$. This means the likelihood is (Section 3.2.3)

$$P(y|\theta) = \binom{10}{7} \theta^7 (1 - \theta)^3,$$

and the highest likelihood in the table is for $\theta = 0.7$. In the third column the prior probability 0.8 for H_1 is uniformly distributed over the five values $0.6, \ldots, 1.0$, while the prior probability of 0.2 for H_0 is uniformly distributed over the six values $0.0, \ldots, 0.5$. The posterior probabilities in the last column are calculated from Equation (12.3) with $\sum P(y|\theta)P(\theta) = 0.1241$. The table shows how the investigator's probability for H_1 has risen from 0.8 to 0.955. This increase can be understood by looking at the fourth column which shows the majority of the likelihood in the H_1 parameter space.

This high level of probability for H_1 is enough for the investigator to stop testing and declare the infection to be endemic in the village. As a comparison, if the investigator was unable to decide whether infection was endemic in a new village, she might give a prior probability of 0.5 to both H_0 and H_1. With 7 infections out of 10 samples, the posterior probability of H_1 is now only 0.84. This level of probability may lead to further testing before either hypothesis can be accepted with sufficient surety. The likely number of further tests needed can be calculated for the prior of 0.5 using the assumption that the percentage of positive stool samples remains the same in further testing (i.e., 70%). To reach the threshold that $P(H_1) > 0.95$, the investigator would need to test 12 more samples (22 in total), assuming 16 positive samples.

In this example, using prior information concerning the conditions for infection has reduced the sample size needed. This could have particular advantages if speed of data collection is an issue or the test is expensive or uncomfortable. The prior information represents the investigator's field experience. If this prior is wrong (e.g., overly optimistic about H_1), then any inference may be wrong. Similarly if the data are wrong (e.g., collected using a biased sample), then any inference may be wrong.

To simplify this example, a discrete parameter space is used for θ with widely spaced values (i.e., $0.0, 0.1, \ldots, 1.0$). It would have been more accurate to use a finer resolution of probability (e.g., $0.00, 0.01, 0.02, \ldots, 1.00$), but this would have increased the size of Table 12.1. This illustrates how the calculations needed for a Bayesian analysis can greatly increase as the likelihood and posterior need to be evaluated over the entire parameter space. See Exercise 12.1 for more details.

12.2 Priors

The clearest difference between Bayesian methods and frequentist methods is that Bayesian methodology combines a **prior** with the observed data to create a **posterior**. Heuristically Equation (12.2) shows that the posterior estimate of θ is dependent on a combination of the data (via the likelihood) and the prior probability. If the prior is constant for all values of θ so that $P(\theta) = c$, then $P(\theta|\mathbf{y}) \propto P(\mathbf{y}|\theta)$. So using a flat or **uninformative prior** gives a posterior that is completely dependent on the data. Many practical Bayesian analyses use uninformative priors, and hence the posterior is exclusively influenced by the data. In this case the results from Bayesian and frequentist analysis are often very similar, although the results from the Bayesian analysis still provide the ideal interpretation for p-values and posterior intervals (Section 12.1.1).

The influence of priors depends on their relative weighting. In the example in the previous section on *Schistosoma japonicum*, the prior probability of H_1 was 4 times that of H_0 ($P(H_1)/P(H_0) = 4$), reflecting the investigator's strong opinion about the village. So the parameter space for θ from 0.6 to 1 was given much greater weight than the parameter space from 0 to 0.5. An uninformative prior would have equally weighted the two hypotheses so that $P(H_1)/P(H_0) = 1$. The further this ratio moves from 1 (in either direction) the greater the influence of the prior on the posterior (see Exercise 12.1).

Many people feel uncomfortable about incorporating prior information into an analysis. There is a feeling that using prior information (especially that based on a personal experience) is unscientific or biased. The reverse to this argument is that most people interpret new results in light of their prior experience, and that Bayesian methodology offers a way to quantify this experience.

Priors do not necessarily reflect an opinion as they can also be used to incorporate information about the model or study design (Berger 1985). An example is shown later in this chapter and in Section 14.2.

Informative and uninformative priors are discussed next.

12.2.1   Informative priors

The example using *Schistosoma japonicum* infection illustrated the use of an informative prior. The prior represented the investigator's belief about the conditions for infection in the village. In a classical statistical analysis the initial belief is that either hypothesis (the null or alternative) could be equally likely. In the *Schistosoma japonicum* example the initial belief was that hypothesis H_1 was much more likely.

In practice the use of informative priors based on opinion is uncommon, although two examples are given below. For both examples the results are also given using an uninformative prior. This is considered to be good practice when using informative priors so that readers who have no prior opinion can interpret the data from their neutral standpoint. Some may also want to view the data from a neutral standpoint so that it can more easily be compared with other studies. Showing the results based on an uninformative prior also helps to quantify the effect of an informative prior.

12.2.2   Example: Sceptical prior

Parmar et al. (1994) describe a clinical trial comparing a new cancer treatment with an existing treatment. The 11 specialists taking part in the trial were asked how effective they thought the new treatment would be compared with the conventional therapy. Historical data on conventional therapy gave an estimated proportion of patients surviving to two years of 0.15. The median response from the specialists was that they expected a 10% absolute improvement over conventional therapy so that the proportion surviving two years would be 0.25. This is an **enthusiastic prior** opinion.

From Equation (10.4) the (cumulative) hazard function is $H_k = -\log(P_k)$, where P_k, the survivor function, is the probability of survival beyond two years for condition k, where $k = 1$ for the conventional therapy and $k = 2$ for the new treatment. The log hazard ratio, LHR, is

$$\text{LHR} = \log\left(\frac{H_1}{H_2}\right) = \log\left(\frac{-\log P_1}{-\log P_2}\right). \tag{12.4}$$

If $P_1 = 0.15$ and $P_2 = 0.25$ then LHR $= 0.3137$. The reason for using the log hazard ratio is that there is good empirical evidence that it is approximately Normally distributed which will make the computation easier.

Many previous new treatments had been only marginally effective which meant that there was a general feeling of scepticism in the wider population of clinicians. For any new treatment to become part of routine practice it would need to show a large enough improvement to convince these sceptics. A prior for the sceptics (**sceptical prior**) would be that there was no improvement and that the probability that two-year survival would be improved by more than 10% (i.e., LHR $= 0.3137$) would be 0.05. This prior can be represented by a Normal distribution with a mean of zero and a standard deviation calculated by setting the 95th percentile 1.645σ equal to 0.3137 to obtain $\sigma = 0.1907$.

After two years there were 78 deaths among 256 patients and the observed log hazard ratio was LHR $= 0.580$ with a 95% confidence interval of $(0.136, 1.024)$ (or 0.580 ± 0.444, so the estimated standard deviation was $\sigma = 0.2266$). Using $P_1 = 0.15$ for the conventional therapy and re-arranging Equation (12.4) gives $P_2 = \exp(\log P_1 / \exp \text{LHR}) = \exp(\log 0.15 / \exp 0.58) = 0.3457$, which represents an improvement of approximately 20% (or double the median for the enthusiastic prior).

The posterior density for the LHR can be calculated using Bayes' Equation (12.3) and Normal distributions for the data (i.e., the likelihood) and the sceptical prior

$$P(\mathbf{y}|\theta)P(\theta) = \frac{1}{c} \times \frac{1}{0.2266\sqrt{2\pi}} \exp\left(-\frac{(\theta - 0.58)^2}{2 \times 0.2266^2}\right)$$
$$\times \frac{1}{0.1907\sqrt{2\pi}} \exp\left(-\frac{\theta^2}{2 \times 0.1907^2}\right),$$

where c is the normalizing constant which can be calculated by integrating the remainder of the expression on the right-hand side of this equation over a suitably wide range, for example, from LHR $= -1.15$ (corresponding to $P_2 =$ 0.0025 or almost no survivors) to LHR $= +1.15$ (corresponding to $P_2 = 0.55$ or a 40% improvement).

The sceptical prior, likelihood and posterior distributions are plotted in Figure 12.1 showing both the LHR and corresponding absolute improvements on the horizontal axis. To contrast the results of using a sceptical prior, the right-hand panel of Figure 12.1 shows an uninformative prior and the likelihood and resulting posterior, which completely overlap—this corresponds to the classical frequentist analysis.

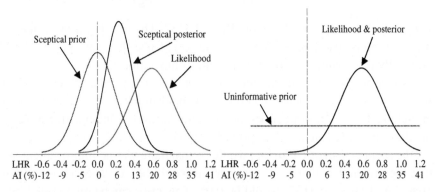

LHR -0.6 -0.4 -0.2 0.0 0.2 0.4 0.6 0.8 1.0 1.2   LHR -0.6 -0.4 -0.2 0.0 0.2 0.4 0.6 0.8 1.0 1.2
AI (%)-12 -9   -5   0   6 13 20 28 35 41   AI (%)-12 -9   -5   0   6 13 20 28 35 41

Figure 12.1 *Differences in new vs. standard treatment in terms of Log Hazard Ratio (LHR) and Absolute Improvement (AI) in survival using sceptical (left panel), and uninformative priors (right panel). (Data from Parmar et al. (1994).)*

The sceptical posterior is between the sceptical prior and the likelihood. The mean is at LHR = 0.240, corresponding to an improvement of 7% for the new treatment. The probability that improvement is greater than 10% (corresponding to LHR = 0.3137) is 0.31 from the sceptical posterior compared with 0.05 from the sceptical prior, but it was still unlikely to provide sufficiently convincing evidence for the new treatment.

Using the posterior distribution from an uninformative prior, the probability that the improvement in survival for the new treatment is greater than 10% is 0.88, and the probability that the improvement is positive is 0.99. Using a classical statistical analysis the new treatment might well be considered superior, but only for those people who started from a neutral standpoint.

In this example the Bayesian analysis focused on the decision to use a new treatment and not on the statistical significance of the trial. Rather than the aim being to demonstrate that one treatment was better than another, the aim was to demonstrate that one treatment could be adopted over another. In practice, decisions to adopt new treatments are often made by combining the results from several trials (using a meta-analysis) or from a weight of evidence, rather than the results of one trial. This is because the results of several trials are often more convincing than a single trial. The aim of this study design was to have a single trial that would lead to a change in practice if the evidence for the treatment was convincing enough.

The posterior distribution using the Normally distributed sceptical prior had a Normal distribution (see Exercise 12.2). This means the prior and posterior are **conjugate distributions** and this applies when the posterior distribution is in the same family as the prior distribution. Conjugacy is useful because it gives posterior results with an exact distribution which can then be used for inference (Gelman et al. 2013). This also avoids the need to estimate the often complicated normalising constant (although this was evaluated in the above example for illustrative purposes).

12.2.3 Example: Overdoses amongst released prisoners

A study in Australia followed a sample of 160 released prisoners (Kinner 2006). An important outcome in this population is drug overdose, but after the first follow-up period of 4 weeks post-release (which is a key risk period), nobody had overdosed out of the 91 subjects who could be contacted.

Assuming independence between prisoners, the number y who overdose could be assumed to have the Binomial distribution $\text{Bin}(n, \theta)$. A frequentist analysis would give an estimate of $\hat{\theta} = y/n = 0/91 = 0$ and the variance of this estimator $\text{var}(\hat{\theta}) = \theta(1 - \theta)/n$ would also be estimated to be zero. These are not realistic estimates as it is well known that overdoses do occur in this population.

Bayesian analysis can cope with such zero responses because Bayesian analysis does not base parameter estimates solely on the data, but is able to incorporate prior information. In this case a prior was used to ensure that $0 < \hat{\theta} < 1$ and hence that $\text{var}(\hat{\theta}) > 0$. This is an example of using a prior to incorporate information on the model rather than prior opinion.

The **Beta distribution** is a useful prior for Binomial data

$$\theta \sim \text{Be}(\alpha, \beta), \text{ for } \theta \text{ in } [0, 1], \alpha > 0, \beta > 0;$$

see Exercise 7.5. The probability density function is

$$P(\theta) \propto \theta^{\alpha-1}(1 - \theta)^{\beta-1}, \tag{12.5}$$

so from Equation (12.2) the posterior distribution has the form

$$
\begin{aligned}
P(\theta|\mathbf{y}) &\propto P(\mathbf{y}|\theta)P(\theta), \\
&\propto \theta^y(1-\theta)^{n-y}\theta^{\alpha-1}(1-\theta)^{\beta-1}, \\
&= \theta^{y+\alpha-1}(1-\theta)^{n-y+\beta-1}, \tag{12.6}
\end{aligned}
$$

which is also a Beta distribution $Be(y+\alpha, n-y+\beta)$. So a Beta prior is conjugate with Binomial data.

Heuristically a $Be(\alpha,\beta)$ prior represents observing $\alpha - 1$ previous "successes" and $\beta - 1$ previous "failures" (although this is only true for relatively large values of α and β; see Exercise 12.4). A $Be(1,1)$ prior therefore corresponds to previously seeing no successes or failures. This corresponds to knowing nothing, which is an ideal uninformative prior. This can be verified by noting that the Beta probability density (12.5) is constant over the range $0 \leq \theta \leq 1$ when $\alpha = \beta = 1$.

If we use the uninformative prior $Be(1,1)$ and the data that of $n = 91$ subjects $y = 0$ overdosed, from Equation (12.6) this gives the posterior distribution $Be(1,92)$. From Exercise 7.5 if $\theta \sim Be(\alpha,\beta)$, then $E(\theta) = \alpha/(\alpha+\beta)$, giving the estimate $\hat{\theta} = 1/93 = 0.01075$. To obtain a 95% posterior interval we need to obtain values corresponding to the 0.025 and 0.975 cumulative probabilities (for example, using the R function qbeta). These are 0.000275 and 0.0393.

The chief investigator expected around one overdose per 200 released prisoners or $E(\theta) = 1/200$, which corresponds to $Be(1,199)$. Combining this prior with the data gives the posterior $Be(1,290)$ and hence $\hat{\theta} = 1/291 = 0.003436$ and 95% posterior limits of 0.0000783 and 0.01264 (Table 12.2).

Table 12.2 *Priors and posteriors for the overdose example.*

Prior		Posterior for overdoses		
Name	Distribution	Distribution	Mean	95% posterior interval
Uninformative	Be(1,1)	Be(1,92)	0.0108	0.0003, 0.0393
Investigator's	Be(1,199)	Be(1,290)	0.0034	0.0001, 0.0126

In frequentist analysis the problem of a zero numerator (or zero cell in a two-by-two table) is often overcome by adding a half to the number of events. For this example using a half as the number of overdoses gives an estimated overdose rate of 0.0055 (or 1 overdose in 182 subjects), with a 95% confidence interval of 0.00014 to 0.03023 (obtained using exact methods, for example, using the Stata command: `cii 182 1, level(95) exact`). These estimates are somewhat similar to the Bayesian estimates, particularly the 95% posterior interval using the uninformative prior. This is not surprising, as by adding a half to the number of overdoses we are saying that we do not believe the zero cell, and there was more data (or data from another prison) it would likely be some small positive value. Bayesian methods incorporate this extra belief in a more formal way, as adding a half is an arbitrary solution.

12.3   Distributions and hierarchies in Bayesian analysis

Bayesian analysis is founded on statistical distributions. Parameters are assigned a prior distribution and because of conjugacy, posterior distributions often follow a known distribution. The flow from prior to posterior is the natural hierarchy of a Bayesian analysis which is illustrated in Figure 12.2. This natural hierarchy makes it is easy to add additional layers as in Figure 12.3. The diagram shows an extra layer from adding a parameter below a prior. An example parameter is a transformation of a prior to a different scale. For example, in Section 12.2.2 the specialists' prior could be transformed from an absolute increase to the log hazard ratio.

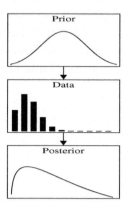

Figure 12.2 *The natural hierarchy of Bayesian analysis.*

12.4   WinBUGS software for Bayesian analysis

WinBUGS is a popular (and free) software package for analysing Bayesian models (Spiegelhalter et al. 2007). For a good practical introduction to Win-

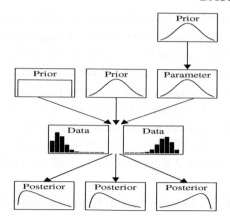

Figure 12.3 *Modelling complex hierarchies in Bayesian analysis.*

BUGS see Lunn et al. (2012). The hierarchies of Bayesian analysis can be visualized by using the DoodleBUGS feature in the WinBUGS language. A *Doodle* is a graphical model of the priors, parameters and data, and the links and hierarchies relating them (Gilks et al. 1996a). Figure 12.4 shows an example *Doodle* using the overdose data. In this graph r[i] is the Bernoulli variable representing overdose (yes/no) for released prisoner i, and p is the probability of overdose. Both these variables are stochastic *nodes*. The overdose node is indexed by each subject and so is within a *plate* (or loop) that goes from *1* to *N* (*N=91* in this example). A Bernoulli distribution is used for r_i with parameter p, and a Beta distribution for p with parameters $\alpha = \beta = 1$ to match the uninformative prior from Table 12.2. So the model in this notation is

$$r_i \sim \mathrm{B}(p), \qquad i = 1, \ldots, n,$$
$$p \sim \mathrm{Be}(1, 1).$$

Building a statistical model via a diagram is an approach that does not appeal to everyone. We can produce standard written code by using the "Write code" function from the *Doodle*, which gives the following:

```
_____ WinBUGS code (prisoner overdose model) _____
model{
   for (i in 1:N){
      r[i] ~ dbern(p);
```

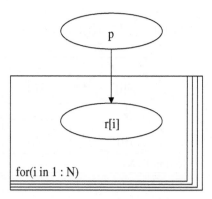

Figure 12.4 *Graphical model (Doodle) for the overdose example.*

```
    }
    p ~ dbeta(1,1);
}
```

In this simple model there is a loop through each Bernoulli response (dbern), and the parameter p has a flat Beta distribution (dbeta).

Figure 12.5 gives a *Doodle* for fitting a regression model to the beetle mortality data (Section 7.3.1). One of the models for these data in Chapter 7 for the number of beetles killed y_i out of n_i at dose x_i was

$$
\begin{aligned}
y_i &\sim \text{Bin}(n_i, \pi_i), \\
\text{logit}(\pi_i) &= \beta_1 + \beta_2 x_i, \qquad i = 1,\ldots,N.
\end{aligned}
$$

The probability π_i, written as pi[i], is a stochastic node and so is oval, whereas the dose x[i] is a fixed node and so is square. The intercept and slope parameters beta[1] and beta[2] are fixed (the same for all i) so they are outside the plate (loop). The number of deaths y[i] is linked to the probability and the number of beetles n[i].

The written code for this model is

— WinBUGS code (logit dose response model, Beetles data) —
```
model{
    for (i in 1:N){
        y[i]~dbin(pi[i],n[i]);
        logit(pi[i])<-beta[1]+beta[2]*x[i];
    }
    beta[1]~dnorm(0.0,1.0E-6);
```

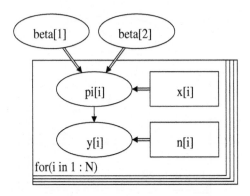

Figure 12.5 *Graphical model (Doodle) for the beetle mortality example.*

```
    beta[2]~dnorm(0.0,1.0E-6);
}
```

Both the beta parameters have a prior which is a Normal distribution (dnorm) with mean of zero and a very large variance (i.e., an uninformative prior). For a Normal distribution WinBUGS uses the accuracy, which is the inverse of the variance. This case uses a small accuracy ($1.0E-6 = 1 \times 10^{-6}$; WinBUGS defaults to scientific notation for small and large numbers).

The data in **S-PLUS format** (which WinBUGS uses) are

```
────────────────── Beetles data in S-PLUS format ──────────────────
list(y=c(6,13,18,28,52,53,61,60),n=c(59,60,62,56,63,59,62,60),
x=c(1.6907,1.7242,1.7552,1.7842,1.8113,1.8369,1.8610,1.8839),N=8)
```

12.5 Exercises

12.1 Reconsider Example 12.1.4 on *Schistosoma japonicum*.

a. Using Table 12.1 calculate the posterior probability for H_1 for the following priors and observed data:

	Observed data	
Prior	5 out of 10 positive	1 out of 10 positive
$P(H_1) = 0.5$		
$P(H_1) = 0.99$		

b. Recalculate the above probabilities using the finer parameter space of $\theta = 0.00, 0.01, 0.02, \ldots, 0.99, 1.00$. Explain the differences in your results.

12.2 Show that the posterior distribution using a Normally distributed prior $N(\mu_0, \sigma_0^2)$ and Normally distributed likelihood $N(\mu_1, \sigma_1^2)$ is also Normally distributed with mean

$$\frac{\mu_0 \sigma_1^2 + \mu_1 \sigma_0^2}{\sigma_0^2 + \sigma_1^2}$$

and variance

$$\frac{\sigma_1^2 \sigma_0^2}{\sigma_0^2 + \sigma_1^2}.$$

12.3 Reconsider Example 12.2.2 about the cancer clinical trial. The 11 specialists taking part in the trial had an enthusiastic prior opinion that the median expected improvement in survival was 10%, which corresponds to an LHR of 0.3137. Assume that their prior opinion can be represented as a Normal distribution with a mean of 0.3137 and a standard deviation of 0.1907 (as per the sceptics' prior).

a. What is their prior probability that the new treatment is effective?

b. What is their posterior probability that the new treatment is effective?

12.4 Reconsider Example 12.2.3 on overdoses among released prisoners. You may find the First Bayes software useful for answering these questions.

a. Use an argument based on $\alpha - 1$ previous successes and $\beta - 1$ previous failures to calculate a heuristic Beta prior. Combine this prior with the data to give a Beta posterior.

b. Calculate the mean of the posterior. Compare this mean to the investigator's prior opinion of 1 overdose in 200 subjects (0.005). Considering that there were no overdoses in the data, what is wrong with this posterior?

Chapter 13

Markov Chain Monte Carlo Methods

13.1 Why standard inference fails

For a continuous parameter space, Bayes' formula for the posterior distribution is

$$P(\theta|\mathbf{y}) = \frac{P(\mathbf{y}|\theta)P(\theta)}{\int P(\mathbf{y}|\theta)P(\theta)d\theta}. \tag{13.1}$$

To use this equation, the normalising constant (the denominator) must be estimated. Unfortunately because of its complexity, the normalising constant cannot be calculated explicitly for all models (Gilks et al. 1996a). If there are m unknown parameters $(\theta_1, \ldots, \theta_m)$, then the denominator involves integration over the m-dimensional parameter space which becomes intractable for large values of m.

In this chapter we introduce a numerical method for calculating complex integrals and hence making inference about $\boldsymbol{\theta}$. The method is called **Markov chain Monte Carlo** (MCMC) and combines two methods: Monte Carlo integration and Markov chain sampling. In the previous chapters the unknown parameters were estimated using the methods of maximum likelihood (Section 1.6.1) and used algorithms such as the Newton–Raphson to find the maximum likelihood (Section 4.2). MCMC has some parallels to these methods (as shown below) but involves many more iterations and so is much more computer intensive. The recent rise in computer power and refinement of MCMC methods has led to an increase in applications using MCMC, including Bayesian methods.

This chapter gives a brief introduction to MCMC methods with a focus on practical application. For a more detailed description of MCMC methods, see Brooks (1998), Gelman et al. (2013, Chapters 10–11), or Gilks et al. (1996b).

13.2 Monte Carlo integration

Monte Carlo integration is a numerical integration method which simplifies a continuous distribution by taking discrete samples. It is useful when a continuous distribution is too complex to integrate explicitly but can readily be sampled (Gelman et al. 2013).

As an example of a relatively complex continuous distribution, a bimodal probability density is shown in Figure 13.1. This bimodal distribution was

(a)   (b)

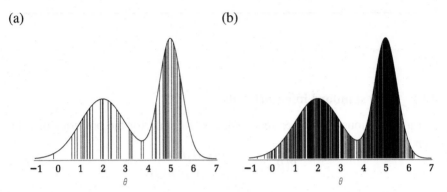

Figure 13.1 *A continuous bimodal distribution and discrete random samples, for sample sizes of (a) 50 and (b) 500.*

generated using two Normal distributions:

$$\theta = Y_1 + Y_2, \quad Y_1 \sim N(2,1), \quad Y_2 \sim N(5,0.5^2).$$

Also shown are discrete random samples from this density (vertical lines), labelled $\theta^{(1)}, \ldots, \theta^{(M)}$. Sample sizes of $M = 50$ and $M = 500$ are shown. Notice how the samples are more frequent at the modes, particularly the higher mode ($\theta = 5$). This is because the values of $\theta^{(i)}$ were selected based on the probability density, $P(\theta)$, so more samples are made where $P(\theta)$ is relatively high.

The samples in Figure 13.2 use a histogram to approximate the continuous distribution of $P(\theta)$. The larger sample more closely approximates the continuous distribution. With the larger sample a narrower bin width of 0.2 was used which gives a smoother looking distribution.

If a histogram is a reasonable approximation to a continuous distribution, then any inferences about $P(\theta)$ can be made by simply using the sampled values. For example, the mean of θ is estimated using the mean of the sampled values

$$\hat{\bar{\theta}} = \frac{1}{M} \sum_{i=1}^{M} \theta^{(i)}. \tag{13.2}$$

Using our sampled data this gives an estimate for the mean of 3.21 for $M = 50$

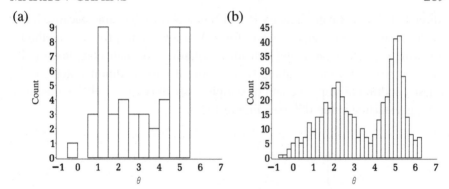

Figure 13.2 *Approximating the continuous bimodal distribution in Figure 13.1 using discrete random samples, for sample sizes (and bin widths) of (a) 50 (0.5) and (b) 500 (0.2).*

and 3.43 for $M = 500$ compared with a true mean of 3.5. Similarly calculating the median of $P(\theta)$ by taking the central value from the samples gives an estimate for the median of 3.28 for $M = 50$ and 3.57 for $M = 500$. The true median of this bimodal distribution is 4.0, and hence our estimates are relatively poor. This is because the true median has a low sample density being between the two modes (Figure 13.1), and hence there were few observed samples in that region. The first quartile of the distribution is 2.0, which is one of the modes. The estimates of the first quartile were 1.44 for $M = 50$ and 2.03 for $M = 500$.

Higher values of M lead to a closer approximation to the continuous distribution. This is easy to see in Figure 13.2. When using Monte Carlo integration, we control the value of M and it may seem best to make M very large. However, higher values of M lead to greater computing time (which can be significant for complex problems). Ideally a value of M is chosen that gives a good enough approximation to the continuous density for the minimum number of samples.

A similar numerical integration technique was used for the *Schistosoma japonicum* example (Section 12.1). In that example the posterior probabilities were estimated by evaluating Equation (13.1) at 11 discrete points (Table 12.1). A regular spaced set of values was used for θ rather than selecting M values at random (see Exercise 13.1).

13.3 Markov chains

In the previous example, samples were drawn from a bimodal probability density $P(\theta)$ and used to make inferences about θ. However, drawing samples directly from the **target density** $P(\theta)$ is not always achievable because it may have a complex, or even unknown, form. Markov chains provide a method of drawing samples from target densities (regardless of their complexity). The method simplifies the sampling by breaking it into conditional steps. Using these conditional steps, a chain of samples is built $(\theta^{(1)}, \ldots, \theta^{(M)})$ after specifying a starting value $\theta^{(0)}$ (see Figure 13.3).

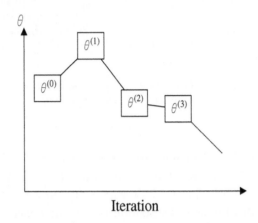

Figure 13.3 *A simple example of a Markov chain.*

A Markov chain is a stochastic process whose future state is only dependent on the current state and is independent of the past, so

$$P(\theta^{(i)} = a|\theta^{(i-1)}, \theta^{(i-2)}, \ldots, \theta^{(0)}) = P(\theta^{(i)} = a|\theta^{(i-1)}).$$

This lack of dependence on the past is called the Markov property (Bartlett 1978). It is this property that allows Markov chains to simplify complex problems, because the next sample in the chain depends only on the previous sample.

An algorithm for creating a Markov chain for a target probability density $P(\theta)$ is

1. Choose an initial value $\theta^{(0)}$. The restriction on the initial value is that it needs to be within the distribution of $P(.)$, so that $P(\theta^{(0)}) > 0$.

2. Create a new sample using $\theta^{(1)} \sim f(\theta^{(1)}|\theta^{(0)}, \mathbf{y})$, where $\mathbf{y}$ is the data.

3. Repeat step 2 M times, each time increasing both indices by 1.

$f(\theta^{(i+1)}|\theta^{(i)})$ is the **transitional density**; it controls the movement, or transition, from $\theta^{(i)}$ to $\theta^{(i+1)}$. The transitional density may have a standard distribution (e.g., Normal) if standard statistical models are used (Brooks 1998).

A *sample* is the generation of an individual value at step 2 (e.g., $\theta^{(1)}$). For problems involving more than one parameter, there will be multiple samples (e.g., $\theta^{(1)}$ and $\beta^{(1)}$). An *iteration* is a complete cycle through every unknown parameter.

The sampling at step 2 is random. There will be many possible values for $\theta^{(i+1)}$, determined by $f(\theta^{(i+1)}|\theta^{(i)})$. An actual value is randomly sampled using pseudo-random numbers. This means that it is possible to obtain many different Markov chains for the same problem. Ideally any chain would give an equally good approximation to the target density $P(\theta)$.

There are a number of different sampling algorithms that can be used for step 2. In the next sections brief introductions are given to two of the most popular algorithms. The algorithms are illustrated using the dose-response model for the beetle mortality data (Section 7.3).

13.3.1 The Metropolis–Hastings sampler

The Metropolis–Hastings sampler works by randomly proposing a new value θ^*. If this proposed value is accepted (according to a criterion below), then the next value in the chain becomes the proposed value $\theta^{(i+1)} = \theta^*$. If the proposal is rejected, then the previous value is retained $\theta^{(i+1)} = \theta^{(i)}$. Another proposal is made and the chain progresses by assessing this new proposal.

One way of creating proposals is to add a random variable to the current value,

$$\theta^* = \theta^{(i)} + Q.$$

Q could be chosen from the standard Normal distribution $Q \sim N(0,1)$ so that proposals closer to the current value are more likely, or from a Uniform distribution $Q \sim U[-1,1]$ so that all proposals within one unit of the current value are equally likely. The probability distribution of Q is called the **proposal density**.

The acceptance criterion is

$$\theta^{(i+1)} = \begin{cases} \theta^*, & \text{if } U < \alpha \\ \theta^{(i)}, & \text{otherwise,} \end{cases}$$

where U is a randomly drawn from a Uniform distribution between 0 and 1 ($U \sim U[0,1]$) and α is the acceptance probability given by

$$\alpha = \min\left\{ \frac{P(\theta^*|\mathbf{y})}{P(\theta^{(i)}|\mathbf{y})} \cdot \frac{Q(\theta^{(i)}|\theta^*)}{Q(\theta^*|\theta^{(i)})}, 1 \right\},$$

where $P(\theta|\mathbf{y})$ is the probability of θ given the data $\mathbf{y}$ (the likelihood). If the proposal density is symmetric (so that $Q(a|b) = Q(b|a)$), then α simplifies to

$$\alpha = \min\left\{ \frac{P(\theta^*|\mathbf{y})}{P(\theta^{(i)}|\mathbf{y})}, 1 \right\}, \tag{13.3}$$

which is the likelihood ratio.

An example of Metropolis–Hastings sampling is shown in Figure 13.4, based on estimating β_2 from the extreme value model for the beetle mortality data. The likelihood $P(\beta_2|\mathbf{D})$ (where $\mathbf{D} = [\mathbf{y}, \mathbf{n}, \mathbf{x}]$) was calculated using Equation (7.3) with $g(\pi_i) = \beta_1 + \beta_2(x_i - \bar{x})$ and fixing $\beta_1 = 0$. The figure shows the current location of the chain, $\beta_2^{(i)} = 22.9$, and two proposals, $\beta_2^* = 23.4$ and $\beta_2^{**} = 22.1$. These proposals were generated using the current location plus a random number from the standard Normal density. As this is a symmetric proposal density the acceptance probability (13.3) can be used. This gives $\alpha = 0.83$ for β_2^* and $\alpha = 1$ for β_2^{**}. So β_2^* is accepted with probability 0.83 and β_2^{**} is always accepted (any proposal is accepted if $P(\theta^*|\mathbf{D}) > P(\theta^{(i)}|\mathbf{D})$). To accept or reject β_2^* a random number is generated $U \sim U[0,1]$. If U is less than 0.83, then the proposal would be accepted; otherwise it would be rejected. Because acceptance or rejection depends on a random number, it is possible to obtain two different chains from this value for $\beta_2^{(i)}$ and this proposal.

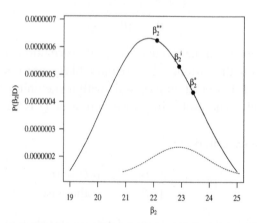

Figure 13.4 *Metropolis–Hastings sampling for β_2 for the beetle mortality example using the extreme value model. The plot shows the likelihood $P(\beta_2|\mathbf{D})$ (solid line), the proposal density $Q(.)$ (dotted line), the current location of the chain $\beta_2^{(i)} = 22.9$, and two different proposals $\beta_2^* = 23.4$ and $\beta_2^{**} = 22.1$.*

A realization of the first 200 Markov chain samples using Metropolis–Hastings sampling for β_1 and β_2 are shown in Figure 13.5. The initial values were $\beta_1^{(0)} = 0$ and $\beta_2^{(0)} = 1$. These samples were created using the MHadaptive library in R (Chivers 2012). Also shown is the log-likelihood.

There are sections of the chain for both β_1 and β_2 that look flat. This is because a number of proposed moves have been rejected, and the chain has remained in the same position for two or more iterations. The proposed values were rejected 105 times for the first 200 iterations.

The log-likelihood shows a big increase with increasing iterations, which then stabilizes after roughly 100 iterations. This is because the initial value for $\beta_2^{(0)} = 1$ was far from a reasonable estimate. From the starting value the chain for β_2 has moved steadily toward larger values. As the estimates for β_2 improve, there is a clear improvement in the log-likelihood.

Such initial periods of poor estimates are not uncommon because the initial values may be far from the maximum likelihood. These early estimates should not be used for any inference but instead discard as a **burn-in**. For this example the first 100 iterations would be discarded.

Reasonable initial estimates can come from summary statistics or non-Bayesian models. For example, for a model using $g(\pi_i) = \beta_1 + \beta_2 x_i$ the initial values of $\beta_1^{(0)} = -39$ and $\beta_2^{(0)} = 22$ from the estimates in Table 7.4 could have been used.

13.3.2   The Gibbs sampler

The Gibbs sampler is another way of generating a Markov chain. It splits the parameters into a number of components and then updates each one in turn (Brooks 1998). For the beetle mortality example, a Gibbs sampler to update the two unknown parameters would be

1. Assign an initial value to the two unknowns: $\beta_1^{(0)}$ and $\beta_2^{(0)}$.

2. a. Generate $\beta_2^{(1)} \sim f(\beta_2 | \mathbf{y}, \mathbf{n}, \mathbf{x}, \beta_1^{(0)})$.

   b. Generate $\beta_1^{(1)} \sim f(\beta_1 | \mathbf{y}, \mathbf{n}, \mathbf{x}, \beta_2^{(1)})$.

3. Repeat the step 2 M times, each time increasing the sample indices by 1.

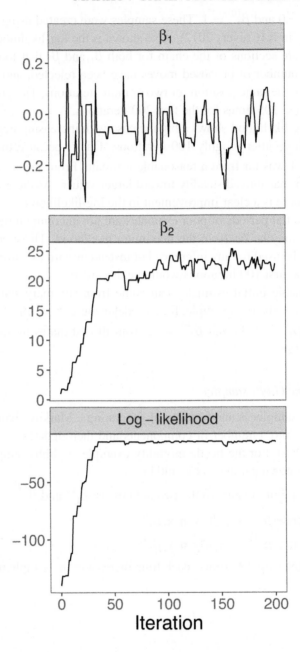

Figure 13.5 *Markov chain samples using Metropolis–Hastings sampling for* β_1, β_2 *and the log-likelihood for the beetle mortality example using the extreme value model with* $g(\pi_i) = \beta_1 + \beta_2(x_i - \bar{x})$.

It is possible to randomize the updating order of the β's in step 2, but this usually makes little difference. Looking at the algorithm it is clear why the Gibbs sampler is also known as alternating conditional sampling.

Two important differences between the Gibbs sampler and the Metropolis–Hastings sampler are (1) the Gibbs sampler always takes a step, whereas the Metropolis–Hastings may remain in the same position, and (2) the Gibbs sampler uses the full conditional distribution, not the marginal for the parameter being considered.

An example of Gibbs sampling using the beetle mortality data is shown in Figure 13.6. These results were generated using the extreme value model with the initial values $\beta_1^{(0)} = 0$ and $\beta_2^{(0)} = 1$ and so are comparable to the Metropolis–Hastings samples in Figure 13.5. Using the Gibbs sampler requires almost no burn-in for this example compared with the burn-in of around 100 for the Metropolis–Hastings samples. In this example the Gibbs sampler took only three iterations to recover from the poor starting values, compared with over 100 for the Metropolis–Hastings sampler. This one example does not mean that the Gibbs sampler is always better than the Metropolis–Hastings sampler.

13.3.3 Comparing a Markov chain to classical maximum likelihood estimation

The iterative process of a Markov chain algorithm is very similar to the iterative process used by some algorithms for obtaining maximum likelihood estimates. For example, in Section 4.2 an iterative Newton–Raphson algorithm is used to obtain maximum likelihood estimates for the scale parameter of a Weibull model. That algorithm started with an initial guess and iteratively worked toward a maximum in the likelihood. The iterations stopped when the ratio of derivatives from one iteration to the next became sufficiently small (this was the convergence criterion).

The Newton–Raphson algorithm can be used to find the maximum likelihood estimates of the model using the logit link for the beetle mortality data. A three-dimensional plot of the log-likelihood against the intercept (β_1) and effect of dose (β_2) is shown in Figure 13.7. The log-likelihood was calculated using Equation (7.4) and $\text{logit}(\pi_i) = \beta_1 + \beta_2(x_i - \bar{x})$. The plots show the first two iterations for two different pairs of initial values. The full iteration history is shown in Table 13.1; the third and fourth iterations were not plotted because they were so close to the second. Both results converged in 4 iterations to the same maximum log-likelihood of -18.715, corresponding to the estimates of $\hat{\beta}_1 = 0.74$ and $\hat{\beta}_2 = 34.27$. The estimates converged because the increase in

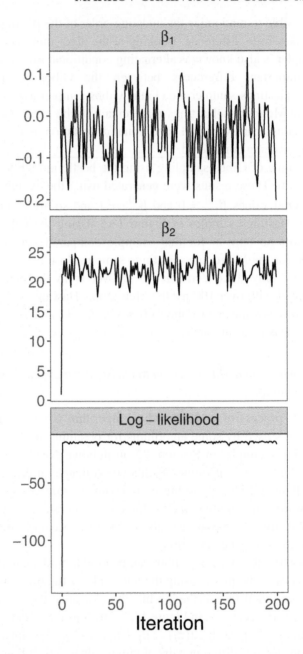

Figure 13.6 *Markov chain samples using Gibbs sampling for β_1, β_2 and the log-likelihood for the beetle mortality example using the extreme value link.*

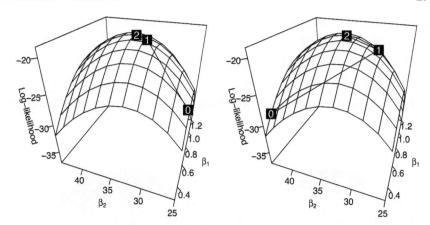

Figure 13.7 *Three-dimensional plot of the log-likelihood function and maximum likelihood iterations using a logit link for the beetle mortality data. Initial values are labelled 0.*

Table 13.1 *Maximum likelihood iterations for the beetle mortality data using the logit link* $logit(\pi_i) = \beta_1 + \beta_2(x_i - \bar{x})$ *for two pairs of starting values.*

Iteration	Intercept (β_1)	Dose (β_2)	Log-likelihood
0	1.20	26.00	−35.059
1	0.70	32.28	−18.961
2	0.74	34.16	−18.716
3	0.74	34.27	−18.715
4	0.74	34.27	−18.715
0	0.40	44.00	−29.225
1	0.81	29.67	−20.652
2	0.74	34.18	−18.716
3	0.74	34.27	−18.715
4	0.74	34.27	−18.715

the log-likelihood from iterations 3 to 4 was sufficiently small. This is akin to saying that the estimates cannot climb any higher up the maximum likelihood "hill."

Stata will show the details of the maximum likelihood iterations and start from a specific set of initial values using the following code:

```
──── Stata code (logit link with ML iteration history) ────
.generate float xdiff = x-1.793425
```

```
.glm y xdiff,family(binomial n) link(logit) trace
from(26 1.2, copy)
```

The iterations using the Metropolis–Hastings sampling algorithm to fit the same model are shown in Figure 13.8. The initial values were $\beta_1^{(0)} = 1.2$ and $\beta_2^{(0)} = 26$. The plots show the chain after 100 iterations and then 1000

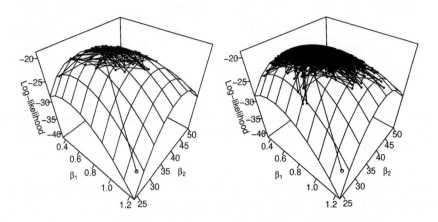

Figure 13.8 *Metropolis–Hastings iterations for the beetle mortality data using the logit link* $logit(\pi_i) = \beta_1 + \beta_2(x_i - \bar{x})$ *after 100 iterations (left panel), 1000 iterations (right panel). The initial value is shown as an open circle and subsequent estimates as closed circles.*

iterations. After 1000 iterations the peak of the log-likelihood has a dense covering of samples which can be used to build a complete picture of the joint distribution of β_1 and β_2. From this joint distribution specific quantities can be estimated, such as the marginal means and posterior intervals. In contrast, the Newton–Raphson algorithm was only concerned with one aspect of the joint distribution: the maximum of the log-likelihood function (which from a frequentist perspective is used to estimate the fixed but unknown parameter means). The cost of the extra information is the computation, 1000 iterations versus just 4.

The extra computation can have other benefits if the likelihood is multimodal, such as that shown in Figure 13.9. The Newton–Raphson algorithm would find only one maximum and the maximum found would depend strongly on the initial value. A reasonably designed Markov chain should find both modes and hence we would be aware that there was more than one solution. Such multiple solutions can occur when the model has been badly parameterized to allow multiple equivalent solutions by alternative linear combinations of the parameters. This can sometimes be solved by reparameterizing such as using a corner-point parameterization or sum-to-zero constraint (Section 2.4).

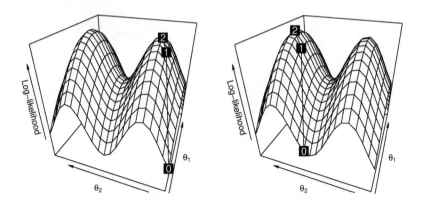

Figure 13.9 *Hypothetical example of a bimodal likelihood demonstrating how the maximum found by the Newton–Raphson algorithm depends on the initial value.*

Another contrast between the Newton–Raphson algorithm and Markov chain sampling is the method used to create the sampling "steps." The Newton–Raphson and other numerical optimization methods take *deterministic* steps. The steps are governed by the fixed likelihood (assuming fixed data) and the initial values. The algorithms vary but heuristically the steps are programmed to always go upwards (until there is no higher step). Given the same starting value, the iterations will always follow the same path to the top.

A Markov chain algorithm takes *random* steps over the likelihood surface, and these steps can go up or down. As shown in Section 13.3.1 the random steps for the Metropolis–Hastings algorithm are designed to favour taking upward steps. This ensures that the chain never moves too far from the maximum (and that the sample density converges to the target density). Also, as the Markov chain steps are random, the chain could be re-started at the same initial value and get a different set of iterations.

13.3.4 Importance of parameterization

The efficiency of the Markov chain sampling is dependent on (amongst other things) the model parameterization (Gilks and Roberts 1996). As an example the iterations of an Metropolis–Hastings sampling algorithm fitted using the extreme value model are shown in Figure 13.10. The only difference between

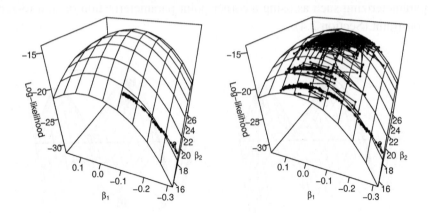

Figure 13.10 *Metropolis–Hastings iterations for the beetle mortality data using the extreme value model with* $g(\pi_i) = \beta_1 + \beta_2(x_i - \bar{x})$ *after 100 iterations (left panel), 1000 iterations (right panel). The initial value is shown as an open circle and subsequent estimates as closed circles.*

this figure and Figure 13.8 is the link function. The plots show the chain after 100 and 1000 iterations. The initial values were $\beta_1^{(0)} = -0.3$ and $\beta_2^{(0)} = 17$.

After 100 Metropolis–Hastings iterations the chain is still nowhere near the maximum, but it has moved closer. This upward movement has occurred only for β_1, whereas β_2 is still close to its initial value. The chain is still burning-in. After 1000 iterations the chain has eventually made its way to the summit and has then densely sampled around that area.

Estimation using the extreme value model converged much more slowly than when the logit link was used. The proposals for β_2 for the extreme value model were consistently rejected in the early stages. This may have been because the proposal distribution was too wide, and so most proposals missed the area of high likelihood (see Exercise 13.2). Proposal distributions are often tuned during the burn-in period (by making them wider or narrower) to give an acceptance rate close to 60%.

Other parameterization can affect chain efficiency, as shown later using centring.

13.4   Bayesian inference

In this section we apply Monte Carlo integration to Markov chains to make Bayesian inferences. Because the Markov chain samples give us the complete distribution (assuming the chain has correctly converged), many forms of inference are possible. For example, we have an estimate of the joint distribution over all parameters, and so can make inferences about joint or conditional probabilities. For the beetle mortality example in the previous section, it would be easy to calculate $P(\beta_1 > 0$ and $\beta_2 > 0)$ or $P(\beta_1 > 0|\beta_2 > 0)$. This ease of making complex inference is one of the major advantages of using MCMC methods.

Figure 13.11 shows the histograms for β_1 and β_2 for the beetle mortality data using Metropolis–Hastings sampling. The first 1000 samples are shown in Figure 13.5. These histograms are based on discarding the first 100 samples (burn-in) and on a sample size of 4900. The bin-width for β_1 is 0.05 and for β_2 is 1. The histogram for β_1 has a bell-shaped distribution, whereas the histogram for β_2 is less symmetric. However, to make inferences about $\boldsymbol{\beta}$ the densities need not be Normal although it is useful to know that they are unimodal.

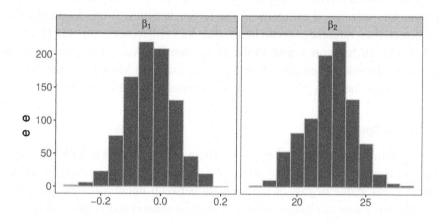

Figure 13.11 *Histograms of Markov chain samples using Metropolis–Hastings sampling for β_1 and β_2 for the beetle mortality example using the extreme value model and $g(\pi_i) = \beta_1 + \beta_2(x_i - \bar{x})$.*

Various summary statistics of $\boldsymbol{\beta}$ can be calculated. The mean is calculated using an equation similar to (13.2) but with an index running from 101 to 5000. The mean for β_1 is -0.03 and the mean for β_2 is 22.36. The estimates for $\boldsymbol{\beta}$ using classical methods are in Table 7.4 and are $\beta_1 = -39.57$ and $\beta_2 = 22.04$. Hence, the slope estimate is almost identical, but the intercept parameter is quite different because the x values were centred around their mean $\bar{x}$. Centring was used so that the MCMC samples converged (see Section 6.7).

A likely range for $\boldsymbol{\beta}$ can be estimated using a **posterior interval**. The $\gamma\%$ posterior interval contains the central $\gamma\%$ of the sampled values, and so the interval contains the true estimate with probability γ. The lower limit of the interval is the $(50 - \gamma/2)$th percentile of the Markov chain samples, and the upper limit is the $(50 + \gamma/2)$th percentile. So a 95% posterior interval goes from the 2.5th to 97.5th percentile.

For the samples from the extreme value model, a 95% posterior interval for β_1 is -0.19 to 0.12, and for β_2 is 18.81 to 25.93. These intervals correspond to the distributions shown in Figure 13.11. Using the estimates in Table 7.4, a 95% confidence interval for β_2 is 18.53 to 25.55, which is very close to the 95% posterior interval.

13.5   Diagnostics of chain convergence

The inferences made in the previous section using the Markov chain samples were based on the assumption that the sample densities for the unknown parameters were good estimates of the target densities. If this assumption is incorrect, then inferences could be invalid. Valid inferences can only be made when a chain has **converged** to the target density. Assessing chain convergence is therefore a key part of any analysis that uses Markov chains. In this section some methods for assessing chain convergence are shown.

13.5.1   Chain history

A simple informal method of assessing chain convergence is to look at the **history** of iterations using a time series plot. A chain that has converged should show a reasonable degree of randomness between iterations, signifying that the Markov chain has found an area of high likelihood and is integrating over the target density (known as **mixing**).

An example of a chain that has not converged and one that has probably converged are shown in Figure 13.12. Such plots are available in WinBUGS using the command `history()`. Both chains show the estimate for β_2 from the logit model but using different parameterizations. Both chains started with an initial value of $\beta_2^{(0)} = 33$. The chain showing poor convergence does not seem to have found a stable area of high likelihood. The regression equation for this model was

$$\text{logit}(\pi_i) = \beta_1 + \beta_2 x_i.$$

The chain showing reasonable convergence used the regression equation

$$\text{logit}(\pi_i) = \beta_1 + \beta_2(x_i - \bar{x}).$$

Centring the dose covariate has greatly improved the convergence because centring reduces the correlation between the parameter estimates β_1 and β_2 (Gilks and Roberts 1996).

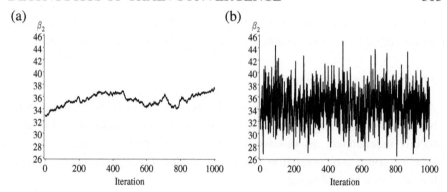

Figure 13.12 *Example of a chain showing (a) poor convergence and (b) reasonable convergence (first 1,000 iterations using Gibbs sampling). Estimate for β_2 using the logit link using two different parameterizations.*

The three-dimensional plots of the likelihood shown in Figure 13.13 show the effect of centring. The likelihood without the uncentered dose has a ridge

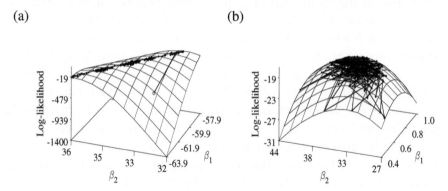

Figure 13.13 *Three-dimensional plots of the log-likelihood and 200 Gibbs samples for the beetle mortality data using the logit link function and (a) uncentered dose or (b) centered dose. The initial value is shown as an open circle and subsequent estimates as closed circles.*

of high likelihood. In contrast the likelihood with a centered dose has a gently curved surface. The first sample for the uncentered dose jumps quickly to the ridge of high likelihood. The subsequent samples become stuck along the ridge creating a narrow and strongly correlated pattern (Gilks and Roberts 1996). The samples with the centered dose are far less correlated as they are freer to move about the parameter space in any direction.

Looking at history plots is a useful check of convergence and can identify obvious problems with the chain. It is also good practice to use more formal testing procedures, which may highlight more subtle problems. In the following sections some examples of these more formal tests are given.

13.5.2   Chain autocorrelation

Autocorrelation is a useful diagnostic because it summarizes the dependence between neighbouring samples. Ideally neighbouring samples would be completely independent, as this would be the most efficient chain possible. In practice some autocorrelation is usually accepted, but large values (greater than 0.4) can be problematic.

For an observed chain $\boldsymbol{\theta} = [\theta^{(1)}, \ldots, \theta^{(i)}]$, the **autocorrelation function** (ACF) at lag k is

$$\rho(k) = \frac{1}{\sigma_\theta^2 (M-k)} \sum_{i=k+1}^{M} \left(\theta^{(i)} - \overline{\theta}\right) \left(\theta^{(i-k)} - \overline{\theta}\right),$$

where σ_θ^2 is the sample standard deviation of $\boldsymbol{\theta}$ and $\overline{\theta}$ is the sample mean.

If there is significant autocorrelation between the θ_i's, then it can be reduced by systematically using every jth sample and discarding the others. This process is known as **thinning**. Assuming the chain follows an autoregressive process of order 1 with parameter γ, then the autocorrelation at lag one is γ. After sampling every jth observation, this is reduced to γ^j. A highly correlated chain ($\gamma = 0.9$) would need to use $j = 9$ to reduce the autocorrelation to under 0.4. This means many iterations are wasted, and the chain would need to be run for a long time to give a reasonable sample size. Chains can be thinned in WinBUGS using the command thin.samples(j).

An example of the reduction in the autocorrelation due to thinning is shown in Figure 13.14. The plot shows the chain histories and autocorrelation from lag zero to twenty ($k = 0, \ldots, 20$) for a chain with no thinning and the same chain thinned by 3 ($j = 3$). The chains are the estimates of β_2 from the extreme value model for the beetle mortality data using Metropolis–Hastings sampling. The first 1000 iterations were discarded as a burn-in. Also shown on the plots are the 95% confidence limits for the autocorrelation assuming that the samples were generated from an uncorrelated and identically distributed process. At lag k these 95% confidence limits are

$$\left[-1.96 \frac{1}{\sqrt{M-k}}, +1.96 \frac{1}{\sqrt{M-k}} \right].$$

Multiple testing is an issue if these limits are used repeatedly to formally reject the hypothesis that the chain has converged. However, we recommend that the limits be used for guidance only.

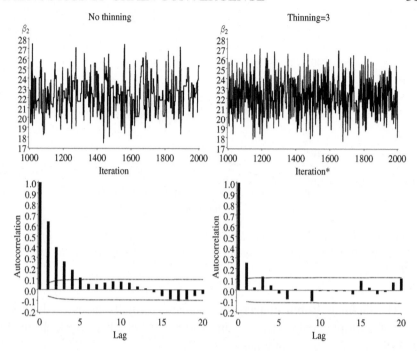

Figure 13.14 *Reduction in autocorrelation of Metropolis–Hastings samples after thinning. Chain history (top row) and ACF (bottom row) for the estimate of β_2 from the beetle mortality example using the extreme value model and a centered dose.*

For the chain with no thinning the autocorrelation at lag 1 is large, $\hat{\rho}(1) = 0.63$. Some of this autocorrelation is due to the Metropolis–Hastings sampling rejecting proposals. Thinning by 3 reduces the autocorrelation at lag 1 to $\hat{\rho}(1) = 0.25$. The cost of this reduction was an extra 2000 samples to give 1000 thinned samples.

13.5.3 Multiple chains

Using multiple chains is a good way to assess convergence. If multiple chains are started at widely varying initial values and each chain converges to the same solution, this would increase our confidence in this solution. This method is particularly good for assessing the influence of initial values.

An example using multiple chains with the extreme value model for the beetle mortality data is given below. An initial value of $\beta_2 = 1$ was used for one chain and $\beta_2 = 40$ for the other. These starting values are on either side of the maximum likelihood estimate (Figure 13.4). Both chains converged toward a common estimate around $\beta_2 \approx 20$ as shown in Figure 13.15.

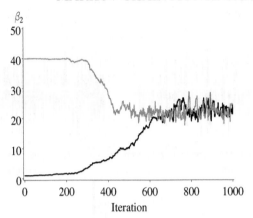

Figure 13.15 *Two chains with different starting values. Estimates of β_2 using Metropolis–Hastings sampling for the extreme value model using the beetle mortality data.*

The Gelman–Rubin convergence statistic formally assesses the confluence of multiple chains (Brooks and Gelman 1998). It is available in Win-BUGS using the `gr()` command.

One drawback to using multiple chains as a diagnostic tool is that it may be difficult to generate suitably varied starting values, particularly for complex problems with many unknown parameters and multidimensional likelihoods.

13.6   Bayesian model fit: the deviance information criterion (DIC)

In previous chapters, the Akaike information criterion (AIC) was used to choose the most parsimonious model (Section 7.5). The deviance information criterion (DIC) is the generalisation of the AIC for Bayesian models fitted using MCMC methods (Spiegelhalter et al. 2002). The deviance is $-2l(\mathbf{y}|\boldsymbol{\theta})$ (defined in Section 5.2) and quantifies the difference between the fitted values and observed data. For observed data $\mathbf{y}$ and parameters $\boldsymbol{\theta}$ the DIC is

$$\text{DIC} = D(\mathbf{y}|\overline{\boldsymbol{\theta}}) + 2p_D, \tag{13.4}$$

where $D(\mathbf{y}|\overline{\boldsymbol{\theta}}) = -2l(\mathbf{y}|\overline{\boldsymbol{\theta}})$ is the deviance at $\overline{\boldsymbol{\theta}}$ and p_D is the *effective* number of parameters. As with the AIC the lower the DIC, the better the model.

The effective number of parameters is estimated as

$$p_D = \overline{D(\mathbf{y}|\boldsymbol{\theta})} - D(\mathbf{y}|\overline{\boldsymbol{\theta}}), \tag{13.5}$$

where $\overline{D(\mathbf{y}|\boldsymbol{\theta})}$ is the average deviance over all values of $\boldsymbol{\theta}$. Thus the effective number of parameters is the difference between the fit of the average model and the fit of the "best" model (assuming the best model comes from using the parameter means).

Substituting the expression for p_D (13.5) into Equation (13.4) gives an alternative form for the DIC:

$$\text{DIC} = \overline{D(y|\boldsymbol{\theta})} + p_D. \tag{13.6}$$

The DIC requires sampled values of the deviance and parameters. Hence, the DIC is easily calculated as part of an MCMC analysis (see Exercise 13.4). In WinBUGS the commands are `dic.set()`, which starts storing deviance values, and `dic.stats()`, which estimates the DIC based on the stored deviances and parameters.

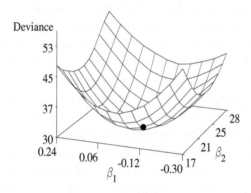

Figure 13.16 *Deviance for the extreme value model fitted to the beetle mortality data. The dot shows the deviance at* $\overline{\boldsymbol{\beta}}$.

An example of calculating p_D and the DIC by plotting the deviance over the range of $\boldsymbol{\beta}$ is shown for the extreme value model for the beetle mortality data (Figure 13.16). The deviance is proportional to the inverted log-likelihood shown in Figure 13.10. The value of the deviance at $\overline{\boldsymbol{\beta}}$ is marked on the figure. This is the value at $(\beta_1, \beta_2) = (-0.046, 22.1)$ and is $D(y|\overline{\boldsymbol{\beta}}) = 29.7$. The model was run for $M = 10,000$ iterations with a burn-in of 5000. The mean deviance using these 5000 samples is $\overline{D(y|\boldsymbol{\beta})} = 31.7$. So the effective number of parameters is $p_D = 31.7 - 29.7 = 2.0$ (which equals the number fitted: β_1 and β_2) and the DIC is 33.7.

There are some important points to note from this example. In this case the deviance was concave with a clear minimum, and $D(y|\overline{\boldsymbol{\theta}})$ was very near this minimum. If the deviance has several minima, then $D(y|\overline{\boldsymbol{\theta}})$ may not be a good estimate of the smallest deviance, and hence, the DIC may not be a good estimate of model fit (Spiegelhalter et al. 2002). Furthermore, the deviances used to calculate the DIC depend on the parameter space. If the parameter space is restricted, for example, by using a prior for β_1 or β_2, then this alters the DIC.

The number of effective parameters is not necessarily an integer. As an example of a non-integer value for p_D consider the random intercept defined as part of a multilevel model (Section 11.5),

$$a_j \sim N(0, \sigma_a^2), \qquad j = 1, \dots, n_j.$$

The number of parameters for the random intercept is not $n_j + 1$, one for each intercept and one for the variance. Rather because a Normal distribution is fitted to the intercepts, strength is gained and each intercept may need less than one parameter; the gain depends on how many parameters are needed to fit the Normal distribution. Hence, the effective number of parameters can be thought of as the amount of information needed to describe the data.

The DIC can be used to assess the value of adding explanatory variables to a model as well as the link function or covariance structure. If a number of different models are fitted (using the same data), the model with the lowest DIC should be selected. However, if the difference in DIC is small, it may be misleading to conclude that one model is better than another. In general differences of less than 5 are considered small, and those above 10 substantial (Lunn et al. 2012, Chapter 8). However, these values are just a guide. Indeed the DIC is a guide to model selection and should not be treated as an absolute decision criterion. A number of examples of selecting optimal models using the DIC are given in the following chapter.

13.7   Exercises

13.1 Reconsider the example on *Schistosoma japonicum* from the previous chapter. In Table 12.1 the posterior probability for H_0 was calculated using an equally spaced set of values for θ. Recalculate the values in Table 12.1 using 11 values for θ generated from the Uniform distribution U[0,1]. Compare the results obtained using the fixed and random values for θ. Should any restrictions be placed on the samples generated from the Uniform distribution?

13.2 The purpose of this exercise is to create a chain of Metropolis–Hastings samples, for a likelihood, $P(\theta|y)$, that is a standard Normal, using symmetric and asymmetric proposal densities. A new value in the chain is proposed by adding a randomly drawn value from the proposal density to the current value of the chain, $\theta^* = \theta^{(i)} + Q$. Use the likelihood ratio, Equation (13.3), to assess the acceptance probability.

If using RStudio, the following commands will be helpful. The sampled values are labelled `theta`.

`theta<-vector(1000,mode='numeric')` creates an empty vector of length 1000

`pnorm(theta)` is the standard Normal probability density

`runif(100,-1,1)` generates 100 random Uniform[-1,1] variables

`hist(theta)` plots a histogram of theta

`plot(theta,type='b')` plots a history of theta

a. Create 1000 samples for θ using a Uniform proposal density, $Q \sim$ U$[-1,1]$. Start the chain at $\theta^{(0)} = 0.5$. Monitor the total number of accepted moves (acceptance rate). Plot a history of the sampled values and a histogram.

b. Try the smaller proposal density of $Q \sim$ U$[-0.1, 0.1]$ and the larger density of $Q \sim$ U$[-10, 10]$. Explain the differences in the acceptance rates and chain histories.

c. It has been suggested that an acceptance rate of around 60% is ideal. Using a proposal density $Q \sim$ U$[-q, q]$, find the value of q that gives an acceptance rate of roughly 60%. Was it the most efficient value for q? How could this be judged?

d. Plot the acceptance rate for $q = 1, \ldots, 20$.

e. Using a Normal proposal density $Q \sim$ N$(0, \sigma^2)$, write an algorithm that "tunes" the values of σ^2 after each iteration to give an acceptance rate of 60%. Base the acceptance rate on the last 30 samples.

13.3 This exercise is an introduction to the R2WinBUGS package that runs WinBUGS from R (Sturtz et al. 2005). R2WinBUGS is an R add-on package which needs to be installed in R. The advantage of using R2WinBUGS rather than WinBUGS directly is that script files can be created to run the entire analysis process of data manipulation, analysis and displaying the results.

a. Open a new script file using RStudio. Load the R2WinBUGS library by typing

`>library(R2WinBUGS)`

Change the working directory to an area where the files created by this exercise can be stored, for example,

`>setwd('C:/Bayes')`

Alternatively create a new project in RStudio which will provide a common place for the files and facilities switching between different projects.

b. WinBUGS accepts data in S-PLUS (i.e., as a list) and rectangular format. Type the following data from the beetle mortality example into a text file (in rectangular format) and save them in the working directory as "BeetlesData.txt"

```
x[] n[] y[]
1.6907 59 6
1.7242 60 13
1.7552 62 18
1.7842 56 28
1.8113 63 52
1.8369 59 53
1.8610 62 61
1.8839 60 60
END
```

c. To conduct some initial investigation of the data, read the data into RStudio using

```
>library(dobson)
>data(beetle)
```

The proportion of deaths is calculated using

```
>beetle$p <- beetle$y/beetle$n
```

A scatter plot can be examined by typing

```
>plot(beetle$x, beetle$p, type='b')
```

What are the major features of the data?

d. Open a new .odc file in WinBUGS and type in the following dose-response model using the extreme value distribution so that $\pi_i = 1 - \exp[-\exp(\beta_1 + \beta_2 x_i)]$:

```
———————— WinBUGS code (dose-response model) ————————
model
{ # Likelihood
    for (i in 1:8){
        y[i]~dbin(pi[i],n[i]); # Binomial
        pi[i]<-1-exp(-exp(pi.r[i])); # Inverse link function
        pi.r[i]<-beta[1]+(beta[2]*x[i]); # Regression
```

```
        fitted[i]<-n[i]*pi[i]; # Fitted values
    }
# Priors
    beta[1]~dnorm(0,1.0E-6); # Intercept
    beta[2]~dnorm(0,1.0E-6); # Slope
}
```

Save the model in the working directory using the filename "BeetlesEx-treme.odc". Although the model will be run from RStudio, errors can be checked for in WinBUGS by clicking on Model ⇒ Specification ⇒ check model. Hopefully the message "model is syntactically correct" appears in the bottom-left corner.

What are the assumptions of this model?

e. The model has two parameters, the intercept and slope labelled beta[1] and beta[2], respectively. Initial values are needed for these parameters. Type the following in RStudio

```
inits = list(list(beta=c(0,0)))
```

What do these initial values translate to in terms of the model?

f. In the R script, type the following (the comments beginning with "#" are optional but can be helpful):

```
———————————— R code (using R2WinBUGS) ————————————
# set up the data
data = list(y=beetle$y, x=beetle$x, n=beetle$n)
parameters = c('beta')
model.file = 'BeetlesExtreme.odc'
# run WinBUGS
bugs.res <- bugs(data, inits=inits, parameters, model.file,
n.chains=1, n.burnin=5000, n.iter=10000, n.thin=1, debug=T,
bugs.directory="c:/Program Files/WinBUGS/")
# display the summary statistics
bugs.res
```

This code sets up the data and runs WinBUGS. The debug=T option keeps WinBUGS open after the model has run which allows you to examine the results in detail. The bugs.directory option will need to be changed to match where WinBUGS is installed.

The following code uses the reshape2 library to arrange the data in long format and the ggplot2 library to plot the chain histories.

———————————— R code (plotting chains) ————————

```
# plot the histories
library(reshape2)
library(ggplot2)
beta.chains = bugs.res$sims.list$beta
to.plot = melt(beta.chains)
names(to.plot)[1:2] = c('iter','beta')
chain.plot = ggplot(data=to.plot, aes(x=iter, y=value))+
geom_line()+
facet_wrap(~beta, scale='free_y')
chain.plot
```

Do these look like good chains?

To examine the autocorrelation type

```
>acf(beta.chains[,1])
>acf(beta.chains[,2])
```

g. The mixing of the chains can be improved by subtracting the mean. Change the regression line in the WinBUGS odc file to

```
pi.r[i]<-beta[1]+(beta[2]*(x[i]-mean.x)); # Regression
```

To add the mean dose (mean.x) to the data file use

```
data$mean.x = mean(data$x)
```

Re-run the RStudio script file. Have the chains improved? Why?

h. The deviance $-2\log p(\mathbf{y}|\boldsymbol{\beta})$ is used to assess model fit; the lower deviance, the better the fit. By default R2WinBUGS monitors the deviance. Plot the deviance for the previous models. What do the deviance plots show?

i. Use the glm command in RStudio to find reasonable starting values for the intercept and slope. Explain the difference.

j. Re-run the model but this time change the RStudio script file so that the fitted values are also monitored. Plot the fitted values against dose and include the observed data.

k. Re-do Exercise 13.3(h) but this time using a logit link. Explain the difference.

13.4 This exercise is about calculating the deviance information criterion (DIC). The two key equations are (13.5) and (13.6) for the number of parameters (pD) and DIC, respectively.

a. R2WinBUGS calculates the DIC automatically, extract the values by typing bugs.res$pD and bugs.res$DIC.

b. Use the chains for the deviance and beta parameters to calculate $\overline{D(\mathbf{y}|\boldsymbol{\beta})}$ and $D(\mathbf{y}|\overline{\boldsymbol{\beta}})$, from which you can estimate p_D and the DIC.

c. Write the results from a. in the first row of the table below and the results from b. in the second row. How well do the calculated results match those calculated by R2WinBUGS?

Version	$\overline{D(\mathbf{y}|\boldsymbol{\beta})}$ (Dbar)	$D(\mathbf{y}|\hat{\boldsymbol{\beta}})$ (Dhat)	p_D	DIC
DIC				
Mean of $\boldsymbol{\beta}$				
Median of $\boldsymbol{\beta}$				
Half variance of $D(.)$				

d. The mean is just one estimate of the "best possible" deviance; the median could be used instead. Calculate the deviance at the medians of $\boldsymbol{\beta}$, then calculate the alternative values for p_D and DIC. Write these values in the third row of the table.

e. Another alternative calculation for p_D is the variance of $D(\mathbf{y}|\hat{\boldsymbol{\beta}})$ divided by 2. Calculate this alternative p_D and alternative DIC and write the results in the last row of the table.

Comment on the differences in the complete table.

f. Re-run exercise 13.4 but this time using a logit link. Which link function gives the best fit? Was the difference consistent regardless of the method used to calculate $D(\mathbf{y}|\hat{\boldsymbol{\beta}})$?

g. Having been through the calculations for the DIC in detail, when might it not work well?

Chapter 14

Example Bayesian Analyses

14.1 Introduction

This chapter comprises examples of Bayesian analyses for generalized linear models. The examples from previous chapters are used and differences in the results between classical (frequentist) and Bayesian methods are discussed. WinBUGS code is given for each example.

A very comprehensive book of Bayesian examples with WinBUGS code is Congdon (2006). There are also many good examples in Gelman et al. (2013) and in the WinBUGS manual of Spiegelhalter et al. (2007).

Reproducing the results of this chapter

The WinBUGS software was used to estimate the models in this chapter (Version 1.4.3, Spiegelhalter et al. 2007). BUGS stands for Bayesian inference Using Gibbs Sampling, although it also uses other sampling methods such as the Metropolis–Hastings algorithm. As discussed in the previous chapter, random numbers are used to generate the Markov chain. Hence, readers trying to reproduce the results shown here may get slightly different answers due to this randomness. Any differences should be small, especially if chains are run for a large number of iterations. Alternatively if readers use the same random number seed, the same number of iterations and burn-in, and the same initial values, they will reproduce the results given here exactly. The random number seed used was 314159 (the first six digits of π and the default WinBUGS starting value). This can be set using the seed command in the WinBUGS model menu.

14.2   Binary variables and logistic regression

In Section 7.3 a number of dose response models for the beetle mortality data were examined including a linear dose response model,

$$\Pr(\text{death}|x) = \beta_1 + \beta_2 x,$$

where x is the dose. For this model there is a constant tolerance defined by the interval $[c_1, c_2]$ (Figure 7.1). This model cannot be fitted using standard methods because the interval must accommodate the observed doses so that $c_1 \leq x \leq c_2$. This condition cannot be satisfied using classical methods, but it can be satisfied in a Bayesian framework by using Uniform priors so that

$$c_1 \quad \sim \quad U[L_1, \min(x)]$$
$$c_2 \quad \sim \quad U[\max(x), L_2].$$

From the data in Table 7.2 $\min(x) = 1.6907$ and $\max(x) = 1.8839$. The value c_1 is the dose at which beetles begin to die (the lowest toxic dose). Hence, the prior for c_1 needs to cover the full range of possible lowest toxic doses. This can easily be achieved by giving L_1 a very small value, say $L_1 = 0.01$. Similarly the value of c_2 is the dose at which all beetles are dead. The range of possibilities can be covered by giving L_2 a large value, say $L_2 = 10$ (which is over 5 times the largest dose in the data). Flat (Uniform) priors are used for c_1 and c_2 because they are not being used to incorporate known information but to satisfy model constraints.

The number of deaths are modelled using the Binomial distribution

$$y_i \quad \sim \quad \text{Bin}(n_i, p_i),$$
$$p_i \quad = \quad \beta_1 + \beta_2 x_i,$$
$$\beta_1 \quad = \quad -c_1/(c_2 - c_1),$$
$$\beta_2 \quad = \quad 1/(c_2 - c_1).$$

The identity link function is used for p_i, whereas the other models in Section 7.3.1 used more complex link functions such as the logit. The Bayesian model above estimates c_1 and c_2, and β_1 and β_2 are simply functions of c_1 and c_2. The classical dose response models in Section 7.3.1 all estimate β_1 and β_2 directly.

The WinBUGS code, data structure and initial values to fit the linear dose response model are given below. The data are in two parts. The number of responses and minimum and maximum dose are in S-PLUS data format, using the `list` command. The doses, observed number of deaths and number of beetles are in **rectangular data format**. Rectangular format is generally better for entering large amounts of data, whereas small amounts of data can be easily added using the S-PLUS format.

The minimum and maximum of the dose are included as data (min.x and max.x). It is possible to let WinBUGS calculate these values, using the commands min.x<-min(x[]) and max.x<-max(x[]). However, every calculation in WinBUGS increases the computation time, so it is more efficient to do such calculations outside WinBUGS.

———————— WinBUGS code (linear dose response model) ————————

```
model{
# Likelihood
    for (i in 1:N){
        y[i]~dbin(pi[i],n[i]);
        pi[i]<-beta[1]+(beta[2]*x[i]);
        fitted[i]<-n[i]*pi[i];
    }
# Priors
    c[1]~dunif(0.01,min.x);
    c[2]~dunif(max.x,10);
# Scalars
    beta[1]<- -c[1]/(c[2]-c[1]);
    beta[2]<- 1/(c[2]-c[1]);
}


# Data
list(N=8,min.x=1.6907,max.x=1.8839)
x[] n[] y[]
1.6907 59 6
1.7242 60 13
1.7552 62 18
1.7842 56 28
1.8113 63 52
1.8369 59 53
1.8610 62 61
1.8839 60 60
END


# Initial values
list(c=c(0.01,10))
```

The model was fitted using 10,000 iterations with a burn-in of 5,000. The parameter estimates are given in Table 14.1. The width of the posterior interval for c_1 is 0.031 (1.679 − 1.648) which is over six times the width of the interval for c_2 which is just 0.005 (1.889 − 1.884). This suggests greater confidence about the highest toxic limit (c_2) than the lowest (c_1). The posterior intervals, fitted linear dose response and the data are shown in Figure 14.1. The plot helps to explain the greater confidence about c_2 than c_1, as there are more data with the high proportions of deaths compared with lower proportions. The lower limit for c_2 is bounded by 1.884, the highest dose at which the proportion killed was 1.

Table 14.1 *Fitting a linear dose response model to the beetle mortality data.*

Parameter	Mean	95% posterior interval
β_1	−7.604	−8.172, −6.929
β_2	4.564	4.203, 4.866
c_1	1.666	1.648, 1.679
c_2	1.885	1.884, 1.889

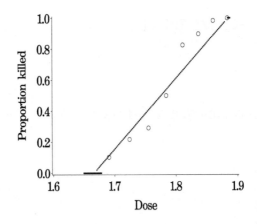

Figure 14.1 *Observed proportion killed (circles), estimated linear dose response model (dotted line) and 95% posterior intervals for c_1 and c_2 (horizontal lines) for the beetle mortality data from Table 7.2.*

Table 14.2 compares the fit of the linear model with the extreme value model, and shows the deviance D, Equation (7.5). It is clear from Figure 14.1 that a linear fit to the data is not appropriate, and this explains the smaller deviance of the S-shaped extreme value model.

The deviance information criterion (DIC) (Section 13.6) can be used to

Table 14.2 *Comparison of observed numbers killed with fitted values obtained from classical extreme value model and Bayesian linear dose response model for the beetle mortality data.*

Observed value of Y	Fitted values	
	Classical extreme value model	Bayesian linear model, mean (95% PI)
6	5.59	6.58 (3.27, 10.46)
13	11.28	15.87 (13.07, 19.11)
18	20.95	25.17 (22.83, 27.88)
28	30.37	30.15 (28.50, 32.05)
52	47.78	41.72 (40.33, 43.29)
53	51.14	45.96 (44.96, 46.97)
61	61.11	55.12 (54.20, 55.76)
60	59.95	59.61 (58.67, 59.99)
Deviance, D	3.45	28.18

assess the models fitted in a Bayesian setting. The classical deviances and Bayesian DIC are shown in Table 14.3. The deviance for the linear model was calculated using the fitted values from the Bayesian model. The DIC results confirm those of the deviance: the extreme value model is the best fit to the data. The logit and probit models have a similar fit as judged by either the deviance or the DIC. The linear model gives the poorest fit to the data.

Table 14.3 *Model fit criteria for the dose response models using the beetle mortality data. Models ordered by DIC.*

Model	Classical	Bayesian	
	Deviance (D)	Estimated number of parameters	DIC
Extreme value	3.45	2.0	33.7
Probit	10.12	2.0	40.4
Logistic	11.23	1.9	41.2
Linear	28.13	0.9	56.2

14.2.1 Prevalence ratios for logistic regression

Section 7.9 compared using odds ratios and prevalence ratios to describe the estimated differences in probability when using logistic regression. Prevalence ratios can be calculated by using the log link in place of the logit link, but estimates often converge using the logit link and fail to converge using the log link because the logit-transformed estimates can take any value.

These convergence issues can be avoided in WinBUGS by using the logit link and constructing prevalence ratios based on the estimated probabilities. The WinBUGS code below is an example using the beetle mortality data. To aid convergence the dose (x) was centred using the mean $(x - \bar{x})$. The prevalence ratios were estimated for two doses relative to the mean, and for comparison the odds ratios were also calculated.

─────────────── WinBUGS code (prevalence ratios) ───────────────

```
model{
# Likelihood
for (i in 1:N){
y[i]~dbin(pi[i],n[i]);
logit(pi[i])<-beta[1]+(beta[2]*x[i]);
}
# Priors
beta[1]~dnorm(0,1.0E-5)
beta[2]~dnorm(0,1.0E-5)
# Scalars
logit(pi.hat[1]) <- beta[1]
logit(pi.hat[2]) <- beta[1] + (0.05*beta[2])
logit(pi.hat[3]) <- beta[1] + (0.1*beta[2])
PR[1] <- pi.hat[2] / pi.hat[1]
PR[2] <- pi.hat[3] / pi.hat[2]
OR[1] <- (pi.hat[2]/(1-pi.hat[2])) / (pi.hat[1]/(1-pi.hat[1]))
OR[2] <- (pi.hat[3]/(1-pi.hat[3])) / (pi.hat[2]/(1-pi.hat[2]))
}
```

The estimated probabilities, odds ratios and prevalence ratios are in Table 14.4. The estimated prevalence ratio for mortality for an increase in dose of 0.05 above the mean is 1.36, whereas the odds ratio for the same increase is a much larger 5.67. Such large odds ratios occur when probabilities approach zero or one. The estimated odds ratio of 5.67 is the same for the increases in dose from $\bar{x}$ to $\bar{x} + 0.05$ and from $\bar{x} + 0.05$ to $\bar{x} + 0.1$ as the log odds ratio is linear for each unit increase in x. However, the prevalence ratios are not consistent for each unit increase and depend on the reference dose, hence any interpretation of prevalence ratios for continuous explanatory variables needs to give both the reference value and the increase. The estimated S-shaped association between dose and probability is shown in Figure 14.2. The S-shaped association is smooth but the prevalence ratio will be a linear interpolation between two selected doses on the S-curve. This shows how the prevalence ratio can change greatly depending on the baseline dose, and also shows how using a large increase in dose, for example the prevalence ratio for the smallest to largest dose, will accurately reflect the overall change in probability but would miss the variation in risk.

Table 14.4 *Estimated probabilities, odds ratios and prevalence ratios for mortality for the beetle mortality data, estimates from a Bayesian model using the logit link.*

Estimate	Mean	95% credible interval
Pr(death $\mid x =$ mean)	0.677	0.614, 0.733
Pr(death $\mid x =$ mean + 0.05)	0.920	0.884, 0.948
Pr(death $\mid x =$ mean + 0.1)	0.984	0.972, 0.993
Odds ratio, mean vs. mean +0.05	5.67	4.30, 7.51
Prevalence ratio, mean vs. mean +0.05	1.36	1.28, 1.46
Odds ratio, mean+0.05 vs. mean +0.1	5.67	4.30, 7.51
Prevalence ratio, mean+0.05 vs. mean +0.1	1.07	1.05, 1.10

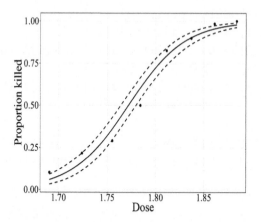

Figure 14.2 *Observed proportion killed (dots), estimated logit dose response model (solid line) and 95% posterior intervals (dotted lines) from a Bayesian model for the beetle mortality data from Table 7.2.*

14.3   Nominal logistic regression

Section 8.3 covered nominal logistic regression using an example based on car preferences. This section shows the equivalent Bayesian model.

The WinBUGS code below uses the car preference example to calculate the nominal logistic regression model (8.11). The calculation of the preference probabilities π, required the use of the intermediary variable ϕ:

$$\log(\phi_j) \;=\; \beta_{0j} + \beta_{1j}x_1 + \beta_{2j}x_2 + \beta_{3j}x_3, \qquad j = 2,3 \qquad (14.1)$$

$$\pi_j \;=\; \frac{\phi_j}{\sum_{j=1}^{3}\phi_j}, \qquad j = 1,2,3.$$

This parameterization ensures that $\pi_1 + \pi_2 + \pi_3 = 1$. To ensure that the parameter estimates are identifiable and that the above model equals (8.11), ϕ_1 is set to 1 (see Exercise 14.1). The use of ϕ is identical to the way the nominal model probabilities were estimated on page 186. The observed number of responses has a multinomial distribution (dmulti).

———— WinBUGS code (nominal logistic regression) ————
```
model{
# Likelihood
 for (i in 1:6){
  freq[i,1:3]~dmulti(pi[i,1:3],n[i]);
  phi[i,1]<-1; # For identifiability
  log(phi[i,2])<-beta0[1]+(beta1[1]*sex[i])+(beta2[1]*age1[i])
              +(beta3[1]*age2[i]);
  log(phi[i,3])<-beta0[2]+(beta1[2]*sex[i])+(beta2[2]*age1[i])
              +(beta3[2]*age2[i]);
  for (j in 1:3){
    pi[i,j]<-phi[i,j]/sum(phi[i,]); # Normalising
    fitted[i,j]<-pi[i,j]*n[i]; # Fitted values
  }
 }
# Priors
 for (k in 1:2){
  beta0[k]~dnorm(0,0.0001);
  beta1[k]~dnorm(0,0.0001);
  beta2[k]~dnorm(0,0.0001);
  beta3[k]~dnorm(0,0.0001);
 }
# Scalars
 for (k in 1:2){
```

```
   or1[k]<-exp(beta1[k]);
   or2[k]<-exp(beta2[k]);
   or3[k]<-exp(beta3[k]);
 }
}
# Data
sex[] age1[] age2[] freq[,1] freq[,2] freq[,3] n[]
0 0 0 26 12 7   45
0 1 0 9   21 15 45
0 0 1 5   14 41 60
1 0 0 40 17 8   65
1 1 0 17 15 12 44
1 0 1 8   15 18 41
END
# Initial values
list(beta0=c(0,0),beta1=c(0,0),beta2=c(0,0),beta3=c(0,0))
```

The estimates from this model are shown in Table 14.5. The estimates were based on 10,000 MCMC iterations followed by 50,000 iterations thinned by 5. The fitted values from WinBUGS were almost identical to those in Table 8.3, and the goodness of fit statistic was $X^2 = 3.923$ (compared with $X^2 = 3.931$ from Table 8.3).

Table 14.5 *Results of fitting the nominal logistic regression model (8.11) to the data in Table 8.1. Results can be compared with Table 8.2.*

Parameter β	Estimate b (std. dev.)	Odds ratio, $OR = e^b$ (95% posterior interval)	
$\log(\pi_2/\pi_1)$: important vs. no/little importance			
β_{02}: constant	−0.602 (0.285)		
β_{12}: men	−0.386 (0.304)	0.71	(0.37, 1.24)
β_{22}: age 24–40	1.139 (0.344)	3.31	(1.61, 6.18)
β_{32}: age > 40	1.619 (0.403)	5.48	(2.32, 11.26)
$\log(\pi_3/\pi_1)$: very important vs. no/little importance			
β_{03}: constant	−1.063 (0.330)		
β_{13}: men	−0.816 (0.320)	0.47	(0.24, 0.82)
β_{23}: age 24–40	1.498 (0.407)	4.86	(2.03, 10.02)
β_{33}: age > 40	2.969 (0.433)	21.43	(8.47, 45.99)

14.4   Latent variable model

In Section 8.4 the use of latent variables for ordinal regression was examined. A latent variable model assumes there is an underlying continuous response (z) which is observed as an ordinal variable (see Figure 8.2).

Such latent variable models can be fitted in WinBUGS by estimating the latent cutpoints (C) and using the cumulative logit model (8.14). The probability that the continuous latent response (z) is greater than the jth cutpoint is

$$Q_j = P(z > C_j).$$

Then the cumulative logit model is

$$\log\left(\frac{Q_j}{1 - Q_j}\right) = \frac{P(z > C_j)}{P(z \leq C_j)} = \mathbf{x}\boldsymbol{\beta} - C_j. \tag{14.2}$$

Note that the j subscript on the parameter estimates $\boldsymbol{\beta}$ has been dropped. This formulation of the latent variable model is equivalent to the proportional odds model (8.17) (see Exercise 14.2).

From this model the probabilities can be estimated using

$$\pi_j = \begin{cases} 1 - Q_j, & j = 1 \\ Q_{j-1} - Q_j, & j = 2, \ldots, J - 1 \\ Q_j, & j = J \end{cases},$$

where J is the total number of latent classes.

The difficultly in fitting such models in standard statistical packages is in estimating the cutpoints. Using Bayesian methods Uniform prior distributions can be used to ensure the cutpoints are ordered: $C_1 < C_2 < \ldots < C_{J-1}$.

$$\begin{aligned} C_1 &= 0, \\ C_j &\sim U[C_{j-1}, C_{j+1}], & j = 2, \ldots, J - 2 \\ C_{J-1} &\sim U[C_{J-2}, \infty]. \end{aligned}$$

The first cutpoint is assigned at an arbitrary fixed value of zero to ensure identifiability. For the last cutpoint, C_{J-1}, a large upper limit is used in place of ∞.

The car preference example has three ratings and so requires two cutpoints C_1 and C_2. In this example the underlying latent variable, z, is the importance rating. Although it is observed as an ordinal variable taking values 1, 2 or 3, it is assumed that the underlying (latent) z value is continuous. The WinBUGS code for this model is below.

```
_____ WinBUGS code (latent variable model) _____
model{
## Likelihood
  for (i in 1:6){
    freq[i,1:ncat]~dmulti(pi[i,1:ncat],n[i]);
    # Cumulative probability of > category k given cutpoint
    for (k in 1:ncat-1){
      logit(Q[i,k])<-(beta[1]+(beta[2]*sex[i])+(beta[3]*age1[i])
                  +(beta[4]*age2[i]))-C[k];
    }
# Calculate probabilities
    pi[i,1] <- 1-Q[i,1]; # Pr(cat=1)=1-Pr(cat>1);
    for (k in 2:ncat-1) {
      pi[i,k] <- Q[i,k-1]-Q[i,k]; # Pr(cat>k-1)-Pr(cat>k);
    }
    pi[i,ncat] <- Q[i,ncat-1]; # Pr(cat=k)=Pr(cat>k-1);
    for (j in 1:3){
      fitted[i,j]<-pi[i,j]*n[i]; # Fitted values
    }
  }
## Priors
  for (k in 1:4){beta[k]~dnorm(0,0.00001);}
# ordered cut-offs
    C[1]<-0; # for identifiability
    C[2]~dunif(C[1],10);
## Scalars
  for (k in 1:3){or[k]<-exp(beta[k+1]);}
}
# Data
list(ncat=3)
sex[] age1[] age2[] freq[,1] freq[,2] freq[,3] n[]
0 0 0 26 12 7  45
0 1 0 9  21 15 45
0 0 1 5  14 41 60
1 0 0 40 17 8  65
1 1 0 17 15 12 44
1 0 1 8  15 18 41
END
# Initial values
list(beta=c(0,0,0,0), C=c(NA,1))
```

The initial value for the first cut-point is missing (NA) because it is set to zero for identifiability and therefore does not need to be sampled.

The results from fitting the latent variable model are shown in Table 14.6. The odds ratios, $\exp(\beta)$, are the odds of being in a higher category. The results are very similar to those from the proportional odds ordinal regression model in Table 8.4.

Table 14.6 *Results of fitting the latent variable ordinal model to the data in Table 8.1.*

Parameter β	Estimate b (std. dev.)	Odds ratio, $OR = e^b$ (95% posterior interval)
β_0: constant	−0.045 (0.232)	
β_1: men	−0.580 (0.227)	0.57 (0.56, 0.88)
β_2: age 24–40	1.162 (0.277)	3.32 (1.88, 5.51)
β_3: age > 40	2.258 (0.291)	9.98 (5.48, 17.2)

The DIC for the nominal model was 66.8 with 8.0 estimated parameters. The DIC for the latent variable model was 61.5 with 5.1 estimated parameters, indicating a reasonable improvement in model fit and reduction in complexity. The goodness of fit statistic for the latent variable model was $X^2 = 4.56$.

14.5   Survival analysis

WinBUGS deals with censored values in a different way to standard software. It uses two values for survival time: the observed time and the minimum time. A right censored observation has a missing observed time and a minimum time equal to the censored time. An uncensored observation has a non-missing observed time and a notional minimum time of zero. This formulation treats censored observations as missing data. A corollary of this parameterization is that the survival times for censored observations are estimated.

The notation for survival time is

$$y = \max\{y_r, y^\star\}$$

where y_r is the observed survival time and $y^\star$ the minimum time. For the observed time a parametric distribution might be used such as the Weibull model, $y_r \sim \text{Wei}(\phi, \lambda)$. The distribution for y is then a truncated Weibull distribution.

A WinBUGS survival analysis is illustrated using the data on remission times (Table 10.1). The data need to have both the observed time and the minimum time. The first four rows of the ordered data are shown below for the treatment group (trt[]=1).

```
WinBUGS code (formatting survival data, ''Remission.txt'')
trt[] y[] y.star[]
1 6 0
1 6 0
1 6 0
1 NA 6
...
```

The first three subjects all died at 6 weeks. The fourth subject was censored at 6 weeks, so the observed time is unknown (NA) and the minimum time (y.star[]) is 6.

The WinBUGS model using a Weibull distribution for survival time is shown below.

```
— WinBUGS code (Weibull survival model, ''Weibull.odc'') —
model{
    for(i in 1:N){
        y[i]~dweib(lambda,phi[i])I(y.star[i],);
        log(phi[i])<-beta_0 + beta_1*trt[i]; # log link
    }
    lambda~dexp(0.001);
    beta_0~dnorm (0,0.001);
    beta_1~dnorm (0,0.001);
    # Median survival in control
    median[1]<-pow(log(2)*exp(-beta_0),1/lambda);
    # Median survival in treatment
    median[2]<-pow(log(2)*exp(-(beta_0+beta_1)),1/lambda);
    diff<-median[1]-median[2];
}
```

Using the Bayesian analysis, the interesting test statistic of the difference in median survival time between the treatment and control groups can be calculated (diff).

A useful option in WinBUGS is to run models using a script file, rather than pointing and clicking on the menu bars. Script files do speed up analysis, but it is recommended that the model first be checked using the menu bars as the error messages from script files are less specific. An example is given

below for the Weibull survival model. The relevant files are assumed to be in
a directory called c:/bayes.

```
_____ WinBUGS code (script file for the remission data) _____
display('log')
check('c:/bayes/Weibull.odc')
data('c:/bayes/Remission.txt')
data('c:/bayes/RemissionN.txt')
compile(2)
inits(1,'c:/bayes/Initials.txt')
inits(2,'c:/bayes/Initials.txt')
gen.inits() # For censored survival times
update(5000) # Burn-in
set(beta_0)
set(beta_1)
set(lambda)
update(10000)
gr(*) # Gelman-Rubin convergence test
history(*)
density(*)
stats(*)
save('c:/bayes/Weibull_results.odc')
```

The file Weibull.odc contains the model file shown above, and the file
Remission.txt contains the data in rectangular format (the first four lines
of which are shown above). The file RemissionN.txt contains the sample
size N in S-PLUS format. The file Initials.txt contains the initial values in S-PLUS format. This script runs two chains and so is able to use the
Gelman–Rubin statistic to assess convergence (Section 13.5.3).

The results of fitting the model are shown in Table 14.7. These results can
be compared with the classical model results in Table 10.3. The estimated
difference (and 95% posterior interval) between the median survival times is
a particular advantage of the Bayesian model.

14.6   Random effects

In this section random effects models are considered. These models are also
known as mixed models because they use a mixture of fixed and random effects. In Chapter 11 a mixed model using classical statistics was applied. The
parameterization of a mixed model in a Bayesian setting is identical to the frequentist approach; hence, mixed models are one of the closest links between

Table 14.7 *Results of fitting the Weibull model to the remission times data.*

Variable	Mean	95% posterior interval
Group β_1	-1.778	$-2.64, -0.978$
Intercept β_0	-3.162	$-4.349, -2.145$
Shape λ	1.39	$1.023, 1.802$
Median survival		
Control	7.538	$5.183, 10.47$
Treatment	27.65	$17.24, 45.33$
Difference	20.11	$9.367, 38.12$

classical and Bayesian methods. The biggest difference is the parameter estimation, which is made using maximum likelihood for classical methods and MCMC for Bayesian methods.

Random intercepts

A random effects model using a random intercepts model for longitudinal data, such as the stroke recovery scores (Table 11.1), is illustrated. Random intercepts are useful for longitudinal data because they can capture the dependence among repeated results from the same subject. For the stroke recovery data for subject j at time t, a random intercepts model is

$$Y_{jt} = \alpha_g + a_j + \beta_g t + e_{jt}, \quad j = 1, \ldots, 24, \ t = 1, \ldots, 8, \ g = 1, 2, 3,$$
$$e_{jt} \sim N(0, \sigma_e^2),$$

where α_g is the intercept in group g and a_j is the departure from this intercept for subject j which is specified using a Normal distribution

$$a_j \sim N(0, \sigma_a^2).$$

The intercept for subject j in treatment group g is $\alpha_g + a_j$. The number of parameters is much larger than the models fitted in Section 11.3 because there is an intercept for each of the 24 subjects together with the variance (σ_a^2).

The WinBUGS code for the random intercepts model is given below. Uniform priors are used for the variance σ_a^2 (var.subject) and the residual variance σ_e^2 (var.resid). The code inverts these priors to create the precisions (tau), which are used by the Normal distribution (dnorm). To improve convergence, the original ability scores were roughly centered by subtracting 50 (Section 13.5). To make the results comparable with those from the classical analysis the intercepts (α's) are then re-scaled by adding 50.

─────────── WinBUGS code (random intercepts) ───────────

```
model{
# likelihood
  for(subject in 1:N){ # loop in subject
    intercept[subject]~dnorm(0,tau.subject);
    for(time in 1:T){ # loop in time
      mu[subject,time]<-intercept[subject] +
        alpha.c[group[subject]] + (beta[group[subject]]*time);
      ability[subject,time]~dnorm(mu[subject,time],tau.resid);
      } # end of time loop
  } # end of subject loop
# priors
  var.subject~dunif(0,1.0E4);
  var.resid~dunif(0,1.0E4);
  beta[1]~dnorm(0,1.0E-4); # Linear effect of time (group=A)
  beta[2]~dnorm(0,1.0E-4); # Linear effect of time (group=B)
  beta[3]~dnorm(0,1.0E-4); # Linear effect of time (group=C)
  alpha.c[1]~dnorm(0,1.0E-4); # Centered intercept (group=A)
  alpha.c[2]~dnorm(0,1.0E-4); # Centered intercept (group=B)
  alpha.c[3]~dnorm(0,1.0E-4); # Centered intercept (group=C)
# scalars
  tau.subject<-1/var.subject;
  tau.resid<-1/var.resid;
  rho<-var.subject/(var.resid+var.subject); # Within-sub corr.
  b.diff[1]<-beta[2]-beta[1];
  b.diff[2]<-beta[3]-beta[1];
  a.diff[1]<-alpha.c[2]-alpha.c[1];
  a.diff[2]<-alpha.c[3]-alpha.c[1];
}
```

The group intercepts α_g (alpha.c in the above code) are fixed effects each having a single fixed value. However, in a Bayesian setting the prior for these fixed effects is given a distribution, which makes them appear like random effects. In the above code a vague Normal prior was used, $\alpha_g \sim N(0, 10000)$.

The estimates of the subject-specific intercepts and re-scaled intercepts in each treatment group ($\hat{\alpha}_g + 50$) are shown in Figure 14.3. There was a reasonably wide variation in intercepts and $\hat{\sigma}_a$ was 22.4. The variation in intercepts is also evident in the plot of the data (Figure 11.2).

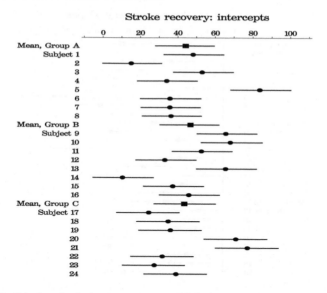

Figure 14.3 *Random intercept estimates and 95% posterior intervals for the stroke recovery data. Squares represent mean treatment intercepts ($\hat{\alpha}_g$); circles represent mean subject intercepts ($\hat{\alpha}_g + \hat{a}_j$); the horizontal lines are the 95% posterior intervals.*

Random slopes

The random intercepts model can be extended by also adding random (or subject-specific) slopes using

$$Y_{jt} = \alpha_g + a_j + (\beta_g + b_j)t + e_{jt},$$
$$b_j \sim N(0, \sigma_b^2),$$

so the slope for subject j is $\beta_g + b_j$. The estimates of the subject-specific slopes and slopes in each treatment group ($\hat{\beta}_g$) are shown in Figure 14.4. There was wide variation in slopes reflected by $\hat{\sigma}_b = 3.3$. In every treatment group there was at least one subject who did not significantly improve over time as the 95% posterior interval for their slope included zero.

Other estimates from these random effect models are shown in the next section.

14.7 Longitudinal data analysis

In this section Bayesian methods to analyse longitudinal data are used. The first model considered uses a multivariate Normal distribution. For observed

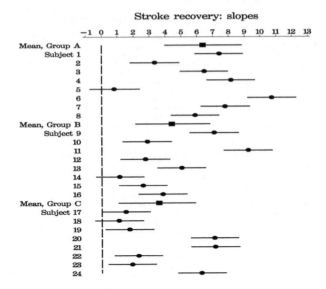

Figure 14.4 *Random slopes and 95% posterior intervals for the stroke recovery data. Squares represent mean treatment slopes ($\hat{\beta}_g$); circles represent mean subject slopes ($\hat{\beta}_g + \hat{b}_j$); the horizontal lines are the 95% posterior intervals.*

data **y** (of length n) and covariates **X**, the equations are

$$\begin{aligned} \mathbf{y} &= \mathbf{X}\boldsymbol{\beta} + \mathbf{e}, \\ \mathbf{e} &\sim \text{MVN}(\mathbf{0}, \mathbf{V}), \end{aligned} \tag{14.3}$$

where **e** has a multivariate Normal distribution (of dimension T) and **V** is a $T \times T$ variance–covariance matrix (with T being the maximum number of time points). The matrix **V** must be symmetric and positive definite (Section 1.5). The covariance between the errors accounts for the longitudinal nature of the data. Fitzmaurice et al. (2012, Chapter 7) discuss these models in detail, and they define them as **covariance pattern models**. The model structure is very similar to the Generalized Estimating Equation (GEE) model from Section 11.3, but in a Bayesian setting the model is fitted using MCMC methods rather than score equations.

Defining the variance–covariance matrix

In the previous example (Section 14.6) the Normal distribution was specified in terms of the precision not the variance. Similarly for multivariate Normal data WinBUGS uses the inverse of the variance–covariance matrix $\mathbf{V}^{-1}$. Examples are given below of inverse variance–covariance matrices for four common residual structures. The individual elements of the matrix are referred to using the indices $\{j,k\} = 1,\ldots,m$. Also given are examples of suitable priors.

1. For an independent covariance the diagonal of $\mathbf{V}^{-1}$ is given an equal precision and the off-diagonals are zero

$$V_{jk}^{-1} = \begin{cases} 1/\sigma^2, & j=k \\ 0, & \text{otherwise} \end{cases},$$

$$\sigma^2 \sim U[0,1000],$$

where σ^2 is the variance of the errors from model (14.3) can be assigned a vague Uniform prior.

2. An exchangeable variance–covariance is defined by Equation (11.7). Using the same notation the inverse variance–covariance is

$$V_{jk}^{-1} = \begin{cases} [1+(T-2)\rho]/\gamma, & j=k=1,\ldots,T \\ -\rho/\gamma, & \text{otherwise} \end{cases},$$

$$\gamma = \sigma^2[1+(T-2)\rho+(T-1)\rho^2],$$

$$\sigma^2 \sim U[0,1000],$$

$$\rho \sim U(-1,1),$$

where ρ is the within-subject correlation.

3. An autoregressive (AR) variance–covariance is defined by Equation (11.9). The inverse covariance matrix is specified using a known matrix inversion:

$$V_{jk}^{-1} = \begin{cases} \tau, & j=k=1,T \\ \tau(1+\rho^2), & j=k=2,\ldots,T-1 \\ -\tau\rho, & j=1,\ldots,T-1, k=j+1 \\ -\tau\rho, & k=1,\ldots,T-1, j=k+1 \\ 0, & \text{otherwise} \end{cases},$$

$$\rho \sim U(-1,1),$$

$$\tau \sim G(0.001,0.001),$$

$$\sigma^2 = [\tau(1-\rho^2)]^{-1}.$$

In this model ρ^d is the correlation between observations that are d time-points apart and σ^2 is the variance of the errors. For this matrix a Gamma prior is used for the precision (τ) and the variance (σ^2) is calculated.

4. An unstructured variance–covariance is defined by Equation (11.10). Rather than giving a structure to the inverse covariance matrix (as for the

above three covariance types), an unstructured matrix can be created using the Wishart distribution

$$\mathbf{V}^{-1} \sim W(\mathbf{R}, v),$$

where $\mathbf{R}$ is a $T \times T$ matrix and v is the degrees of freedom. The inverse Wishart is the conjugate prior for the multivariate Normal and gives covariance matrices that are symmetric and positive definite. The conjugacy means that this posterior estimate for $\mathbf{V}^{-1}$ also follows a Wishart distribution with degrees of freedom $n + v$. The posterior mean for $\mathbf{V}$ has the simple weighted form (Leonard and Hsu 1992):

$$\mathbf{V} = \left(n\mathbf{S} + v\mathbf{R}^{-1}\right)/(n+v)$$

where $\mathbf{S} = \sum \mathbf{y}\mathbf{y}^T /n$ is the sample covariance matrix. It is clear that a large value for v gives more weight to the prior matrix $\mathbf{R}^{-1}$, and larger sample sizes give greater weight to the sample covariance $\mathbf{S}$.

Example results using the stroke recovery data

The multivariate Normal model (14.3) and the above four covariance structures are illustrated using the stroke recovery data from Chapter 11. The following multivariate normal was fitted:

$$\begin{aligned}
\mathbf{Y}_i &= \alpha_g + \beta_g t_i + \mathbf{e}_i, & i = 1, \ldots, 27,\ g = 1, 2, 3, \\
\mathbf{e}_i &\sim \text{MVN}(\mathbf{0}, \mathbf{V}).
\end{aligned}$$

As in the last section a different intercept (α) and slope (β) was used for each of the three treatment groups ($g = 1, 2, 3$). These parameters were given vague Normal priors

$$\alpha_g \sim N(0, 1000), \qquad \beta_g \sim N(0, 1000), \qquad g = 1, 2, 3.$$

Vague priors are also used for the covariance matrix $\mathbf{V}$.

.2The intercept and slope estimates from fitting the different covariance structures are given in Table 14.8. For comparison the results of a random effects model are also included, using a random intercept for each subject (Section 14.6). The results can be compared with those using classical methods in Table 11.7. The intercepts ($\hat{\alpha}$) are slightly different between the Bayesian and classical results. However, both sets of results indicate that a common intercept may provide an adequate fit to the data. A common intercept also seems plausible when examining Figure 11.3. Despite the difference in intercepts, the estimates for the slopes were similar for the Bayesian and classical methods.

Table 14.8 *Estimates of intercepts and changes over time for the stroke recovery data. Results can be compared with Table 11.7.*

	Intercept estimates		
	$\hat{\alpha}_1$ (s.d.)	$\hat{\alpha}_2 - \hat{\alpha}_1$ (s.d.)	$\hat{\alpha}_3 - \hat{\alpha}_1$ (s.d.)
Covariance pattern models			
Independent	29.55 (5.864)	3.496 (8.318)	−0.013 (8.222)
Exchangeable	29.71 (8.154)	3.282 (11.59)	−0.084 (11.59)
AR(1)	33.17 (7.889)	−0.043 (11.21)	−6.139 (11.23)
Unstructured	34.62 (6.830)	−5.022 (9.918)	−8.946 (9.966)
Random effects models			
Intercepts	29.98 (8.051)	2.682 (11.74)	−0.407 (11.95)
Intercepts + slopes	30.38 (7.988)	3.675 (11.96)	−0.941 (11.48)

	Slope estimates		
	$\hat{\beta}_1$ (s.d.)	$\hat{\beta}_2 - \hat{\beta}_1$ (s.d.)	$\hat{\beta}_3 - \hat{\beta}_1$ (s.d.)
Covariance pattern models			
Independent	6.341 (1.159)	−2.027 (1.645)	−2.696 (1.635)
Exchangeable	6.320 (0.472)	−1.985 (0.665)	−2.680 (0.666)
AR(1)	6.062 (0.843)	−2.128 (1.199)	−2.236 (1.187)
Unstructured	6.436 (1.042)	−2.533 (1.507)	−3.023 (1.526)
Random effects models			
Intercepts	6.325 (0.476)	−1.991 (0.674)	−2.694 (0.668)
Intercepts + slopes	6.378 (1.207)	−1.966 (1.718)	−2.784 (1.658)

For the random effects model the estimate of the within-subject correlation was 0.858 with a 95% posterior interval (PI) of 0.766 to 0.930. For the exchangeable correlation the within-subject correlation was 0.856 with a 95% PI of 0.764 to 0.923. For the AR(1) correlation the estimate of $\hat{\rho}$ was 0.946 with a 95% PI of 0.915 to 0.971. The posterior intervals for these correlations are easily calculated in WinBUGS, but are not given by any standard statistical package as they are difficult to calculate in a classical paradigm.

The WinBUGS code to calculate the AR(1) model is given below. To improve the convergence of the model, 50 was subtracted from every stroke recovery score (Table 11.1).

——————— WinBUGS code (AR(1) covariance pattern) ———————

```
model{
# likelihood
 for(subject in 1:N) { # loop in subject
  ability[subject,1:T]~dmnorm(mu[subject,1:T],omega.obs[1:T,1:T]);
```

```
for(time in 1:T) { # loop in time
  mu[subject,time]<- alpha.c[group[subject]]
                    + (beta[group[subject]]*time);
 } # end of time loop
} # end of subject loop
# variance--covariance matrix
omega.obs[1,1]<- tau.obs; omega.obs[T,T]<- tau.obs;
for (j in 2:T-1){omega.obs[j, j]<- tau.obs*(1+pow(rho,2));}
for (j in 1:T-1){omega.obs[j, j+1]<- -tau.obs*rho;
  omega.obs[j+1, j] <-omega.obs[j, j+1];} # symmetry
for (i in 1:T-1) {
  for (j in 2+i:T) {
   omega.obs[i, j] <-0; omega.obs[j, i] <-0;
  }
}
# priors
tau.obs~dgamma(0.001,0.001);
rho~dunif(-0.99,0.99); # correlation parameter
beta[1]~dnorm(0,1.0E-4); # Linear effect of time (group=A)
beta[2]~dnorm(0,1.0E-4); # Linear effect of time (group=B)
beta[3]~dnorm(0,1.0E-4); # Linear effect of time (group=C)
alpha.c[1]~dnorm(0,1.0E-4); # Centered intercept (group=A)
alpha.c[2]~dnorm(0,1.0E-4); # Centered intercept (group=B)
alpha.c[3]~dnorm(0,1.0E-4); # Centered intercept (group=C)
# scalars
b.diff[1]<-beta[2]-beta[1];
b.diff[2]<-beta[3]-beta[1];
a.diff[1]<-alpha.c[2]-alpha.c[1];
a.diff[2]<-alpha.c[3]-alpha.c[1];
var.obs <-1 / (tau.obs*(1-pow(rho,2))); # sigma^2
}
```

The estimated unstructured covariance and correlation matrices are plotted in Figure 14.5. The correlation matrix was calculated from the covariance matrix using the following formula:

$$\rho_{jk} = \frac{V_{jk}}{\sqrt{V_{jj}V_{kk}}}, \qquad j,k = 1,\dots,8.$$

The covariance matrix in Figure 14.5 shows a clear increase in variance over time. The correlation matrix shows a steady decay with increasing distance between observations. The correlations shown in Figure 14.5 are

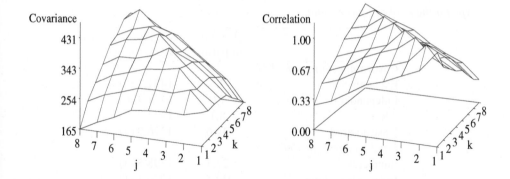

Figure 14.5 *Estimated unstructured covariance and correlation matrices for the stroke recovery data using a vague prior.*

slightly smaller than those in Table 11.2, which are based on the raw data. The correlations in Figure 14.5 are for the residuals after fitting terms for treatment and treatment-by-time. So some of the correlation between the raw results has been explained by this model.

Covariance selection

There are a number of different covariance structures available, but which one gives the best fit to the data? The model fit can be assessed using the DIC (Section 13.6). Table 14.9 shows the DIC values for the four covariance pattern models and two random effects models.

Using an independent correlation the DIC calculation shows the estimated number of parameters as 7.0. For this model the DIC calculation has given the exact number of parameters (three intercepts + three slopes + one variance). Based on the DIC the independent model is the worst fit. An AR(1) correlation uses only one more parameter but gives a much better fit than the independent model (change in DIC: $1721.3 - 1342.0 = 379.3$). An AR(1) correlation seems sensible when considering the pattern shown by the unstructured correlations in Figure 14.5.

The DIC suggests that the best fitting model overall was a model with random slopes and intercepts. This had a substantially better fit than the AR(1) covariance pattern model, change in DIC: $1342.0 - 1226.1 = 115.9$. This better fit was achieved using the largest number of parameters. As there are 24 subjects a random intercept and slope for each would be 48 parameters, plus 2

Table 14.9 *Using DIC values for choosing the optimal covariance structure or model type for the stroke recovery data.*

	Est. no. of parms.	DIC (rank[†])
Covariance pattern models		
Exchangeable	7.5	1471.1 (5)
Independent	7.0	1721.3 (6)
AR(1)	8.1	1342.0 (2)
Unstructured	34.6	1376.9 (3)
Random effects models		
Intercepts	27.4	1398.5 (4)
Intercepts + slopes	46.7	1226.1 (1)

[†] lower values indicate a better model

more for the Normal variances. Subtracting the 7 regression parameters from the estimated number of parameters gives 39.7 random parameters (only 8.3 short of 48). This suggests that many subjects had different intercepts and slopes. The variance in the intercepts and slopes is clear in Figures 14.3 and 14.4.

Table 14.9 highlights a major strength of the DIC: it can be used to compare very different models (e.g., random effects and covariance pattern models). This is because WinBUGS uses the same algorithm to fit both models (i.e., MCMC), whereas classical statistics uses different algorithms depending on the problem (e.g., score equations for GEE models, and maximum likelihood estimation for random effects models). Comparing such models in a classical setting is only possible using different approaches such as cross-validation.

The DIC fit statistics for the covariance pattern models in Table 14.9 are comparable with the AIC statistics in Table 11.8. To make the number of parameters comparable six was added to the parameter numbers to allow for the three intercepts and three slopes in Table 11.8 as these are only for the variance–covariance parameters. The parameter numbers differ by less than half and the model fits have the same interpretation with the autoregressive model as a clear winner followed by the unstructured with the independent model as the worst choice.

14.8   Bayesian model averaging

Uncertainty in inference was previously considered using standard errors, confidence intervals or credible intervals, with larger standard errors and wider intervals indicating greater uncertainty. Another source of uncertainty is **model uncertainty** which arises when there is more than one model that could be used to describe the data. For example, in the previous section six different models were fitted to the stroke data (Table 14.9), but the parameter estimates were only given for the best fitting model (e.g., the difference in slopes) this could lead to over-confident inferences that ignore other potential explanations of the data.

This model uncertainty can be considered using **Bayesian model averaging** which combines inferences from multiple models. For a detailed introduction to Bayesian model averaging see Hoeting et al. (1999). Using Bayesian model averaging the posterior distribution for the parameter of interest θ given the data $\mathbf{y}$ is averaged over K models:

$$P(\theta|\mathbf{y}) = \sum_{k=1}^{K} P(\theta|M_k, \mathbf{y}) P(M_k|\mathbf{y}). \tag{14.4}$$

and

$$P(M_k|\mathbf{y}) = \frac{P(\mathbf{y}|M_k) P(M_k)}{\sum_{j=1}^{K} P(\mathbf{y}|M_j) P(M_j)}, \tag{14.5}$$

so that over all K models considered

$$\sum_{k=1}^{K} P(M_k|\mathbf{y}) = 1.$$

The posterior distribution (14.4) uses the parameter estimates given a model M_k, multiplied by the probability of that model. This approach means that K possible models are considered and not just a model with the best fit to the data, although models that have a better fit will have a greater influence as their probability will be larger.

The posterior mean for the parameter θ is simply:

$$E(\theta|\mathbf{y}) = \sum_{k=1}^{K} E(\theta|M_k, \mathbf{y}) P(M_k|\mathbf{y}).$$

which is a weighted mean of the estimates from all models. The posterior variance for the parameter θ is

$$Var(\theta|\mathbf{y}) = \sum_{k=1}^{K} \left\{ \left(Var(\theta|M_k, \mathbf{y}) + E(\theta|M_k, \mathbf{y})^2 \right) P(M_k|\mathbf{y}) \right\} - E(\theta|\mathbf{y})^2.$$

If all models are assumed equally likely, so that $P(M_k) = 1/K$, then

the posterior probability of each model (14.5) can be estimated using the
Bayesian information criterion (Section 7.5) as:

$$P(M_k|\mathbf{y}) \approx \frac{-0.5\exp\left[\text{BIC}_k - \max(\text{BIC})\right]}{\sum_{j=1}^{K} -0.5\exp\left[\text{BIC}_j - \max(\text{BIC})\right]}$$

where $\max(\text{BIC})$ is the largest BIC (poorest fit) among all the models fitted.

14.8.1   Example: Stroke recovery

For the stroke recovery data we previously compared four covariance pattern
models and two mixed models (Table 14.9). The BIC values and posterior
probabilities for these six models are in Table 14.10. The AR(1) model is
a far superior with a posterior probability close to one and so there will be
no point in using Bayesian model averaging as the results will be virtually
identical to the AR(1) model. There may still exist some model uncertainty
because other plausible variance–covariance structures have not been tested
(see Exercise 14.6 for an example).

Table 14.10 *BIC values and estimated posterior probabilities for the six models fitted
to the recovery from stroke data.*

Model	BIC	Δ BIC	Posterior probability
Independent	1726	0	< 0.001
Exchangeable	1479	−248	< 0.001
AR(1)	1346	−380	> 0.999
Unstructured	1451	−275	< 0.001
Random intercept	1479	−247	< 0.001
Random intercept and slope	1378	−348	< 0.001

14.8.2   Example: PLOS Medicine *journal data*

This next example does not have one dominant model. It uses the *PLOS
Medicine* data of 875 journal articles with a dependent variable of the number
of online page views for each article (see Section 6.7.1). Because page views
are strongly positively skewed a base 2 logarithmic transform is used to reduce the skewness, so every one-unit increase represents a doubling of page
views. Four explanatory variables are used which are: the number of authors,
the number of characters in the title, and a sinusoidal seasonal pattern using

sine and cosine transformations of the online publication date. The full model including all four variables is

$$
\begin{aligned}
\log(Y_i, 2) \;=\; & \beta_0 + \beta_1 x_{i,1}/5 + \beta_2 x_{i,2}/10 + \beta_3 \sin\left[2\pi(x_{i,3} - 1)/52\right] \\
& + \beta_4 \cos\left[2\pi(x_{i,3} - 1)/52\right], \qquad i = 1,\ldots,875,
\end{aligned}
$$

where $\mathbf{x}$ is an $n \times 3$ matrix with $x_{i,1}$ equal to the number of authors, $x_{i,2}$ equal to the number of characters in the title, and $x_{i,3}$ equal to the week of the year that the article was published (between 1 and 52). To show meaningful sized changes, the number of authors is scaled by 5 and the number of title characters by 10 (see Section 6.7).

With $K = 4$ explanatory variables there are 15 different models to choose from using the combinations of all 4 variables, any 3 variables (4 models), any 2 variables (6 models) and any 1 variable (4 models) (assuming the intercept is included in every model).

The parameter estimates for the models with the four highest posterior probabilities are in Table 14.11, together with the model averaged posterior means and standard deviations. The model averaged standard deviations are larger for each variable, reflecting the greater uncertainty by accounting for multiple models. Apart from the intercept, the posterior means are closer to zero compared with the means from any of these four models, this occurs because a parameter estimate is zero when the variable is not included in the model. These zeros also increase in the posterior standard deviation.

There is large model uncertainty as the largest model posterior probability $(P(M_k|\mathbf{y}))$ is only 0.244 and the top four models combined have a posterior probability of 0.758. A combined posterior probability greater than 0.95 requires the top seven models. There is also uncertainty for most variables as reflected in the variable posterior probability which is the probability that that variable is not zero, calculated as the sum of the model posterior probabilities over all models which contain that variable. The cosine seasonal variable was not included in any of the top four models and had the lowest variable posterior probability of just 0.08, making it unlikely that this is an important explanatory variable.

In this example with $K = 4$ there were 15 different models to compare. However, the number of combinations increases rapidly with K, and $K = 10$ variables gives 1023 model combinations which could be computationally prohibitive depending on the size of the dataset. The paper by Hoeting et al. (1999) discusses ways to reduce the number of models considered prior to using Bayesian model averaging.

Bayesian model averaging can be applied in R using the BMA package (Raftery et al. 2005), and the code below gives the estimates in Table 14.11.

Table 14.11 *Posterior means and standard deviations using Bayesian model averaging for the PLOS Medicine data, together with the means and standard deviations of the four models with the highest model posterior probabilities.*

Variable	Model 1 (SD)	Model 2 (SD)	Model 3 (SD)	Model 4 (SD)	Posterior mean (SD)	Variable posterior probability
$\hat{\beta}_0$, intercept	13.33 (0.031)	13.31 (0.032)	13.32 (0.032)	13.33 (0.031)	13.32 (0.033)	1
$\hat{\beta}_1$, authors	0.058 (0.016)	0.045 (0.017)		0.058 (0.016)	0.040 (0.027)	0.76
$\hat{\beta}_2$, characters		0.022 (0.009)	0.029 (0.008)		0.014 (0.014)	0.58
$\hat{\beta}_3$, sin date				0.097 (0.041)	0.030 (0.050)	0.31
$\hat{\beta}_4$, cos date					0.006 (0.022)	0.08
No. of variables	2	3	2	3		
$P(M_k|\mathbf{y})$	0.244	0.221	0.160	0.133		

```
_____ R code (Bayesian model averaging) _____
>data$cosw = cos((2*pi*data$week-1)/52)
>data$sinw = sin((2*pi*data$week-1)/52)
>data$nchar = (data$nchar - 100)/ 10
>data$authors = (data$authors - 6)/ 5
>xmat = subset(data, select=c(nchar, authors, cosw, sinw))
>bma.out <- bicreg(x=xmat, y=log2(data$views))
>summary(bma.out, n.models=4)
```

14.9   Some practical tips for WinBUGS

WinBUGS is quite different from other statistical packages. New users have to devote time to learning a different style of programming. It can sometimes be frustrating as WinBUGS's error messages tend to be non-specific. One of the most common problems is failing to get a model started. That is, the model may compile but fail when sampling starts. If this happens it is recommended to try the following strategies (in rough order of importance):

1. Look at the current state of the chain as this often indicates the problem. In WinBUGS click on Model⇒Save State to see all the unknown parameters and their current value. Very large or very small values will often be the source of the problem (as they can throw the chain to a distant region of the

likelihood). Try giving any parameters with large values a smaller initial value or tighter prior. The current samples (with or without editing) can also be used as a complete set of initial values.

2. Use a dataset of standardized covariates (i.e., with zero mean and variance of one). WinBUGS finds it much easier to deal with covariates with a zero mean (as shown in Figure 13.12). It also prefers covariates with a reasonably sized variance, as very narrow or wide distributions can be difficult to sample from. The downside to such standardization is that the parameters estimates are now on a standard deviation scale. This scale may not be meaningful for every dataset, but it is easy to re-scale estimates by suitably transforming the final results.

3. Fit a similar model in a classical statistical analysis and use the estimated parameters as initial values in the Bayesian model. A good example of this is given in Figure 13.10. This Metropolis–Hastings chain was generated using initial values of $\beta_1^{(0)} = -0.2$ and $\beta_2^{(0)} = 18$. Using initial values equal to the maximum likelihood estimates of $\beta_1^{(0)} = 0.0$ and $\beta_2^{(0)} = 22$ (Table 7.4) would have saved many burn-in iterations.

4. If the model involves sampling from a Binomial (or Multinomial) distribution where the probability is very close to zero or one, then it may help to truncate the estimated probability. For example,

```
y[i]~Bin(n[i],p[i])
p[i]<-max(0.00001,p.reg[i])
logit(p.reg[i])<-alpha+beta*x[i]
```

Alternatively use the complementary log-log link instead of the logit.

5. Try reducing the variance of the uninformative priors (e.g., use a tolerance of 0.001 for a Normal distribution, rather than the WinBUGS default of 0.00001). This is a good way to get some estimates from the model. The variances can be relaxed again if these estimates reveal a problem with the model or the initial values.

6. Use the gen.inits() to generate initial values. This is especially helpful for numerous random effects; for example,

$$\alpha_j \sim N(\mu_\alpha, \sigma_\alpha^2), \qquad j = 1, \ldots, 100$$

would require creating 102 initial values. It is a good idea to first give initial values to the mean and variance (or precision). The generated initial values then come from this distribution. Use a relatively small initial variance, so that the initial random effects are close to the mean and can then fan-out as the Markov chain progresses.

7. Try simplifying the model, for example, using a fixed effect in place of a random effect. If that model converges, then use the mean estimates as the starting values in a more complex model.

14.10   Exercises

14.1 Confirm that setting $\phi_1 = 1$ in model (14.1) gives model (8.11).

14.2 Prove that the latent variable model (14.2) is equal to the proportional odds model (8.17).

14.3 a. Run the Weibull model for the remission times survival data, but this time monitor the DIC.

b. Create an exponential model for the remission times in WinBUGS. Monitor the DIC. Is the exponential model a better model than the Weibull? Compare the results with Section 10.7.

14.4 a. Fit the random intercepts and slopes model to the stroke recovery data in WinBUGS. Create a scatter plot of the estimated mean random slopes against the random intercepts. Calculate the Pearson correlation between the intercepts and slopes.

b. Model the random slopes and intercepts so that each subject's intercept is correlated with his or her slope (using a multivariate Normal distribution). Find the mean and 95% posterior interval for the correlation between the intercept and slope. Interpret the correlation and give reasons for the somewhat surprising value.

Hint: To start the MCMC sampling, it may be necessary to use initial values based on the means of the random intercepts and slopes model from part (a).

14.5 This exercise illustrates the effect of the choice of the Wishart prior for the unstructured covariance matrix using the stroke recovery data.

a. Fit a simple linear regression model to the stroke recovery data with terms for treatment and treatment by time. Store the residuals. Use either Bayesian or classical methods to fit the model.

b. Adapt the WinBUGS code for the AR(1) covariance pattern model to an unstructured covariance.

c. Run the model using a vague Wishart prior defined by $\mathbf{R} = \hat{\sigma}^2 \mathbf{I}$ and $v = 9$. $\mathbf{I}$ is the 8×8 identity matrix and $\hat{\sigma}^2$ is the variance of the residuals from part (a).

d. Run the model using a strong Wishart prior defined by an $\mathbf{R}$ equal to the covariances of the residuals from part (a) and $v = 500$. Monitor the intercepts, slopes and covariance and the DIC. Use the same sized burn-in and samples as part (c).

What does the prior value of v of 500 imply?

e. Compare the results from the vague and strong priors. What similarities and differences do you notice for the parameter estimates and covariance matrix? Explain the large difference in the DIC.

14.6 a. Find the inverse of the variance–covariance matrix

$$
\mathbf{V} =
\begin{bmatrix}
\sigma^2 & \rho\sigma^2 & 0 & 0 & 0 & 0 & 0 \\
\rho\sigma^2 & \sigma^2 & \rho\sigma^2 & 0 & 0 & 0 & 0 \\
0 & \rho\sigma^2 & \sigma^2 & \rho\sigma^2 & 0 & 0 & 0 \\
0 & 0 & \rho\sigma^2 & \sigma^2 & \rho\sigma^2 & 0 & 0 \\
0 & 0 & 0 & \rho\sigma^2 & \sigma^2 & \rho\sigma^2 & 0 \\
0 & 0 & 0 & 0 & \rho\sigma^2 & \sigma^2 & \rho\sigma^2 \\
0 & 0 & 0 & 0 & 0 & \rho\sigma^2 & \sigma^2
\end{bmatrix}
$$

b. Fit the model to the stroke recovery data as a covariance pattern model. Compare the parameter estimates and overall fit (using the DIC) to the results in Tables 14.8 and 14.9.

c. What does the matrix assume about the correlation between responses from the same subject?

Postface

Far too often mistakes are made in data analysis (Altman 1994; Chalmers and Glasziou 2009). Many analyses are well conducted, but lack detail in their reporting so readers cannot understand what methods were used or replicate the results. Real harm can occur to people and societies when erroneous or incomplete results are used to change policy or practice. Poor scientific practice also harms science as it undermines the public's confidence in the ability of science to discover truths. Widespread failings in data analysis and reporting have been documented across most scientific fields and it has been called the "**reproducibility crisis**" because so few studies can be reproduced (Baker 2016).

Poor statistical practices are an important part of the reproducibility crisis, and problems include using the wrong methods, excluding data that do not fit with prior beliefs, re-analyzing data until a desired result is achieved, and a preoccupation with achieving "statistical significance" (usually a p-value less than 0.05) (Ziliak and McCloskey 2008). There are well established good statistical practices that avoid these problems and help give evidence that best reflects the data (National Academy of Sciences and National Academy of Engineering and Institute of Medicine 2009). In this Postface steps are discussed that should be taken at the analysis and reporting stages to help ensure reproducible research. These steps can be applied to any type of study and are pertinent to all fields of science and all the methods presented in this book.

1. Write a **protocol** prior to collecting data. A protocol pre-specifies key information such as the research questions, data collection methods and sample size. Protocols help avoid data dredging, where multiple analyses are tried until an "exciting" or "statistically significant" result is found and then reported. Such findings have a much higher risk of being spurious associations that are not reproducible.

    Data mining is a legitimate technique that can uncover true associations and is particularly useful for large and complex data. However, if data mining or multiple analyses are used to generate the results, then this should be reported so that readers can understand how the results were generated and if they are robust.

Thought should be given to which variables need to be collected and the protocol can include a complete list of variables and data sources. It is usually too late at the analysis stage to realize that an important variable has not been collected. Conversely many researchers make the mistake of collecting too many variables especially where there are minimal costs to collecting data (e.g., automated measurements). However, collecting too many variables can harm the overall quality of the data. An example of this is where data become hard to manage because of the size meaning important details might be missed. Researchers also often make the mistake of using long surveys which people may partially complete or throw away. Having too many variables can also slow research because it requires additional data management and can greatly complicate the analysis, especially where multiple highly correlated variables have been collected.

Protocols are more commonly used for prospectively collected data such as planned experiments or randomized trials. They are less common for the retrospective analysis of existing data, but such studies still require a statistical analysis plan (see next point).

The World Health Organization has a useful guideline on what to include in a protocol: www.who.int/rpc/research_ethics/guide_rp/en/.

2. Write a **statistical analysis plan** prior to analyzing the data. This plan goes into more depth than the protocol and ideally gives the details on all the planned tables, figures and statistical models. It is even possible to create a dummy report of tables and figures with no data so that your colleagues can better understand what the final results will look like and improvements or additions can be made prior to any analysis.

3. Clean and verify the data. Errors in the data are likely to produce erroneous results. Errors can occur during data recording or data entry, and also when the data are moved between programs (e.g., reading data from Excel into R). Data recording errors can be minimized by using data collection tools such as REDCap that report logic errors when the data are entered (Harris et al. 2009).

All data should be "cleaned" prior to analysis, meaning that impossible values are removed, e.g., a negative age. Data checking should include verifying that continuous and categorical variables are within their possible ranges. Logical cross-checks can also be made across two or more variables, e.g., checking for men who record being pregnant. Missing data need to be carefully coded to distinguish between data that are actually missing and data that were never meant to be collected.

4. Use **exploratory analyses** to better understand the data before conducting any modelling. The exploratory analysis should include summary statistics. For continuous variables the most useful summary statistics are often the mean, standard deviation, minimum and maximum, but for data with a strong skew or bimodal distribution the most useful summary statistics are often the mode, median, inter-quartile range, minimum and maximum. Histograms of continuous variables are useful for identifying features that may need to be factored into modelling, such as skewness and a large proportion of zeros, which may mean that a variable is best transformed or categorized before being used as an explanatory variable in a regression model. For categorical variables use tables and bar charts for exploratory analyses.

Scatter plots of pairs of continuous variables can be useful for showing strong associations and potentially highlighting non-linear associations. For pairs of categorical variables use cross-tabulations to highlight associations. Box plots can be used to examine how a continuous variable is associated with a categorical variable.

Where there are lots of related variables it may be useful to combine them into a **composite variable**. For example, data on multiple long-term health conditions such as arthritis, asthma, diabetes, etc, may have been collected using a binary yes or no response, but rather than fitting many binary variables it would be more parsimonious to combine the binary variables into a single count of long-term health conditions. The decision to use a composite variable should be based on the goals of the model and how important each individual variable is, bearing in mind that potentially useful information is lost when variables are combined in this way.

Where there are lots of highly correlated variables it may be useful to reduce the number of variables prior to modelling using a data reduction technique such as principal components analysis or factor analysis. This can avoid multicollinearity (see Section 6.3.4) by creating a reduced set of explanatory variables that are more orthogonal. A potential downside to this approach is that these orthogonal variables may not have a clear interpretation as they are weighted combinations of the original variables.

5. Use scrambled and blinded reporting (Järvinen et al. 2014). Scrambling involves randomizing the key information, for example in a clinical trial a random treatment is created in place of the actual treatment. Blinding involves concealing the key information, for example in a clinical trial the

real treatment is used but with dummy labels. Preliminary reports can be created that show the results but with scrambled or blinded data. This can be useful for finding data or programming errors which can then be corrected before the actual data are used. This reduces the chances of **confirmation bias**, where results are only checked and corrected if they disagree with the researchers' a priori opinions.

An example of scrambled reporting is shown using the hepatitis trial data in Table 10.5. A random treatment group was created by randomly reordering the actual treatment group as shown in the R code below, with Figure P.1 showing the key comparison between groups using Kaplan–Meier survival estimates.

```
- R code (survival curves with random treatment groups) -
>hepatitis$rgroup = sample(hepatitis$group, size=
  nrow(hepatitis), replace=F)
>s = survfit(Surv('survival time', event=censor=='died')
  ~ rgroup, data=hepatitis)
>plot(s, xlab='Time, months', ylab='Survival probability',
  lty=c(1,2), mark.time = T)
```

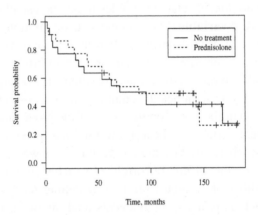

Figure P.1 *Kaplan–Meier estimates for the hepatitis trial data using a randomly generated treatment group. Censored times are shown using a vertical line.*

As expected, the empirical survival estimates in Figure P.1 show no clear difference between the two groups. The plot shows a steady accumulation of deaths over time and that there is an appreciable drop in survival, which indicates that the plot using the observed data should provide a useful comparison between the two groups. The plot also provides an example of what the results will look like if the data are consistent with the null hypothesis. An example of blinded reporting using the hepatitis trial data are shown Figure P.2. This shows a substantial difference between the groups as deaths occur more rapidly in one group, but it is not known if the treatment (Prednisolone) is helping or harming patients. Researchers and clinicians can be shown this graph and asked whether they believe the difference in survival is clinically meaningful. Gathering opinions at this stage reduces the potential biases from conflicts of interest that could influence how the results are interpreted, for example, a desire to favor a company's product. Other statistics using blinded labels could have been added to help interpretation of the results, such as the median survival times.

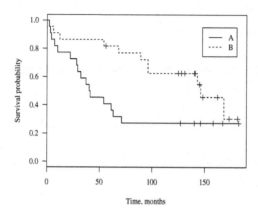

Figure P.2 *Kaplan–Meier estimates for the hepatitis trial data using blinded labels.*

6. If regression models are used, then it is necessary to thoroughly check their adequacy. Examine the residuals using histograms and scatter plots to look for non-linearity, heteroscedasity and outliers. Check for influential observations using delta-beta statistics and Cook's distance (see Section 6.2.7). Check for multicollinearity using the variance inflation factor (see Section 6.3.4).

7. Keep records of the data management and analyses. Results generated by "pointing and clicking" drop-down menus in statistical software do not create an accurate record of the decisions made which makes it difficult to replicate an analysis. There are textbooks that describe how to create reproducible workflows for Stata (Long 2009) and R (Gandrud 2013). Keeping detailed and complete records is good research practice because it allows repetition or refinement of the analysis if needed months or even

years later. Analysis records can also be updated for related projects or shared with colleagues. Syntax code can be written for most statistical packages such as Stata, SPSS and SAS. These packages have options for keeping a log of all the code used. RStudio, which runs R, includes the option of mixing text and analyses to produce a complete report using Rmarkdown. An example is shown below with the results in Figure P.3.

————————— R markdown example code —————————

```
---
title: "An R markdown example"
author: "Dobson & Barnett"
output: pdf_document
---


## R Markdown

This simple example shows how text and code can be
combined in R Markdown.

Summary statistics for the cyclones data.
```{r}
library(dobson)
summary(cyclones)
```
```

It is also good practice to keep the original data as this allows the results to be reproduced from scratch. It is recommend that the "raw" original data and a cleaned version that is ready for analysis are kept, as well as the code that transforms the data from the raw to the cleaned version. Ideally investigators should be willing to share their data and code with other interested researchers, although this is not always possible if sharing the data raises ethical or legal concerns.

8. Use **reporting guidelines** to write up the results. Reporting guidelines describe which aspects of the study should be reported and why, and include a checklist of what to report (Simera and Altman 2013). Using a checklist helps ensure that all the important details are included, this means the results are more likely to be correctly interpreted and used in practice. There are guidelines for common designs such as Randomized Controlled Trials (Moher et al. 2001) and for more specialist designs such as multivariable

An R markdown example

Dobson & Barnett

R Markdown

This simple example shows how text and code can be combined in R Markdown.

Summary statistics for the cyclones data.

```
library(dobson)
summary(cyclones)
```

```
##      years                season          number
##  Length:13            Min.    : 1     Min.    : 2.000
##  Class :character     1st Qu.: 4     1st Qu.: 4.000
##  Mode  :character     Median : 7     Median : 6.000
##                       Mean    : 7     Mean    : 5.538
##                       3rd Qu.:10     3rd Qu.: 6.000
##                       Max.    :13     Max.    :12.000
```

Figure P.3 *Example output using R markdown to create well documented statistical analysis.*

prediction models (Moons et al. 2015). There are even guidelines on the number of decimal places to report (Cole 2015).

When reporting **p-values** give the actual value and do not write "non-significant" or "$p > 0.05$" if the p-value is above the commonly used 0.05 threshold. Using "non-significant" implies that a p-value of 0.06 is equivalent to a value of 0.99, whereas these provide very different levels of support for the null hypothesis (Sterne et al. 2001). The exception is for very small p-values, e.g., 1×10^{-8}, which can be reported using a sensible threshold such as < 0.001, although small values are often reported exactly in genetics studies due to the great number of hypotheses tested. Also, be careful to give the correct interpretation of p-values (see Section 12.1) as they are often misreported or misconstrued (Goodman 2008).

9. **Continuing professional development**. Learning about the latest developments in statistics helps you to remain up to date with the best methods. This can be done by joining a statistical society, reading relevant journals and books, attending conferences, or attending courses, and there are now many good courses online.

Appendix

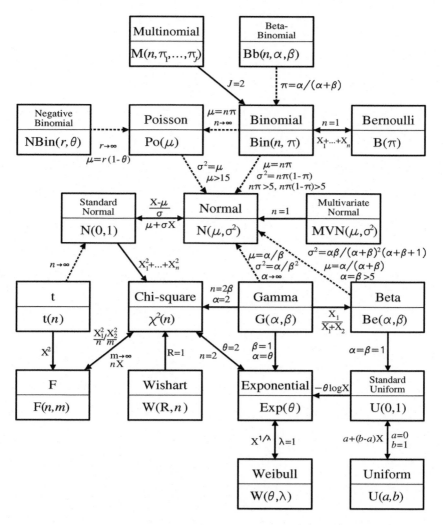

Some relationships between common distributions, adapted from Leemis (1986). Dotted lines indicate an asymptotic relationship and solid lines a transformation.

Software

First Bayes (free)

- Tony O'Hagan, Department of Probability and Statistics, University of Sheffield, Hicks Building, Sheffield, S3 7RH, United Kingdom
- www.firstbayes.co.uk

Genstat

- VSN International Ltd, 5 The Waterhouse, Waterhouse Street, Hemel Hempstead, HP1 1ES, United Kingdom
- www.vsni.co.uk/software/genstat

Minitab

- Minitab Inc., Quality Plaza, 1829 Pine Hall Rd, State College, PA 16801-3008, U.S.A.
- www.minitab.com

MLwiN

- Centre for Multilevel Modelling, Graduate School of Education, Bristol Institute of Public Affairs, University of Bristol, 2 Priory Road, Bristol, BS8 1TX, United Kingdom
- www.cmm.bristol.ac.uk/MLwiN/index.shtml

Python (free)

- Python Software Foundation, Delaware, United States
- https://www.python.org/
- Python solutions to the examples in this book are available at: github.com/thomas-haslwanter/dobson

R (free)

- R Foundation for Statistical Computing, Vienna, Austria
- www.r-project.org

- The example data sets used in this book are available in the *dobson* package. To install it use the R command `install.packages('dobson')`.

Rstudio (free) graphical user interface for running R

- Rstudio Inc., 250 Northern Ave, Boston, MA 02210, U.S.A.
- www.rstudio.com

Stata

- StataCorp LP, 4905 Lakeway Drive, College Station, TX 77845, U.S.A.
- www.stata.com

SAS

- SAS Institute Inc., 100 SAS Campus Drive, Cary, NC 27513-2414, U.S.A.
- www.sas.com

S-PLUS

- Insightful Corporation, 1700 Westlake Avenue North, Suite 500, Seattle, WA 98109-3044, U.S.A.
- www.spotfire.tibco.com

StatXact and LogXact

- Cytel Inc., 675 Massachusetts Avenue, 3rd Floor, Cambridge, MA 02139-3309, U.S.A.
- www.cytel.com/software

WinBUGS (free)

- MRC Biostatistics Unit, Cambridge, United Kingdom & Imperial College School of Medicine at St Mary's, London, United Kingdom
- www.mrc-bsu.cam.ac.uk/bugs

References

Agresti, A. (2007). *An Introduction to Categorical Data Analysis* (2nd ed.). Hoboken, N.J.: Wiley.

Agresti, A. (2010). *Analysis of Ordinal Categorical Data* (2nd ed.). Hoboken, N.J.: Wiley.

Agresti, A. (2013). *Categorical Data Analysis* (3rd ed.). Hoboken, N.J.: Wiley.

Aitkin, M., D. Anderson, B. Francis, and J. Hinde (2005). *Statistical Modelling in GLIM* (2nd ed.). Oxford: Oxford University Press.

Aitkin, M. and D. Clayton (1980). The fitting of exponential, Weibull and extreme value distributions to complex censored survival data using GLIM. *Applied Statistics 29*, 156–163.

Alfons, A. (2012). *cvTools: Cross-validation tools for regression models*. R package version 0.3.2.

Altman, D. G. (1994). The scandal of poor medical research. *BMJ 308*(6924), 283–284.

Altman, D. G. and J. M. Bland (1998). Statistical notes: times to event (survival) data. *British Medical Journal 317*, 468–469.

Altman, D. G., D. Machin, T. N. Bryant, and M. J. Gardner (Eds.) (2000). *Statistics with Confidence* (2nd ed.). London: Blackwell BMJ Books.

Ananth, C. V. and D. G. Kleinbaum (1997). Regression models for ordinal responses: a review of methods and applications. *International Journal of Epidemiology 26*, 1323–1333.

Andrews, D. F. and A. M. Herzberg (1985). *Data: A Collection of Problems from Many Fields for the Student and Research Worker*. New York: Springer Verlag.

Aranda-Ordaz, F. J. (1981). On two families of transformations to additivity for binary response data. *Biometrika 68*, 357–363.

Ashby, D., C. R. West, and D. Ames (1989). The ordered logistic regression model in psychiatry: rising prevalence of dementia in old people's homes.

Statistics in Medicine 8, 1317–1326.

Australian Bureau of Statistics (1998). Apparent consumption of foodstuffs 1996–97. Technical Report Publication 4306.0, Canberra.

Baker, M. (2016). 1,500 scientists lift the lid on reproducibility. *Nature 533*, 452–454.

Barnett, A. G., N. Koper, A. J. Dobson, F. Schmiegelow, and M. Manseau (2010). Using information criteria to select the correct variance–covariance structure for longitudinal data in ecology. *Methods in Ecology and Evolution 1*, 15–24.

Bartlett, M. S. (1978). *An Introduction to Stochastic Processes: with Special Reference to Methods and Applications* (3rd ed.). Cambridge; New York: Cambridge University Press.

Baxter, L. A., S. M. Coutts, and G. A. F. Ross (1980). Applications of linear models in motor insurance. Zurich, pp. 11–29. Proceedings of the 21st International Congress of Actuaries.

Belsley, D. A., E. Kuh, and R. E. Welsch (2004). *Regression Diagnostics: Identifying Influential Observations and Sources of Collinearity*. New York: Wiley.

Berger, J. O. (1985). *Statistical Decision Theory and Bayesian Analysis* (2nd ed.). New York: Springer-Verlag.

Birch, M. W. (1963). Maximum likelihood in three-way contingency tables. *Journal of the Royal Statistical Society, Series B 25*, 220–233.

Bliss, C. I. (1935). The calculation of the dose-mortality curve. *Annals of Applied Biology 22*, 134–167.

Boltwood, C. M., R. Appleyard, and S. A. Glantz (1989). Left ventricular volume measurement by conductance catheter in intact dogs: the parallel conductance volume increases with end-systolic volume. *Circulation 80*, 1360–1377.

Box, G. E. P. and D. R. Cox (1964). An analysis of transformations. *Journal of the Royal Statistical Society, Series B 26*, 211–234.

Breslow, N. E. and N. E. Day (1987). *Statistical Methods in Cancer Research, Volume 2: The Design and Analysis of Cohort Studies*. Lyon: International Agency for Research on Cancer.

Brooks, S. P. (1998). Markov chain Monte Carlo method and its application. *The Statistician 47*, 69–100.

Brooks, S. P. and A. Gelman (1998). Alternative methods for monitoring convergence of iterative simulations. *Journal of Computational and Graph-*

ical Statistics 7, 434–455.

Brown, C. C. (1982). On a goodness of fit test for the logistic model based on score statistics. *Communications in Statistics—Theory and Methods 11*, 1087–1105.

Burton, P., L. Gurrin, and P. Sly (1998). Tutorial in biostatistics—extending the simple linear regression model to account for correlated responses: an introduction to generalized estimating equations and multi-level modelling. *Statistics in Medicine 17*, 1261–1291.

Cai, Z. and C.-L. Tsai (1999). Diagnostics for non-linearity in generalized linear models. *Computational Statistics and Data Analysis 29*, 445–469.

Carlin, J. B., R. Wolfe, C. Coffey, and G. Patton (1999). Tutorial in biostatistics—analysis of binary outcomes in longitudinal studies using weighted estimating equations and discrete-time survival methods: prevalence and incidence of smoking in an adolescent cohort. *Statistics in Medicine 18*, 2655–2679.

Chalmers, I. and P. Glasziou (2009). Avoidable waste in the production and reporting of research evidence. *The Lancet 374*(9683), 86–89.

Charnes, A., E. L. Frome, and P. L. Yu (1976). The equivalence of generalized least squares and maximum likelihood estimates in the exponential family. *Journal of the American Statistical Association 71*, 169–171.

Chivers, C. (2012). *MHadaptive: General Markov Chain Monte Carlo for Bayesian Inference using adaptive Metropolis–Hastings sampling*. R package version 1.1-8.

Cnaan, A., N. M. Laird, and P. Slasor (1997). Tutorial in biostatistics—using the generalized linear model to analyze unbalanced repeated measures and longitudinal data. *Statistics in Medicine 16*, 2349–2380.

Cole, T. J. (2015). Too many digits: the presentation of numerical data. *Archives of Disease in Childhood 100*(7), 608–609.

Collett, D. (2003). *Modelling Binary Data* (2nd ed.). London: CRC Press.

Collett, D. (2014). *Modelling Survival Data in Medical Research* (3rd ed.). London: CRC Press.

Congdon, P. (2006). *Bayesian Statistical Modelling* (2nd ed.). Chichester; New York: Wiley.

Cook, R. D. and S. Weisberg (1999). *Applied Regression including Computing and Graphics*. New York: Wiley.

Cox, D. R. (1972). Regression models and life tables (with discussion). *Journal of the Royal Statistical Society, Series B 34*, 187–220.

Cox, D. R. and D. V. Hinkley (1974). *Theoretical Statistics*. London: Chapman & Hall.

Cox, D. R. and E. J. Snell (1968). A general definition of residuals. *Journal of the Royal Statistical Society, Series B 30*, 248–275.

Cox, D. R. and E. J. Snell (1981). *Applied Statistics: Principles and Examples*. London: Chapman & Hall.

Cox, D. R. and E. J. Snell (1989). *Analysis of Binary Data* (2nd ed.). London: Chapman & Hall.

Cressie, N. and T. R. C. Read (1989). Pearson's χ^2 and the loglikelihood ratio statistic g^2: a comparative review. *International Statistical Review 57*, 19–43.

Crowley, J. and M. Hu (1977). Covariance analysis of heart transplant survival data. *Journal of the American Statistical Association 72*, 27–36.

Deddens, J. A. and M. R. Petersen (2008). Approaches for estimating prevalence ratios. *Occupational and Environmental Medicine 65*(7), 501–506.

Dickman, P. W., A. Sloggett, M. Hills, and T. Hakulinen (2004). Regression models for relative survival. *Statistics in Medicine 23*, 51–64.

Diggle, P. J., P. Heagerty, K.-Y. Liang, and S. L. Zeger (2002). *Analysis of Longitudinal Data* (2nd ed.). Oxford: Oxford University Press.

Dobson, A. J. and J. Stewart (1974). Frequencies of tropical cyclones in the northeastern Australian area. *Australian Meteorological Magazine 22*, 27–36.

Draper, N. R. and H. Smith (1998). *Applied Regression Analysis* (3rd ed.). New York: Wiley-Interscience.

Duggan, J. M., A. J. Dobson, H. Johnson, and P. P. Fahey (1986). Peptic ulcer and non-steroidal anti-inflammatory agents. *Gut 27*, 929–933.

Egger, G., G. Fisher, S. Piers, K. Bedford, G. Morseau, S. Sabasio, B. Taipim, G. Bani, M. Assan, and P. Mills (1999). Abdominal obesity reduction in Indigenous men. *International Journal of Obesity 23*, 564–569.

Evans, M., N. Hastings, and B. Peacock (2000). *Statistical Distributions* (3rd ed.). New York: Wiley.

Fahrmeir, L. and H. Kaufmann (1985). Consistency and asymptotic normality of the maximum likelihood estimator in generalized linear models. *Annals of Statistics 13*, 342–368.

Feigl, P. and M. Zelen (1965). Estimation of exponential probabilities with concomitant information. *Biometrics 21*, 826–838.

Finney, D. J. (1973). *Statistical Methods in Bioassay* (2nd ed.). New York: Hafner.

Fitzmaurice, G. M., N. M. Laird, and J. H. Ware (2012). *Applied Longitudinal Analysis* (2nd ed.). Hoboken, N.J.: Wiley-Interscience.

Fleming, T. R. and D. P. Harrington (2005). *Counting Processes and Survival Analysis* (2nd ed.). New York: Wiley.

Forbes, C., M. Evans, N. Hastings, and B. Peacock (2010). *Statistical Distributions* (4th ed.). New York: Wiley.

Fox, J. and S. Weisberg (2011). *An R Companion to Applied Regression* (Second ed.). Thousand Oaks CA: Sage.

Friedman, J., T. Hastie, and R. Tibshirani (2010). Regularization paths for generalized linear models via coordinate descent. *Journal of Statistical Software 33*, 1–22.

Gandrud, C. (2013). *Reproducible Research with R and R Studio*. London: Chapman & Hall.

Gasparrini, A., Y. Guo, M. Hashizume, et al. (2015). Mortality risk attributable to high and low ambient temperature: a multicountry observational study. *Lancet 386*(9991), 369–375.

Gehan, E. A. (1965). A generalized Wilcoxon test for comparing arbitrarily singly-censored samples. *Biometrika 52*, 203–223.

Gelman, A., J. B. Carlin, H. S. Stern, D. B. Dunson, A. Vehtari, and D. B. Rubin (2013). *Bayesian Data Analysis* (3rd ed.). Boca Raton, Fla.: Chapman & Hall/CRC.

Gilks, W. R., S. Richardson, and D. J. Spiegelhalter (1996a). Introducing Markov chain Monte Carlo. In W. R. Gilks, S. Richardson, and D. J. Spiegelhalter (Eds.), *Markov chain Monte Carlo in practice*. London: Chapman & Hall.

Gilks, W. R., S. Richardson, and D. J. Spiegelhalter (1996b). *Markov chain Monte Carlo in practice*. London: Chapman & Hall.

Gilks, W. R. and G. O. Roberts (1996). Strategies for improving MCMC. In W. R. Gilks, S. Richardson, and D. J. Spiegelhalter (Eds.), *Markov chain Monte Carlo in practice*. London: Chapman & Hall.

Glantz, S. A. and B. K. Slinker (1990). *Primer of Applied Regression and Analysis of Variance*. New York: McGraw Hill.

Glymour, M. M. and S. Greenland (2008). Causal diagrams. In K. Rothman, S. Greenland, and T. Lash (Eds.), *Modern Epidemiology*. Wolters Kluwer Health/Lippincott Williams & Wilkins.

Goldstein, H. (2011). *Multilevel Statistical Models* (4th ed.). London: Arnold.

Goodman, S. (2008). A dirty dozen: Twelve p-value misconceptions. *Seminars in Hematology 45*(3), 135–140.

Harris, P. A., R. Taylor, R. Thielke, J. Payne, N. Gonzalez, and J. G. Conde (2009). Research electronic data capture (REDCap) - A metadata-driven methodology and workflow process for providing translational research informatics support. *Journal of Biomedical Informatics 42*, 377–381.

Hastie, T. J. and R. J. Tibshirani (1990). *Generalized Additive Models*. London: Chapman & Hall.

Hastie, T. J., R. J. Tibshirani, and J. Friedman (2009). *The Elements of Statistical Learning: Data Mining, Inference, and Prediction* (2nd ed.). New York: Springer-Verlag.

Hebbali, A. (2017). *olsrr: Tools for Teaching and Learning OLS Regression.* R package version 0.3.0.

Hilbe, J. M. (2014). *Modeling Count Data*. Cambridge: Cambridge University Press.

Hilbe, J. M. (2015). *Practical Guide to Logistic Regression*. Baton Rouge: Chapman & Hall/CRC.

Højsgaard, S. and U. Halekoh (2016). *doBy: Groupwise Statistics, LSmeans, Linear Contrasts, Utilities*. R package version 4.5-15.

Højsgaard, S., U. Halekoh, and J. Yan (2005). The R package geepack for generalized estimating equations. *Journal of Statistical Software 15*, 1–11.

Hoeting, J. A., D. Madigan, A. E. Raftery, and C. T. Volinsky (1999). Bayesian model averaging: a tutorial. *Statistical Science 14*(4), 382–417.

Holtbrugger, W. and M. Schumacher (1991). A comparison of regression models for the analysis of ordered categorical data. *Applied Statistics 40*, 249–259.

Hosmer, D. W. and S. Lemeshow (1980). Goodness of fit tests for the multiple logistic model. *Communications in Statistics—Theory and Methods A9*, 1043–1069.

Hosmer, D. W. and S. Lemeshow (2008). *Applied Survival Analysis: Regression Modeling of Time to Event Data* (2nd ed.). New York: Wiley.

Hosmer, D. W., S. Lemeshow, and R. X. Sturdivant (2013). *Applied Logistic Regression* (3rd ed.). New York: Wiley.

Järvinen, T. L., R. Sihvonen, M. Bhandari, et al. (2014). Blinded interpretation of study results can feasibly and effectively diminish interpretation bias. *Journal of Clinical Epidemiology 67*, 769–772.

Jones, R. H. (1987). Serial correlation in unbalanced mixed models. *Bulletin of the International Statistical Institute 52*, 105–122.

Kalbfleisch, J. D. and R. L. Prentice (2002). *The Statistical Analysis of Failure Time Data* (2nd ed.). New York: Wiley.

Kinner, S. A. (2006). Continuity of health impairment and substance misuse among adult prisoners in Queensland, Australia. *International Journal of Prisoner Health 2*, 101–113.

Klein, J., H. C. van Houwelingen, J. G. Ibrahim, and T. H. Scheike (2013). *Handbook of Survival Analysis*. London: Chapman & Hall/CRC.

Kleinbaum, D. G. and M. Klein (2012). *Survival Analysis - A Self-Learning Text* (3rd ed.). New York: Springer.

Kleinbaum, D. G., L. L. Kupper, K. E. Muller, and A. Nizam (2007). *Applied Regression Analysis and Multivariable Methods* (4th ed.). Pacific Grove, Calif.: Duxbury.

Krzanowski, W. J. (1998). *An Introduction to Statistical Modelling*. London: Arnold.

Kutner, M. H., C. J. Nachtsheim, J. Neter, and W. Li (2005). *Applied Linear Statistical Models* (5th ed.). Chicago: McGraw-Hill/Irwin.

Lawless, J. F. (2002). *Statistical Models and Methods for Lifetime Data* (2nd ed.). New York: Wiley.

Lee, C., A. J. Dobson, W. J. Brown, L. Bryson, J. Byles, P. Warner-Smith, and A. F. Young (2005). Cohort profile: The Australian longitudinal study on women's health. *International Journal of Epidemiology 34*, 987–991.

Lee, E. T. and J. W. Wang (2003). *Statistical Methods for Survival Data Analysis* (3rd ed.). New York: Wiley.

Leemis, L. M. (1986). Relationships among common univariate distributions. *The American Statistician 40*, 143–146.

Leonard, T. and J. S. J. Hsu (1992). Bayesian inference for a covariance matrix. *The Annals of Statistics 20*, 1669–1696.

Lewis, T. (1987). Uneven sex ratios in the light brown apple moth: a problem in outlier allocation. In D. J. Hand and B. S. Everitt (Eds.), *The Statistical Consultant in Action*. Cambridge: Cambridge University Press.

Liang, K.-Y. and S. L. Zeger (1986). Longitudinal data analysis using generalized linear models. *Biometrika 73*, 13–22.

Liao, J. and D. McGee (2003). Adjusted coefficients of determination for logistic regression. *The American Statistician 57*, 161–165.

Lipsitz, S. R., N. M. Laird, and D. P. Harrington (1991). Generalized estimating equations for correlated binary data: using the odds ratio as a measure of association. *Biometrika 78*, 153–160.

Long, J. S. (2009). *The Workflow of Data Analysis Using Stata*. Texas: Stata Press.

Lunn, D., C. Jackson, N. Best, A. Thomas, and D. Spiegelhalter (2012). *The BUGS Book: A Practical Introduction to Bayesian Analysis*. Chapman & Hall/CRC Texts in Statistical Science. London: Taylor & Francis.

Madsen, M. (1971). Statistical analysis of multiple contingency tables. two examples. *Scandinavian Journal of Statistics 3*, 97–106.

McCullagh, P. (1980). Regression models for ordinal data (with discussion). *Journal of the Royal Statistical Society, Series B 42*, 109–142.

McCullagh, P. and J. A. Nelder (1989). *Generalized Linear Models* (2nd ed.). London: Chapman & Hall.

McFadden, M., J. Powers, W. Brown, and M. Walker (2000). Vehicle and driver attributes affecting distance from the steering wheel in motor vehicles. *Human Factors 42*, 676–682.

McKinlay, S. M. (1978). The effect of nonzero second-order interaction on combined estimators of the odds ratio. *Biometrika 65*, 191–202.

Mittlbock, M. and H. Heinzl (2001). A note on R^2 measures for Poisson and logistic regression models when both are applicable. *Journal of Clinical Epidemiology 54*, 99–103.

Moher, D., K. F. Schulz, D. Altman, and for the CONSORT Group (2001). The CONSORT statement: Revised recommendations for improving the quality of reports of parallel-group randomized trials. *JAMA 285*, 1987–1991.

Molenberghs, G. and G. Verbeke (2005). *Models for Discrete Longitudinal Data*. Berlin: Springer.

Montgomery, D. C., E. A. Peck, and G. G. Vining (2006). *Introduction to Linear Regression Analysis* (4th ed.). New York: Wiley.

Moons, K. M., D. G. Altman, J. B. Reitsma, et al. (2015). Transparent reporting of a multivariable prediction model for individual prognosis or diagnosis (TRIPOD): Explanation and elaboration. *Annals of Internal Medicine 162*, W1–W73.

Myers, R. H., D. C. Montgomery, G. G. Vining, and T. J. Robinson (2010).

Generalized Linear Models: Applications in Engineering and the Sciences (2nd ed.). Hoboken, N.J.: Wiley.

National Academy of Sciences and National Academy of Engineering and Institute of Medicine (2009). *On Being a Scientist: A Guide to Responsible Conduct in Research* (3rd ed.). Washington, DC: The National Academies Press.

National Centre for HIV Epidemiology and Clinical Research (1994). Australian HIV surveillance report. Technical Report 10.

Nelder, J. A. and R. W. M. Wedderburn (1972). Generalized linear models. *Journal of the Royal Statistical Society, Series A 135*, 370–384.

Otake, M. (1979). Comparison of time risks based on a multinomial logistic response model in longitudinal studies. Technical Report No. 5, RERF, Hiroshima, Japan.

Parmar, M. K., D. J. Spiegelhalter, L. S. Freedman, and the CHART steering committee (1994). The CHART trials: Bayesian design and monitoring in practice. *Statistics in Medicine 13*, 1297–1312.

Pierce, D. A. and D. W. Schafer (1986). Residuals in generalized linear models. *Journal of the American Statistical Association 81*, 977–986.

Pinheiro, J., D. Bates, S. DebRoy, D. Sarkar, and R Core Team (2017). *nlme: Linear and Nonlinear Mixed Effects Models*. R package version 3.1-131.

Pregibon, D. (1981). Logistic regression diagnostics. *Annals of Statistics 9*, 705–724.

Rabe-Hesketh, S. and A. Skrondal (2012). *Multilevel and Longitudinal Modeling Using Stata, Volumes I and II, Third Edition*. Boca Raton, Fla.: Taylor & Francis.

Raftery, A. E., I. S. Painter, and C. T. Volinsky (2005). BMA: An R package for Bayesian model averaging. *R News 5/2*, 2–8.

Roberts, G., A. L. Martyn, A. J. Dobson, and W. H. McCarthy (1981). Tumour thickness and histological type in malignant melanoma in New South Wales, Australia, 1970–76. *Pathology 13*, 763–770.

Rosner, B. (1989). Multivariate methods for clustered binary data with more than one level of nesting. *Journal of the American Statistical Association 84*, 373–380.

Royston, P., G. Ambler, and W. Sauerbrei (1999). The use of fractional polynomials to model continuous risk variables in epidemiology. *International Journal of Epidemiology 28*, 964.

Sangwan-Norrell, B. S. (1977). Androgenic stimulating factor in the anther and isolated pollen grain culture of *Datura innoxia* mill. *Journal of Experimental Biology 28*, 843–852.

Senn, S., L. Stevens, and N. Chaturvedi (2000). Tutorial in biostatistics—repeated measures in clinical trials: simple strategies for analysis using summary measures. *Statistics in Medicine 19*, 861–877.

Simera, I. and D. G. Altman (2013). Reporting medical research. *International Journal of Clinical Practice 67*, 710–716.

Sinclair, D. F. and M. E. Probert (1986). A fertilizer response model for a mixed pasture system. In I. S. Francis, B. Manly, and F. Lam (Eds.), *Pacific Statistical Congress*, pp. 470–474. Amsterdam: Elsevier.

Spiegelhalter, D. J., N. G. Best, B. P. Carlin, and A. van der Linde (2002). Bayesian measures of model complexity and fit (with discussion). *Journal of the Royal Statistical Society Series B 64*, 583–640.

Spiegelhalter, D. J., A. Thomas, N. G. Best, and D. Lunn (2007). WinBUGS version 1.4.3 user manual, MRC Biostatitics Unit, Cambridge.

Stephen, D. M. and A. G. Barnett (2016). Effect of temperature and precipitation on salmonellosis cases in south-east Queensland, Australia: an observational study. *BMJ Open 6*, e010204.

Sterne, J. A. C., D. R. Cox, and G. D. Smith (2001). Sifting the evidence—what's wrong with significance tests? another comment on the role of statistical methods. *BMJ 322*(7280), 226–231.

Stroup, W. (2012). *Generalized Linear Mixed Models: Modern Concepts, Methods and Applications*. Boca Raton, Fla.: Taylor & Francis.

Sturtz, S., U. Ligges, and A. Gelman (2005). R2WinBUGS: A package for running WinBUGS from R. *Journal of Statistical Software 12*(3), 1–16.

Tibshirani, R. (1994). Regression shrinkage and selection via the lasso. *Journal of the Royal Statistical Society, Series B 58*, 267–288.

Twisk, J. W. R. (2006). *Applied Multilevel Analysis: A Practical Guide*. Cambridge: Cambridge University Press.

Twisk, J. W. R. (2013). *Applied Longitudinal Data Analysis for Epidemiology—A Practical Guide* (2nd ed.). Cambridge: Cambridge University Press.

Venables, W. N. and B. D. Ripley (2002). *Modern Applied Statistics with S* (4th ed.). New York: Springer.

Verbeke, G. and G. Molenberghs (2000). *Linear Mixed Models for Longitudinal Data*. New York: Springer.

Wei, L. J. (1992). The accelerated failure time model: a useful alternative to the Cox regression model in survival analysis. *Statistics in Medicine 11*, 1871–1879.

Weisberg, S. (2014). *Applied Linear Regression*. Hoboken, N.J.: Wiley.

Winer, B. J. (1971). *Statistical Principles in Experimental Design* (2nd ed.). New York: McGraw-Hill.

Wood, C. L. (1978). Comparison of linear trends in binomial proportions. *Biometrics 34*(3), 496–504.

Zeger, S. L., K.-Y. Liang, and P. Albert (1988). Models for longitudinal data: a generalized estimating equation approach. *Biometrics 44*, 1049–1060.

Ziliak, S. T. and D. N. McCloskey (2008). *The Cult of Statistical Significance: How the Standard Error Costs Us Jobs, Justice, and Lives*. Ann Arbor: University of Michigan Press.

Index